D0394612

Theory and Design
of Adaptive Filters

TOPICS IN DIGITAL SIGNAL PROCESSING

Theory and Design of Adaptive Filters

John R. Treichler, Applied Signal Technology

C. Richard Johnson, Jr., Cornell University

Michael G. Larimore, Applied Signal Technology

With an Appendix by Texas Instruments
and Teknekron Communications Systems

A WILEY-INTERSCIENCE PUBLICATION

JOHN WILEY & SONS

New York   Chichester   Brisbane   Toronto   Singapore

Copyright © 1987 by Texas Instruments, Inc.
Published by John Wiley & Sons, Inc.

Library of Congress Cataloging in Publication Data:

Treichler, John R.
  Theory and design of adaptive filters.

  "A Wiley-Interscience publication."
  Bibliography: p.
  1. Adaptive filters. I. Johnson, C. Richard, Jr.
II. Larimore, Michael G. III. Title.
TK7872.F5T65 1987   621.3815'324   87-6062
ISBN 0-471-83220-0

Printed in the United States of America

10 9 8 7 6 5 4 3 2

To Sally, Betty, and Mary Anne

Preface

This book is a pedagogical compilation of fundamental adaptive filtering concepts, algorithm forms, behavioral insights, and application guidelines. The analysis and design of three basic classes of adaptive filters are presented: (1) adaptive finite-impulse-response (FIR) filters, (2) adaptive infinite-impulse-response (IIR) filters, and (3) adaptive property-restoral filters.

For the widely studied (and utilized) class of adaptive FIR filters, we develop the most popular analytical tools and distill a tutorial collection of insightful design guidelines of proven utility. For more recently developed adaptive IIR and adaptive property-restoral filters, we focus on emerging theoretical foundations and suggested applications. Our presentation is augmented by listings of FORTRAN code for the more basic of these algorithms. We also detail a real-time solution—using a Texas Instruments signal-processing chip (TMS320)—to one adaptive FIR filter problem. Thus, both practicing engineers interested in designing appropriate adaptive filters for various applications and graduate students interested in acquiring a cohesive pedagogy for initiation of basic research in adaptive filter theory can benefit from this book.

The first chapter contains a detailed description of the purpose and motivation of adaptive filtering in three practical signal-processing problems. Basically, this chapter is for those readers who do not yet know of the practical need for and utility of adaptive filters. An understanding of nonadaptive digital signal processing is assumed of the reader. More specifically, this book was written with the expectation that the reader would have a rudimentary understanding of digital filter theory (e.g., discrete Fourier transform, analog-to-digital filter transformation, and frequency-response shaping techniques of FIR and IIR digital filter design), stochastic processes (e.g., correlation functions), linear algebra (e.g., eigenvalues and eigenvectors), and systems theory (e.g., state representations, z-transforms, transfer functions, and stability theory). If the reader does not possess such a background, the following texts are recommended as possible sources: A. V. Oppenheim and A. S. Willsky, *Signals and Systems* (Prentice-Hall, 1983) and L. R. Rabiner and B. Gold, *Theory and Application of Digital Signal Processing* (Prentice-Hall, 1975). Given such a background, the reader should find this book a self-contained primer and reference source on adaptive filtering.

Chapter 2 provides an intuitive illustration of the basic principles underlying the conversion of (mis)performance information into digital filter adaptation. A set of number guessing games with logically obvious algorithmic solutions is examined and recognized to strongly parallel digital filter adaptation where the filter parameters are effectively the numbers being guessed. These examples of adaptation are simple enough to be readily understood yet complex enough to illustrate several of the basic attributes of adaptive filters. The reader will also find these number guessing games helpful in interpreting the purpose and usefulness of algorithmic elaborations that are introduced in subsequent chapters. Essentially, this chapter is intended for those readers who believe adaptive filters are useful but do not yet have a firm idea as to the source and form of filter parameter adaptation.

Chapters 3, 4, 5, and 6 present the fundamentals of adaptive filter theory. Chapters 3 and 4 focus on adaptive FIR filters, which have been widely analyzed and applied since their conception in the late 1950s. Chapter 3 develops popular adaptive filter analysis tools for dynamic parameter estimate moment analysis. Chapter 4 is based on selections of fundamental material from the large body of useful work on adaptive FIR filter design. Chapters 3 and 4 are widely acknowledged as the traditional domain of adaptive filtering. Chapters 5 and 6 provide rudimentary descriptions of emerging theory and proposed applications of more recently "discovered" adaptive filter algorithms. Adaptive IIR filters, first proposed in the open literature in the mid 1970s, are examined in Chapter 5 from the two conceptually distinct approaches of gradient descent minimization and stability theory. The resulting algorithm forms are interpreted as logical extensions on the basic adaptive FIR filter algorithm. Chapter 5 is the most heavily theoretical of the book. This is due in part to the fact that the pedagogy of adaptive IIR filters is still in its infancy. This more intense theoretical tone is also due in part to the increase in complexity of adaptive IIR filters relative to adaptive FIR filters. Chapter 6 introduces the class of adaptive property-restoral filters, initially formulated in the early 1980s. This property-restoral class is a conceptual enlargement of the basis of the adaptive FIR and IIR filters of Chapters 4 and 5. Property-restoral adaptive filters adjust their coefficients to restore some known property of the desired output sequence without (necessarily) requiring a full sample-by-sample description of the desired output sequence, as do the more traditional adaptive filter algorithms of Chapters 4 and 5. Application of the property-restoral concept has proceeded faster than development of an appropriate comprehensive theory. Thus, Chapter 6 focuses more on the particular engineering successes of this emerging concept rather than its abstract theoretical analysis.

Where Chapters 4 through 6 focus on theoretically derived algorithms valid under certain idealizations, practice dictates certain modifications to help ensure cost-effective, robust performance in the more harsh environment of real applications. Such issues and related implementation concerns are summarized in Chapter 7. Since the study of adaptive filters is driven by the need for (and success of) their use, Chapter 8 returns to the applications cited in Chapter 1, specifically noise cancelling, adaptive differential pulse code modulation, and channel equalization, as illustrations of the use of the adaptive filter algorithms studied in the intervening chapters.

In this book the stress is on fundamentals. Admittedly, those selected are a reflection of the biases of the authors. The possible tunnel vision imposed by such a procedure is mitigated by the citation of references examining more advanced issues.

An attribute of this book that enhances its reference text quality is the inclusion in Appendix A of FORTRAN listings of some basic algorithms. One of the major hurdles for new "students" of adaptive filtering is the conversion of algorithm equations into implementable code. In fact, our teaching experience indicates that the construction of such code is a major learning experience. The listings included are intended to aid this effort. We have chosen only the simpler versions of the algorithms examined due to their pedagogical clarity. The high data rate and minimal computation cost objectives of the majority of adaptive filtering applications support this focus on the simplest of algorithms with proven attractive properties. However, high-level language code is still somewhat removed from the coding (and hardware) necessities of actual applications. Thus, in Appendix B we detail a solution to the tone canceller problem of Section 1.1 using a Texas Instruments TMS320 general-purpose signal-processing chip.

A number of people deserve our thanks for their part in this book. Sid Burrus and Tom Parks, both at Rice University at the time, suggested this project to us as the second entry in a series of books on topics in digital signal processing sponsored by Texas Instruments. (Incidentally, their book *DFT/FFT and Convolution Algorithms: Theory and Implementation* (Wiley-Interscience, 1985) is the first in this series.) Linda Struzinsky of Cornell University typed our entire manuscript several times. Fran Costanzo and Evelyn Liebgold of Applied Signal Technology drafted the figures. George Troullinos of Texas Instruments drew the schematics and wrote the TMS320 code in Appendix B. Dennis Morgan of AT&T Bell Laboratories carefully reviewed our "final" draft and provided a number of suggestions that we subsequently adopted. We are very grateful to each of you.

This book is the culmination of our collaborations in adaptive filter research begun in the office we shared, fondly known then as The Zoo, while we were all struggling towards PhDs. In the intervening years our contacts with and employment by ARGOSystems (Sunnyvale, CA), Applied Signal Technology

(Sunnyvale, CA), and Tellabs Research Laboratory (South Bend, IN) proved indispensable to our understanding of the design aspects of adaptive filters. Also, this book would not be in its present form without the direct financial support of another company, Texas Instruments, that commissioned this manuscript. These interactions are responsible for the increased emphasis this book places on applications relative to more traditional texts.

It is the widespread success of adaptive filter applications that makes this book possible. We acknowledge our indebtedness to the efforts of the community of adaptive filtering theorists and practitioners who are responsible for these successes.

JOHN TREICHLER
RICK JOHNSON
MIKE LARIMORE

Chatham, Massachusetts
October 1986

Contents

Theory and Design
of Adaptive Filters

1

The Need for Adaptive Filtering

● *PRECIS   Many practical signal-processing applications call for an analog or digital filter whose characteristics can be automatically modified to accommodate incomplete knowledge or time variation in the nature of the input signals.*

An adaptive filter is very generally defined as a filter whose characteristics can be modified to achieve some end or objective, and is usually assumed to accomplish this modification (or "adaptation") automatically, without the need for substantial intervention by the user. While not necessarily required, it is also usually assumed that the time scale of the modification is very slow compared to the bandwidth of the signal being filtered. Implicit in this assumption is that the system designer could in fact use a time-invariant, nonadaptive filter if only he knew enough about the input signals to design the filter before its use. This lack of knowledge may spring from true uncertainty about the characteristics of the signal when the filter is turned on, or because the characteristics of the input signal can slowly change during the filter's operation. Lacking this knowledge, the designer then turns to an "adaptive" filter which can "learn" the signal characteristics when first turned on and thereafter can "track" slow changes in these characteristics.

As later chapters will show, adaptive filters can be implemented in a variety of ways, allowing an even wider variety of practical problems to be solved. We find that many aspects of adaptive filter design, and even the development of some of the adaptive algorithms, are governed by the nature of the applications themselves. Because of this, and the desire to impart some perspective about how adaptive filters may be employed, this chapter examines three of the many

1

possible applications. In each case we will examine no details about the adaptive filters themselves—this is done in later chapters—but instead will take three practical signal-processing problems and formulate a solution for each that employs some form of "smart" filter. How to impart this "intelligence" to such a filter will be discussed in Chapters 2 through 6. A variety of implementation issues are discussed in Chapter 7, and the three specific problems examined in this chapter form the basis for the design examples in Chapter 8. Software examples in FORTRAN for many adaptive algorithms for digital filters are listed in Appendix A. An example of assembly language for the Texas Instruments TMS32020 Signal Processor is shown in Appendix B as part of a design example.

1.1 REMOVAL OF POWER LINE HUM FROM MEDICAL INSTRUMENTS

We first consider the problem shown in Figure 1.1. A remote sensor is connected to an amplifier via a length of cable. The amplifier output contains not only the sensor signal but also a 50- or 60-Hz signal component due to stray pickup from power mains. Such hum may prove to be merely bothersome in a high-fidelity sound reproduction system, but it can be overwhelming when attempting to measure very small voltages, such as are found when trying to obtain a patient's

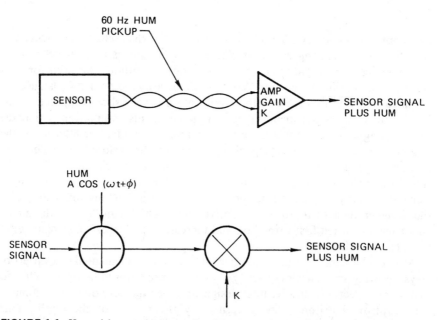

FIGURE 1.1. Hum pickup at the input of an instrumentation amplifier: physical problem and simplified model.

electrocardiogram (EKG). If the spectrum of the sensor signal does not include 60 Hz, then the hum component can be removed via bandpass filtering, but many applications, including EKG measurement, cannot tolerate exclusion of signal energy below 100 Hz. A very narrow notch filter might be appropriate, as long as the frequency of the electric power is steady enough to remain within the notch of the filter. Figure 1.2*a* illustrates this method.

Another approach to solving this problem is shown in Figure 1.2*b*. Suppose we measure the power main itself and attempt to use that signal to subtract off the 60-Hz component entering the amplifier. Clearly if that can be done, then the amplifier input and output will contain only the sensor signal. Moreover, if the power line frequency changes, then both the hum and the "reference" measurement will change together, allowing good cancellation to continue.

The problem with this cancellation approach is that the amplitude and phase of the 60-Hz reference signal must be carefully adjusted to make it accurately cancel the hum at the input of the amplifier. In some applications the required amplitude and phase settings might be determined once and then remain fixed. In most cases, however, the amount of stray 60-Hz pickup will change with the exact lead placement, the amount of power being drawn in the room, and any number of other factors which either change with time or over which

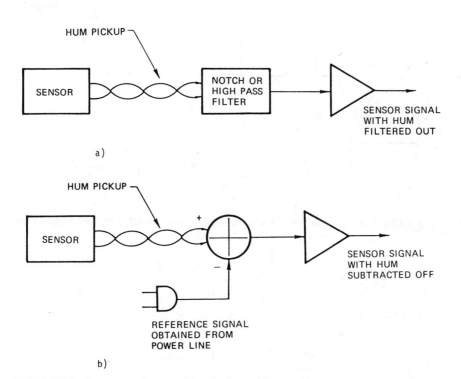

FIGURE 1.2. Two approaches to solving the hum pickup problem, (a) hum removal by inline filtering (b) hum removal by subtraction of a reference signal.

the user has no control. As a practical matter, then, the gain and phase rotation applied to the reference 60-Hz waveform must be variable, and some automatic technique should be available to adjust them in real time to assure good canceliation. Figure 1.3a shows a simplified version of how such a canceller might be designed. The modified version of Figure 1.3a shown in Figure 1.3b is one step closer to practicality since it uses two adjustable coefficients w_0 and w_1 to control the gain and phase of the filter [Widrow et al., 1975b]. The variable phase shifter shown in Figure 1.3a is difficult to implement at low frequencies like 60 Hz.

What we have done here is to develop an adaptive filter to solve the hum problem. The filter accepts a 60-Hz reference signal (from the wall plug) and changes its characteristics (gain and phase) to produce a signal which cancels the

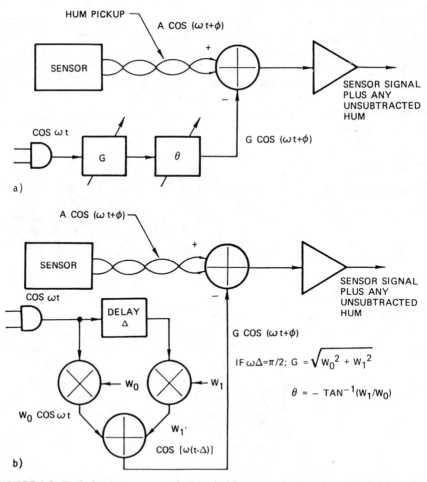

FIGURE 1.3. Evolutionary steps toward a practical hum canceller (a) using adjustable gain and phase to remove hum (b) using a two-coefficient filter to remove hum.

hum to a great extent. How to build adaptation into this filter will be left to later chapters but this example shows that an automatically adjustable or adaptive filter can be the basis of an elegant solution to the hum removal problem.

1.2 ADAPTIVE DIFFERENTIAL PULSE CODE MODULATION

Digital transmission of analog signals has been rapidly evolving over the past 40 years, principally due to the error immunity which digital transmission offers and the rapid decrease in the cost of digital hardware compared to analog systems. The dominant technique for transforming an analog signal to a form suitable for digital transmission is pulse code modulation (PCM) [Owen, 1982]. The analog waveform (e.g., voice or television) is lowpass-filtered and then digitized with a B-bit analog-to-digital (A/D) converter. The samples from the A/D converter are then sent via the transmission system to the receiver. This transmission is usually done in bit-serial format, resulting in a required bit rate of $B \cdot f_s$ bits/sec to convey the analog signal.

The sampling rate is determined by the signal bandwidth. For example, voice signals are typically lowpass-filtered to confine all energy to less than 3300 Hz, and the filter output is sampled at an 8-kHz rate. The number of bits used to represent each sample, B, is typically chosen by compromising among several considerations:

(a) Lower B implies a lower bit rate for the transmission of each signal, thus ultimately lowering the cost of transmission;

(b) B must be large enough to represent the analog signal over all or most of its voltage range; and

(c) B must be large enough to reduce "quantization noise" to a tolerable level.

In most practical systems, considerations (b) and (c) are used to determine the minimum tolerable value of B. For example, in digital telephony the combination of dynamic range requirements and signal-to-noise requirements force B to be on the order of 12 or 13 if linear A/D conversion is done. Given an 8-kHz sampling rate and a digital sample of 12 bits, a 96-kbps transmission link must be employed for a single voice signal with "toll quality." For many reasons, most of them ultimately economic, it is desirable to reduce this data rate. Lowering the bit rate usually leads to a decrease in the price of the circuit.

As a first step, virtually all commercial digital telephone systems use a nonlinear compression of the samples to reduce B from 12 to 8 bits. This compression tends to reduce the signal-to-noise ratio (SNR) of strong signals, which have more than adequate quality, while preserving the SNR of weak signals. This compromise yields acceptable quality over a broad range of signal amplitudes, and, as a result, 64 kbps is the standard for voice telephone transmission.

Substantial research has gone into reducing the value of B still more. One of the many techniques [Gibson, 1980] is shown in Figure 1.4. Assume that the input is a voice signal. Suppose further that we can somehow predict the next sample of the signal and that we subtract this predicted value from the signal itself. If our prediction is relatively accurate, then the difference signal should be substantially smaller in magnitude than the signal itself. Since the difference is smaller, fewer bits, say $\bar{B}$, are needed to represent it. As the quality of prediction worsens, the required value of $\bar{B}$ grows.

Figure 1.4b shows how a communication system can be developed based on the idea of signal prediction. The signal is predicted at the transmitter, the estimate is subtracted from the signal itself, the difference is quantized to $\bar{B}$ bits, and those bits are transmitted. At the receiver the input signal is again predicted, but here it is added to the received signal. The sum of the two is easily seen to be the original input signal. We conclude then that good prediction of the input signal could theoretically produce a reduction in the bit rate required to quantize and transmit the "smaller" difference signal rather than the original signal itself.

In general, the techniques used to predict a signal will depend a great deal on a priori knowledge about how the signal was generated. We will continue to use

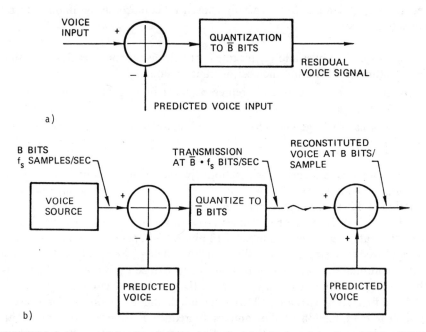

FIGURE 1.4. The use of signal prediction to lower the data rate of a pulse code modulation (PCM) voice signal (a) redundancy reduction given predictable component of voice input (b) residual signal transmission and source signal reconstruction.

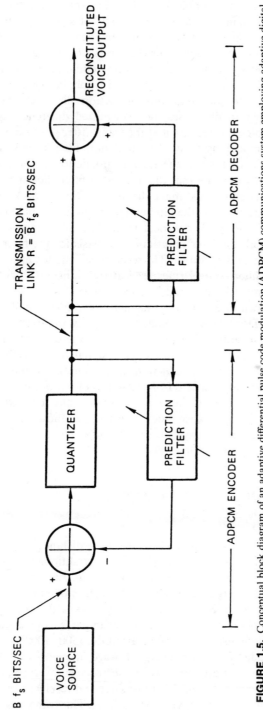
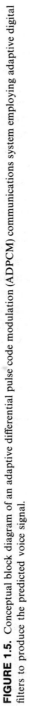

FIGURE 1.5. Conceptual block diagram of an adaptive differential pulse code modulation (ADPCM) communications system employing adaptive digital filters to produce the predicted voice signal.

7

voice transmission as our example here. Considerable work has gone into characterizing speech signals, and several models of speech generation have been developed. Many use the fact that the human vocal tract can be reasonably modeled as a filter whose poles and zeros change as the speaker moves his jaws and tongue. This implies that if suitable excitation could be found, a filter with properly adjusted poles and zeros could be used to predict speech.

Figure 1.5 shows one way that this prediction filter scheme could be employed. This prediction filter uses the transmitted difference signal as its input. The difference signal is also used at the transmitter to sense how well the current vocal tract model used by the prediction filter is working. If the difference signal is small on the average, then the prediction is working well, and large differences usually indicate that the prediction filter should be improved. Suitable rules or algorithms would make the modifications. This scheme would then constantly work to improve the filter and in the process would continuously track the state of the speaker's vocal tract. The advantage of using the difference signal as both the filter input and as a basis for the filter parameter adjustment can be seen at the receiver. The receiver can employ exactly the same prediction filter and filter control rules used at the transmitter with no other signal than the transmitted/received difference signal.

The technique shown in Figure 1.5, plus many variants, is called "adaptive differential PCM" (ADPCM), differential because only the difference is transmitted and adaptive since the filter (and sometimes the quantizer step size) is adapted based on the characteristics of the input signal. Such ADPCM systems can usually preduce a significant reduction in the number of bits required to transmit a signal, and reduction of B from 12 to a $\bar{B}$ of 4 is not unusual in voice telephony. The use of ADPCM in voice telephony has allowed the recent standardization of a 32-kbps transmission rate scheme, allowing a doubling of circuits on existing transmission systems. Refer to section 6.5.3 of [Jayant and Noll, 1984] for a description of this standard.

1.3 EQUALIZATION OF TROPOSCATTER COMMUNICATION SIGNALS

Many techniques exist for communicating over long distances. High-frequency (HF) radio and line-of-sight microwave systems are traditional approaches, and lately the use of communications satellites has spread. Each technique has its own merits and disadvantages, and each one has developed a niche where its mixture of economic and technical advantages makes it superior to other approaches. The technique we examine in this example is troposcatter communications. This technique uses a powerful transmitter (500–1000 W) and a large antenna to beam radio frequency energy toward the horizon. Some small portion of this energy is scattered toward the receiver by turbulence in the troposphere. The distance between the transmitter and receiver can be as much as 500 km, with no intervening repeaters or satellites. This transmission distance makes troposcatter communications systems very cost-effective for communicating across large bodies of water or otherwise inhospitable routes (e.g., the Arctic).

The economic value of a communication path usually depends on its capacity and on its reliability. In the case of troposcatter communications the scattering path limits both. A typical troposcatter link uses two transmitters and two receivers in some form of spatial and/or frequency diversity combining scheme to ensure adequate reliability. Signal capacity is also typically limited to 120 voice channels. We now explore how these limitations come about.

The effect the troposphere has on a radio signal can be modeled in a simplified fashion as shown in Figure 1.6. As a portion of the transmitted beam hits a particular region of turbulence it is refracted because of differences in the atmosphere's refractive index. The amount of energy refracted in any given direction depends on the exact behavior of the turbulent region and cannot be reliably predicted. However, the transmitted beam illuminates many such regions, and many of those are visible at the receiver. The resulting received signal is the sum of the individual components and, depending on the exact phase relationship of these components, they can add either constructively or destructively. Consequently, the instantaneous received power level is effectively a random variable. Even so, its average value can be reasonably predicted given the transmitter and receiver locations.

Using Figure 1.6, we can develop an expression for the received signal $r(t)$,

$$r(t) = \sum_{i=1}^{I} s(t - \Delta_i)g_i(t) \tag{1.3.1}$$

where $s(t)$ is the transmitted signal, $g_i(t)$ is the gain of the propagation path through refractor i, and Δ_i is the added delay caused by taking the ith path compared to the shortest path. The received signal is the sum of I delayed and

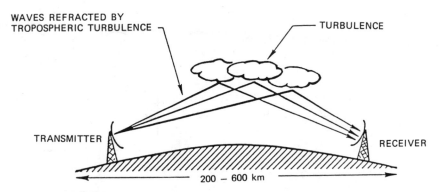

WAVES REFRACTED BY
TROPOSPHERIC TURBULENCE

TURBULENCE

TRANSMITTER

RECEIVER

200 — 600 km

FIGURE 1.6. Cross-section view of an over-the-horizon communication system employing troposcatter propagation.

scaled versions of the transmitted signal. We can rewrite $r(t)$ in (1.3.1) as follows:

$$r(t) = h(t, \tau) * s(t), \qquad (1.3.2)$$

where the asterisk denotes convolution and

$$h(t, \tau) = \sum_{i=1}^{I} g_i(t)\delta(t - \tau - \Delta_i), \qquad (1.3.3)$$

with $\delta(t)$ the Dirac impulse function, that is, $\delta(t)$ is infinite for $t = 0$ and 0 for $t \neq 0$. The description of $r(t)$ in (1.3.2) implies that the received signal can be written as a convolution of the transmitted signal with a time-varying impulse response $h(t, \tau)$ which represents the effect of the propagation channel. [Bello, 1963] and others have shown that after reception, down-conversion, and digitization the received discrete-time signal can be represented by the discrete-time version of (1.3.2)

$$r(k) = h(k, \tau) * s(k), \qquad (1.3.4)$$

where $s(k)$ is a sampled down-converted version of the transmitted signal and $h(k, \tau)$ is the impulse response of a discrete-time, time-varying finite-impulse-response filter. Such a filter is shown in Figure 1.7. The signal passes through a delay line with M taps, each separated by T seconds. Each of the M delayed versions of $s(k)$ is multiplied by a time-varying gain and all are summed to produce $r(k)$.

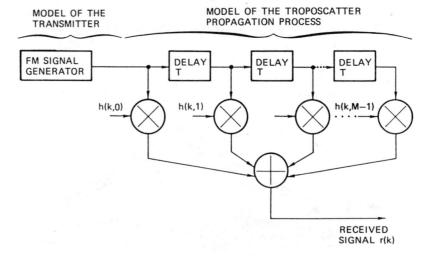

FIGURE 1.7. A practical model of the process which generates the signal received in a troposcatter communication system.

The filtering which $h(k, \tau)$ imposes on the transmitted signal can be roughly characterized by two parameters. The first, the "delay spread" of the propagation channel, is given by M, the number of nonzero filter taps required to represent the channel. The received signal is "spread" by MT seconds by passing through the troposphere. This delay spread is about $0.5\,\mu\text{sec}$ for a typical troposcatter channel. The second parameter is called the "Doppler spread" and describes the output spectrum when a pure sinusoid (with zero bandwidth) is the input. A time-invariant filter would produce a zero bandwidth sinusoid as its output. Any spreading of the tone is caused by time variation of the filter. Thus, the Doppler spread indicates the degree to which the tap gains vary with time. A typical troposcatter channel has a Doppler spread of about 20 Hz.

The impact of this time-varying filter on a transmitted signal depends on the nature of the signal itself. The effect is most easily observed for data signals. Such signals are typically composed of a train of symbols, each one of which carries one or several bits of information in its phase and (sometimes) amplitude. The first-order problem caused by the scatter channel is that each data symbol is spread into the next by an amount equal to the delay spread. If the spread is small, e.g., 10 percent of a symbol's interval, then little deleterious effect is noticed at the receiver. However, when more than 50 percent of the symbol is spread into the next, then serious degradation occurs. Thus, symbol rates for data signals can not exceed about 1 megabaud if good quality is to be retained (given a spread of approximately $0.5\,\mu\text{sec}$). Analog waveforms are sent over tropo channels using frequency modulation (FM). Though not obvious, it turns out that the delay spread also adversely affect FM signals, particularly when the delay spread equals or exceeds the signal's inverse bandwidth. Practical FM tropo systems are limited to maximum bandwidths of about 1 MHz.

The power received through the troposcatter channel is usually enough to sustain low-error transmission at much higher data rates and bandwidths than the channel's delay spread will allow. Thus, we wish to develop a signal-processing technique which would allow some of that unused potential to be tapped by finding some way to reduce the delay spread seen by the receiver. Suppose we employ the scheme shown in Figure 1.8. The received signal $r(k)$, already filtered by the troposcatter channel, is passed through yet another filter before going to the demodulator. This filter has an impulse response $c(k, \tau)$

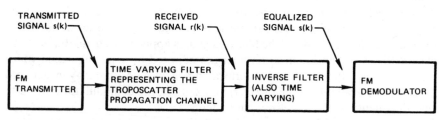

FIGURE 1.8. Employment of an "inverse filter" to equalize the effects of the troposcatter channel on the signal.

which is chosen so that

$$h(k, \tau) * c(k, \tau) = \delta(\tau - D). \tag{1.3.5}$$

With (1.3.5) satisfied, the correction filter $c(k, \tau)$ cancels the effects of the propagation channel to the extent that the two in tandem act like only a flat delay of D sec. The correction filter must be time-variable, hence dependent on k, just as the channel $h(k, \tau)$. Obviously, if we find $c(k, \tau)$ and use those coefficients in the correction filter seen in Figure 1.8, then the demodulator sees the transmitted signal, delayed but not spread, and can thus attain signal quality dependent only on other factors, such as received power level.

The correction filter is usually called an "equalizer" since its function is to equalize the effects of the propagation channel. Suppose for a moment that the channel were time-invariant. It would then be possible to represent the channel by a transfer function with poles and zeros, i.e.,

$$H(z) = \frac{B(z)}{A(z)}. \tag{1.3.6}$$

The equalizer transfer function, now also time-invariant, must satisfy the equation

$$H(z)C(z)z^{-D} = 1 \tag{1.3.7}$$

implying that

$$C(z) = \frac{z^{-D}}{H(z)} = \frac{A(z)z^{-D}}{B(z)}. \tag{1.3.8}$$

The equalizer would then attempt to cancel the spectral effects of the channel by placing poles on the channel's zeros and zeros on the channel's poles. Several practical considerations preclude direct use of this formula for $C(z)$. For example, if $B(z)$ has roots outside the unit circle, then the equalizer is an unstable filter. But still this time-invariant singularity cancellation approach embodied in (1.3.8) serves to illustrate the principal objective of the equalizer, which is to compensate for the channel's effect on the signal.

The block diagram of an equalizer for troposcattered signals is shown in Figure 1.9. The received radio signal is processed to yield a form, analog or digital, suitable for filtering. The filtered signal is applied to a demodulator. The characteristics of the equalization filter are adjusted by some algorithm which attempts to gain the best possible performance. This algorithm may in fact need other signals available in the receiver, for example, the demodulator output, so that the quality of the equalizer can be judged.

If, as assumed earlier, the channel is time-invariant, then the equalizer could be adjusted once and then left fixed. The channel impulse response does vary,

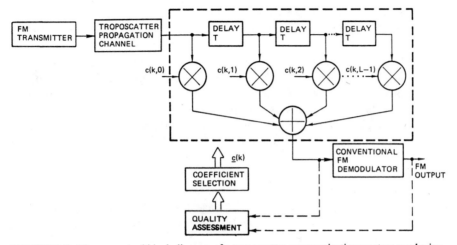

FIGURE 1.9. The conceptual block diagram of a troposcatter communications system employing a tapped-delay-line adaptive filter to equalize the effects of the propagation channel.

however, and the equalizer must be adjusted rapidly enough and accurately enough to maintain good "cancellation" as the channel changes. The Doppler spread, about 20 Hz for tropo channels, indicates the amount of time variation to be expected and the rate at which the adaptive equalizer must track. The delay spread of the channel also impacts the equalizer's design. The delay spread is basically a measure of the time duration of the channel's impulse response. To equalize this, the impulse response of the equalizer must be at least as long and often many times longer. The issue will be illustrated in Chapter 8.

1.4 GENERALITY AND COMMONALITY

Rather than superficially examining a more extensive list of the possible applications benefitting from adaptive filter use, we have looked at three such problems in some detail. In some ways they seem very different. The bandwidths used in removing hum from EKG waveforms are only a few hundred hertz, while a troposcatter communications signal can have a bandwidth of greater than a megahertz. The best filter configuration for the ADPCM problem might be a digital infinite impulse response (IIR) form, while the troposcatter equalizer might be best realized with an analog tapped-delay-line filter, the parameters of which are only periodically adjusted. Even so, all three examples have a few common attributes:

(1) In each case the problem at hand could be solved by a circuit employing just the right filter. For example, in hum removal a reference 60-Hz waveform is provided, but it must be scaled and phase-shifted properly to allow effective subtraction of the hum from the amplifier input.

(2) In all cases the proper choice of filter characteristics (e.g., in hum removal the desired amplitude and phase at 60 Hz) is unknown when the system is initialized.

(3) In all cases the proper choice of coefficients changes with time, usually due to changes occurring in some physical process. In the ADPCM problem, for example, the proper filter for predicting a vocal utterance will change as rapidly as the lips and jaws can move. In the troposcatter equalization problem of communicating through the turbulent troposphere, the Doppler spread is indicative of the rate of desired filter variation.

(4) In most cases, the system into which the adaptive filter was imbedded had one or more points at which the general performance of the adaptive filter could be judged. In the ADPCM case, for example, perfect prediction of the speech would be immediately observed by the continued transmission of zero prediction error.

As we shall see, these common features happen to also be shared by the myriad other applications of adaptive filters. Even so it leaves open a wide variety of important issues, such as:

(a) How can some measure of signal quality be turned into beneficial adjustment of the coefficients of an adaptive filter?

(b) What sort of filter should be used? For example, analog or digital? If digital, which structure? For example, tapped-delay-line, lattice, or direct-form infinite-impulse-response (IIR)?

(c) How can the performance of a particular adaptive filter be analyzed or predicted and thus enhanced by designer choices in particular applications?

(d) How can adaptive filtering concepts be applied to new problems?

Each of these questions will be addressed in later chapters. In the next chapter we begin by looking at the first one: How can one develop rules (or algorithms) for adjusting the coefficients of a filter in response to feedback about how well the filter is performing?

2

Basic Principles
of Adaptive Filtering

● *PRECIS* *The correction term common to adaptive filter parameter update algorithms has an intuitively reasonable form: (a bounded step size) times (a function of the signal multiplying the corresponding parameter in the adaptive filter output computation) times (a function of the measured quality of the adaptive filter output).*

The practical examples in Chapter 1 all suggest the need for an adaptive filter that can alter its frequency response as required by the particular application. The basic structure of such an adaptive filter is shown in Figure 2.1. The input signal is filtered to produce an output which is typically passed on for subsequent processing. The output of this filter is also observed by a circuit which assesses the quality of the output in light of its particular application. This measure of quality, or some function of it, is then passed on to a circuit which uses it to decide how to modify the parameters of the filter to improve the quality of the output. In principle, this processing loop continues to operate until the parameters of the filter are adjusted so that the quality of the filter output is as good as possible. Also, in principle, if the characteristics of the input signal or quality assessment change with time, then this assessment/adjustment loop should readjust the filter's parameters until the new "optimum" output quality is attained.

The functional blocks in Figure 2.1 are quite general and can be chosen in different ways to solve different practical problems. The filter, for example, could be implemented in analog or digital form and could have a tapped-delay-line, pole-zero, or lattice structure. The parameters available for adjustment might be the impulse-response sequence values or more complicated functions of the

15

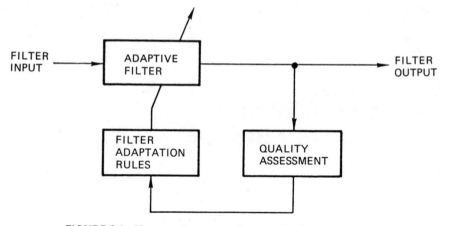

FIGURE 2.1. The general structure of an adaptive filter.

filter's frequency response. Similarly, the circuit which assesses the quality of the filter output can take several forms depending on the adaptive filter's application. The way in which the quality assessment is converted into parameter adjustments, which we will term the adaptive algorithm, can also vary. Variations in the filter structure, the quality assessment mechanism, and the

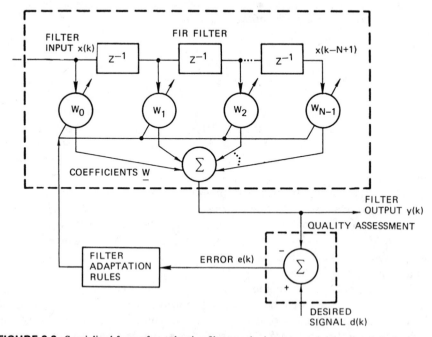

FIGURE 2.2. Specialized form of an adaptive filter employing a tapped-delay-line finite impulse response (FIR) filter and a reference-matching quality assessment.

adaptive update are commonly used to catalog the behavior characteristics, and thus applicability, of various adaptive filters. In fact, these distinctions can be used to characterize the coverage of the following chapters of this book.

The purpose of this chapter is to focus on the adaptive algorithm portion of the adaptive filter, given a particular generic combination of filter structure and quality assessment. We will develop logical, simple algorithms that display significant features shared by the range of adaptive algorithms examined in this book. Thus, while the generality of Figure 2.1 is desirable and will be exploited in later chapters, we focus now on the adaptive filter shown in Figure 2.2.

2.1 THE FIR LINEAR COMBINER

The filter in Figure 2.2 has a discrete-time, finite-impulse-response (FIR) structure, based on a tapped delay line and a set of N adjustable gain coefficients w_0 through w_{N-1}. The filter output $y(k)$ is simply the sum of delayed and scaled inputs, given by

$$y(k) = \sum_{i=0}^{N-1} w_i x(k - i). \tag{2.1.1}$$

For convenience, $y(k)$ can also be written as the dot (or inner) product of two vectors, i.e.,

$$y(k) = \mathbf{X}^t(k)\mathbf{W} \tag{2.1.2}$$

where

$$\mathbf{W} = [w_0 \quad w_1 \quad w_2 \cdots w_{N-1}]^t \tag{2.1.3}$$

and

$$\mathbf{X}(k) = [x(k) \quad x(k - 1) \cdots x(k - N + 1)]^t. \tag{2.1.4}$$

This digital filter structure is one of the simplest. It is stable (for any bounded $\mathbf{W}$) and the impulse response is given directly by the elements of the vector $\mathbf{W}$.

The adaptive filter shown in Figure 2.2 also uses a simple form of quality assessment. The filter output $y(k)$ is compared directly to a desired waveform $d(k)$, and any difference between the two constitutes an error and hence a degradation in quality. The waveform $d(k)$ is sometimes called the desired signal, the reference, or the template waveform. The objective in choosing the coefficients of the adaptive filter is to force $y(k)$ to equal $d(k)$ exactly. One might question why $d(k)$, rather than $y(k)$, is not simply passed on to the next stage of processing, since $d(k)$ is the desired filter output. For now we will cite two circumstances where transmission of $d(k)$ rather than $y(k)$ is infeasible. In certain applications $d(k)$ is only available during a "training" phase in conjunction with a preselected x sequence. In such cases adaptation is ceased during processing of the actual x sequence, for which $d(k)$ is unavailable. In other applications $d(k)$ is "measurable" only after we have computed $y(k)$. In other words, our quality assessment arises from "perfect" hindsight.

Any difference between $d(k)$ and $y(k)$ results in a nonzero error signal $e(k)$, which we intend to use in filter parameter adaptation rules. As we have noted, the objective of this adjustment is to find a set of coefficients, a vector $\mathbf{W}^o$, which reduces $e(k)$, or some function of $e(k)$, to its smallest possible value. When this parametrization is reached (and maintained), the filter is said to have "converged" to the optimal set of filter coefficients, represented by the vector $\mathbf{W}^o$. Note that if $d(k) = \mathbf{X}^t(k)\mathbf{W}^o$, then $\mathbf{W} = \mathbf{W}^o$ implies that $e(k) = 0$ for all k, clearly its smallest possible value. Our filter parameter adjustment rules are termed adaptive if, when the previously optimal solution $\mathbf{W}^o$ changes to a new value, $\mathbf{W}$ tracks it. Such adaptive algorithms are necessarily recursive; as new inputs are processed and their quality assessed, the filter is updated, if necessary.

Even though Figure 2.2 represents a significant simplification and specialization of Figure 2.1, it nonetheless is the structure assumed for most of the practical adaptive filtering work which has been done to date, and it is the focus for more than half of this book. For the remainder of this chapter we turn our attention to the block that contains the filter adaptation rules. Given each $e(k)$, we wish to appropriately revise our choice of $\mathbf{W}$. Development of such rules is the key to adaptive filtering.

2.2 NUMBER GUESSING GAMES

To illustrate an intuitive basis for the development of adaptive parameter estimation algorithms applicable to Figure 2.2, we play a succession of number guessing games. Though at first only vaguely related to the task of Figure 2.2, our number guessing games will be progressively elaborated, along with our respective adaptive algorithm solutions, until the problem structure matches that of Figure 2.2. We shall then be in a position to note certain general features of adaptive filter algorithms that will be elaborated on in subsequent chapters.

2.2.1 Single-Integer Guessing Games

We begin with a trivial integer guessing game with an obvious solution. We pick an integer, between -100 and 100, which you are to guess. The only information we give you for improving your guess is whether it is high or low. Your strategy is immediate. If your guess is too high, you decrease it. If it is too low, you increase your guess. A typical "game" might begin with our picking, unknown to you, 38 as the magic number. You respond by guessing -74, which we indicate is too low. You would simply increase your guess, perhaps cautiously at first. After guessing -65, -30, 0, and 15, you jump to 50 and the sign of your error changes. Ultimately you would converge on the correct value of 38.

Let us complicate the rules of the game. Again we pick an integer, which we label n°, between -100 and 100. But this time we construct a more elaborate procedure for providing you with an indication of the error in your guess. After you indicate your latest (kth) guess, labeled $n(k)$, we will pick another integer, $x(k)$, but we will only tell you if $x(k)$ is positive or negative. We will form the product

$$d(k) = n^\circ x(k). \qquad (2.2.1)$$

We will also compute

$$y(k) = n(k)x(k), \qquad (2.2.2)$$

using your guess $n(k)$, and

$$e(k) = d(k) - y(k) = [n^\circ - n(k)]x(k) \qquad (2.2.3)$$

and then tell you whether e is positive or negative, i.e., whether your estimate of d is too low or too high. If e is positive, this indicates that d is greater than y, i.e.,

$$e(k) = d(k) - y(k) > 0 \rightarrow d(k) > y(k). \qquad (2.2.4)$$

Similarly, if e is negative, then d is less than y. The objective of this game is the same as the preceding one: to ultimately guess the value n°. You will need to choose a strategy that updates $n(k)$ for the next "time" to $n(k + 1)$, using the correction information suggested by the sign of $e(k)$ and the sign of $x(k)$. With a successful strategy your guesses $n(k)$ will converge to n° as k increases. Mathematically, the task is to choose the function f such that

$$n(k + 1) = f(n(k), \text{sgn}(e(k)), \text{sgn}(x(k))) \qquad (2.2.5)$$

causes

$$n(k) \rightarrow n^\circ \quad \text{as} \quad k \rightarrow \infty, \qquad (2.2.6)$$

where

$$\text{sgn}(x) = \begin{cases} 1 & \text{if } x > 0 \\ 0 & \text{if } x = 0 \\ -1 & \text{if } x < 0. \end{cases} \qquad (2.2.7)$$

Now (2.2.5)–(2.2.7) may seem unreasonably abstract for such a simple game. But these equations precisely describe how your kth guess $n(k)$ is "improved" to your next guess $n(k + 1)$ given the signs of $e(k)$ and $x(k)$.

A reasonable strategy is relatively easy to formulate. To begin developing an intuitive strategy that is quite similar to that of the previous integer guessing game, imagine that $\text{sgn}(e(k)) = 1$. In other words, you have been told that $e(k)$ is positive or, from (2.2.4), that $d(k)$ is greater than $y(k)$. Your natural response is to update n so that y is increased. How this is accomplished depends on the sign of $x(k)$. If $x(k)$ is positive, then increasing n increases $y(k)$ in (2.2.2). Conversely, if $x(k)$ is negative, then decreasing n increases $y(k)$ in (2.2.2). So if e is positive, a reasonable strategy is to increase n when x is positive or decrease n when x is negative. Similarly, if e is negative, y is greater than d and n should be corrected to decrease y. So, if e is negative, you should decrease n if x is positive or increase n if x is negative.

To summarize, we want to develop an f for (2.2.5) such that

$$
\begin{aligned}
e > 0 \quad \text{and} \quad x > 0 \quad &\rightarrow \quad \text{increase } n \\[1ex]
e > 0 \quad \text{and} \quad x < 0 \quad &\rightarrow \quad \text{decrease } n \\[1ex]
e < 0 \quad \text{and} \quad x > 0 \quad &\rightarrow \quad \text{decrease } n \\[1ex]
e < 0 \quad \text{and} \quad x < 0 \quad &\rightarrow \quad \text{increase } n.
\end{aligned}
\tag{2.2.8}
$$

Note that, if $e > 0$ and $x > 0$, then $\text{sgn}(e)$ times $\text{sgn}(x)$ is positive. If we want to increase n as indicated in (2.2.8), then we want to add a positive quantity to n. For the first line in (2.2.8) we could use

$$
n(k + 1) = n(k) + \text{sgn}(e(k)) \cdot \text{sgn}(x(k)).
\tag{2.2.9}
$$

Note that (2.2.9) works for all of the other lines of (2.2.8). Essentially, if e and x have the same sign, from (2.2.8) we want to increase n. In (2.2.9) we have chosen to increase n by the addition of $+1$, that is, $\text{sgn}(e(k)) = \text{sgn}(x(k)) \rightarrow \text{sgn}(e(k)) \cdot \text{sgn}(x(k)) = 1$. Conversely, if e and x have opposite signs, then from (2.2.8) we want to decrease n, which (2.2.9) accomplishes by subtracting 1. Note that if $x = 0$, then $e = 0$ from (2.2.3) despite the difference between n° and n. The strategy in (2.2.9) indicates that we appropriately leave our guess for n unchanged because we have not received any indication via e of the error in n. If $n(k) = n^\circ$, then from (2.2.3) $e(k) = 0$ and (2.2.9) appropriately retains $n(k + 1) = n(k) = n^\circ$. Thus, once the correct value is obtained, it is retained.

Now, let's play a sample game. We secretly pick $n^\circ = -2$. Your initial guess $n(1) = 0$ is conservative. With $x(1) = 2$, $d(1) = -4$, $y(1) = 0$, and $e(1) = -4$. So you are told that $\text{sgn}(x(1)) = 1$ and $\text{sgn}(e(1)) = -1$. Using (2.2.9), you correct $n(1)$ to $n(2) = -1$. Now $x(2)$ is selected as -3 and, thus, $d(2) = 6$, $y(2) = 3$, and $e(2) = 3$. So you are told that $\text{sgn}(x(2)) = -1$ and $\text{sgn}(e(2)) = 1$. Again, using (2.2.9), you decrease n by 1 to $n(3) = -2$. With two iterations the strategy of (2.2.9) has converged to the correct answer.

2.2.2 Multiple-Integer Guessing Game

Once again we will increase the complexity of the game. You will now be simultaneously guessing two integers. We will pick two integers n_1^o and n_2^o and form

$$d(k) = n_1^o x_1(k) + n_2^o x_2(k) \tag{2.2.10}$$

given two "inputs" $x_1(k)$ and $x_2(k)$ to this linear combination of terms such as on the right of (2.1.1). Similarly, the prediction of d based on your guesses n_1 and n_2 will be formed via

$$y(k) = n_1(k)x_1(k) + n_2(k)x_2(k) \tag{2.2.11}$$

in order to compute the error

$$e(k) = d(k) - y(k). \tag{2.2.12}$$

You will be told only the sign of e, the sign of x_1, and the sign of x_2. You should update your estimates n_1 and n_2 so that eventually $n_1 \to n_1^o$ and $n_2 \to n_2^o$. Note that the basic parameter estimate correction strategy of the preceding "game" still makes sense: Change n_1 and n_2 to make y closer to d. Thus, the strategy in (2.2.9) could be applied term by term since (2.2.10) simply sums these effects. Consider the following case. Assume $e > 0$, so that you wish to increase y. If x_1 is positive, increasing n_1 will increase y. Similarly, if x_2 is positive, increasing n_2 will also increase y. Repeating all of the cases examined with the preceding game will yield the following update candidate for each n_i

$$n_i(k+1) = n_i(k) + \mathrm{sgn}(e(k)) \cdot \mathrm{sgn}(x_i(k)), \qquad i = 1, 2. \tag{2.2.13}$$

To gain some confidence that (2.2.13) is a reasonable strategy we will consider a sample game as summarized in Table 2.1. The path taken by the sequence of number guesses (or parameter estimates) in the space formed by Cartesian coordinates n_1 and n_2 is shown in Figure 2.3. As we shall discover, a useful translation (actually a rotation and a translation) of Figure 2.4 considers the path of successive estimates drawn instead in the parameter error space with

TABLE 2.1. Two-Integer Guessing Game Example ($n_1^o = 3$, $n_2^o = -1$)

k	$n_1(k)$	$n_2(k)$	$x_1(k)$	$x_2(k)$	$e(k)$
1	0	0	1	−1	4
2	1	−1	2	−1	4
3	2	−2	−1	−2	−3
4	3	−1	—	—	—

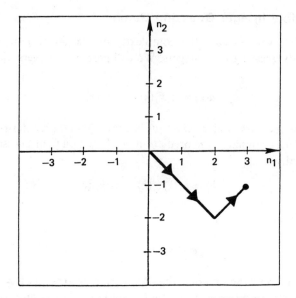

FIGURE 2.3. Estimate path in parameter space (of Table 2.1 example).

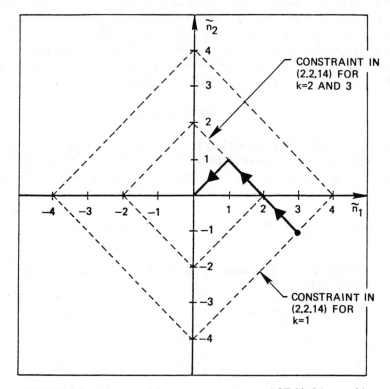

CONSTRAINT IN (2.2.14) FOR k=2 AND 3

CONSTRAINT IN (2.2.14) FOR k=1

FIGURE 2.4. Estimate path in parameter error space (of Table 2.1 example).

coordinates $\tilde{n}_1 = n_1^o - n_1$ and $\tilde{n}_2 = n_2^o - n_2$. Thus, the origin in parameter estimate error space corresponds to $\tilde{n}_1 = \tilde{n}_2 = 0$ or $n_1 = n_1^o$ and $n_2 = n_2^o$. In other words, once the path in parameter estimate error space reaches the origin the number guessing game has been successfully concluded. Therefore, if we could verify the property that the strategy in (2.2.13) never moves the estimate error point "away" from the origin, e.g.,

$$|\tilde{n}_1(k+1)| + |\tilde{n}_2(k+1)| \leqslant |\tilde{n}_1(k)| + |\tilde{n}_2(k)|, \qquad (2.2.14)$$

we would be encouraged to recommend (2.2.13) for playing the game of (2.2.10)–(2.2.12). The inequality in (2.2.14) describes a sense in which the parameter estimator of (2.2.13) never degrades the collective accuracy of the estimates. Note that the example in Table 2.1 satisfies (2.2.14) with $|\tilde{n}_1(k)| + |\tilde{n}_2(k)|$ equal to 4, 2, 2, and 0, respectively, for $k = 1, 2, 3, 4$. In fact, our observation of numerous sample "games" is likely to have been the source of our suggestion of the general validity of (2.2.14). Fortunately, we can *prove* rather directly that (2.2.14) is true for all possible "games."

To help us validate (2.2.14), note that from any point $(n_1(k), n_2(k))$ in the two-integer guessing game parameter space, only nine values are possible for $(n_1(k+1), n_2(k+1))$. These include all the permutations of $+1$, -1, or zero increment in each n_1 and n_2. Note that (2.2.14) only claims that the sum of the magnitudes of the errors in the parameters never increases. This is equivalent to requiring in the two-dimensional error space that the values of $\tilde{n}_1(k+1)$ and $\tilde{n}_2(k+1)$ must remain within a diamond determined by the error in the "initial" guess $(\tilde{n}_1(k), \tilde{n}_2(k))$. For the example in Table 2.1, this diamond is drawn for each k with a dashed line in Figure 2.4. Recognize that of the nine possible moves from any $(\tilde{n}_1(k), \tilde{n}_2(k))$, where both are nonzero, only three are forbidden in order to satisfy (2.2.14). The particular three moves that are forbidden depend on the quadrant in which $(\tilde{n}_1(k), \tilde{n}_2(k))$ is located. When one of the two parameter errors is zero, only four of the nine moves are possible in order for (2.2.14) to be satisfied.

With $\tilde{n}_i = n_i^o - n_i$ and (2.2.13),

$$\Delta\tilde{n}_i(k) \equiv \tilde{n}_i(k+1) - \tilde{n}_i(k) = n_i(k) - n_i(k+1)$$

$$= -\text{sgn}(e(k)) \cdot \text{sgn}(x_i(k)). \qquad (2.2.15)$$

We refer now to Figure 2.5. For example, if $(\tilde{n}_1, \tilde{n}_2)$ exists in the first quadrant, i.e., both $\tilde{n}_1$ and $\tilde{n}_2$ are positive, then moves 1 through 3 should be impossible if (2.2.14) is to be true. Recall from (2.2.10)–(2.2.12) that

$$e(k) = \tilde{n}_1(k)x_1(k) + \tilde{n}_2(k)x_2(k). \qquad (2.2.16)$$

Thus when $\tilde{n}_1(k) > 0$ and $\tilde{n}_2(k) > 0$, for $e(k)$ to be positive at least one of $x_1(k)$ or $x_2(k)$ must be positive. Similarly, for $e(k)$ to be negative at least one of $x_1(k)$ or

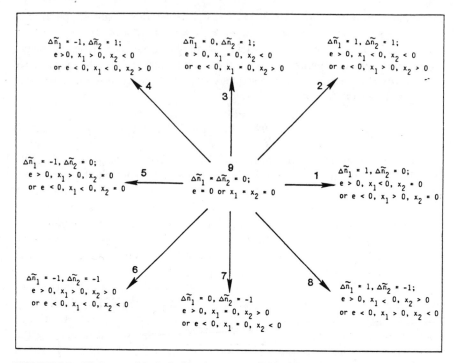

FIGURE 2.5. All nine possible single-iteration moves of (2.2.13) in parameter error space and their cause (in terms of e and x_i).

$x_2(k)$ must be negative. Note that these requirements are excluded from moves 1 through 3. In other words, the three moves forbidden by (2.2.14) for positive $\tilde{n}_1(k)$ and $\tilde{n}_2(k)$ are, in fact, not possible in (2.2.13) or (2.2.15). Such an examination could be continued for the other three quadrants, and points on the axes, with the same result. Due to the restriction of the $\{n_i(k)\}$, $\{n_i^o(k)\}$, and $\{x_i(k)\}$ to integer values, we cannot move from one quadrant to another without first landing on an axis. Thus we have proven (2.2.14) to be valid.

At this stage some observations are pertinent. To begin, note that, since (2.2.14) is not a strict inequality, the sum of the absolute values of the parameter errors ($|\tilde{n}_1| + |\tilde{n}_2|$) need not decrease at every iteration. In fact, the equality implies that the parameter estimator may be stuck at some "distance" and not move any closer to the correct parameter values. Such a failure to further reduce the "distance" to the correct parameters n_i^o is due to the (unsatisfactory) character of the "inputs" x_i. One possibility is that all of the $x_i(k)$ are zero such that $n_i(k + 1) = n_i(k)$ from (2.2.13). In this case $d(k)$ and $y(k)$ in (2.2.10) and (2.2.11), respectively, are also zero and $e(k)$ in (2.2.12) is zero. In other words, the error measure of our parameter estimator is zero despite a nonzero parameter error.

The use of zero $x_i(k)$ for all k is foolish in terms of the stated game. But such a stall could be encountered in a less obvious manner due to a peculiar input sequence. Note, from (2.2.16), that for the game in Table 2.1 the error is

$$e = (3 - n_1)x_1 - (1 + n_2)x_2. \tag{2.2.17}$$

For example, consider a sequence of $x_i(k)$ such that $x_1(k) = -x_2(k) \neq 0$. Thus the error in (2.2.17) is zeroed, i.e.,

$$(3 - n_1) + (1 + n_2) = 0, \tag{2.2.18}$$

by any parameter estimates that satisfy the solution of (2.2.18) as

$$n_1 - n_2 = 4. \tag{2.2.19}$$

With e zeroed, the parameter estimates will not be corrected by (2.2.13) despite their possible inaccuracy. For example, with the inaccurate guesses $n_1 = -n_2 = 2$ for the game in Table 2.1, the sum of the absolute values of the parameter errors is $|3 - 2| + |-1 + 2| = 2$. With $x_1 = -x_2$ it will remain at this value indefinitely while satisfying (2.2.14). To verify this stall due to $e = 0$ despite nonzero $\tilde{n}_i$, form (2.2.17) with n_1 and n_2 satisfying (2.2.19) and any $x_1 = -x_2 \neq 0$. However, if the objective of the "game" is really to zero the prediction error (rather than the parameter estimate error), as is true in the typical adaptive filtering application, this stall is acceptable.

Unfortunately, the freezing of the parameter estimates, despite their inaccuracy, due to the zeroing of the prediction error is not the only way a peculiar x_i sequence can cause (2.2.14) to remain an equality with both sides nonzero. Consider the sequence shown in Table 2.2 and the corresponding parameter error space trajectory shown in Figure 2.6. Note that the sum of the absolute values of the parameter errors is always unity *and* the prediction error is nonzero. Actually, in Table 2.2 any values for x_i with the signs shown would cause the counterclockwise rotation about the origin. Changing the signs on the x_i which are *not* associated with nonzero $\tilde{n}_i$ would alter the direction of the rotation to clockwise. The conclusion is that "peculiar" sequences of the x_i can

TABLE 2.2. Another Two-Integer Guessing Game Example ($n_1^0 = 3$, $n_2^0 = -1$)

k	$n_1(k)$	$n_2(k)$	$x_1(k)$	$x_2(k)$	$e(k)$
1	4	−1	−1	+1	+1
2	3	0	−1	−1	+1
3	2	−1	+1	−1	+1
4	3	−2	+1	+1	+1
5	4	−1	—	—	—

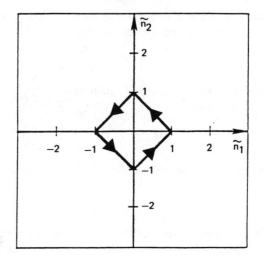

FIGURE 2.6. Nonzero prediction error limit cycle (of Table 2.2 example).

result in nonzero, nondecreasing summed parameter error magnitudes, i.e., equality in (2.2.14), even with nonzero prediction error. In order to successfully conclude the number guessing game, such x_i sequences must be avoided.

One might be tempted to conclude that, since the summed absolute parameter error $\sum_{i=1}^{N} |\tilde{n}_i|$ of the integer guessing game is never-increasing for $N = 2$, the same would be true for $N > 2$. Such an assertion would be false. Consider the three-integer guessing game where $n_1^o = n_2^o = n_3^o = 1$, $x_1(1) = 3$, $x_2(1) = x_3(1) = -1$, and $x_1(k + 1) = x_3(k)$, $x_2(k + 1) = x_1(k)$, $x_3(k + 1) = x_2(k)$ for $k > 0$. With the initial guess of $n_1(0) = n_2(0) = n_3(0) = 0$, the progression of $(n_1(k), n_2(k), n_3(k))$ is $(1, -1, -1)$, $(0, 0, -2)$, $(-1, -1, -1)$ for $k = 1, 2, 3$, respectively. Thus, $\sum_{i=1}^{3} |\tilde{n}_i(3)| = 6$, while $\sum_{i=1}^{3} |\tilde{n}_i(0)| = 3$. Continuing shows that every third iteration $n_i(k) = -k/3$ and $\sum_{i=1}^{3} |\tilde{n}_i(k)| = 3 + k$ for $k = 3, 6, 9, \ldots$. Thus, the summed absolute parameter error is diverging as the parameter estimates diverge. However, for the same situation, but with different initial x_i, that is, $x_1(1) = 1$ and $x_2(1) = x_3(1) = -1$, after three iterations $n_i(3) = n_i^o$ and $\sum_{i=1}^{3} |\tilde{n}_i(k)| = 0$ for $k \geq 3$. Thus, with a different $\{x_i\}$ sequence, convergence rather than divergence results. This again emphasizes the importance of avoiding certain $\{x_i\}$ sequences to attain acceptable behavior.

This (false) conjecture of the nondivergence of multiple (three or more) integer guessing via (2.2.13) due to its nondivergence in guessing two integers also serves as a warning against overextending algorithm behavior predictions from a set of examples. Unfortunately, this tendency is all too common when dealing with complicated time-varying (and often nonlinear) adaptive algorithms that defy analysis. Consider yourself warned.

2.2.3 Multiple-Real-Numbers Guessing Game

Again we alter the rules of the number guessing game. We no longer limit n_i^o or x_i to be integers. Instead, they can be any real numbers. Consider using the adaptive algorithm in (2.2.13). Since the correction term, $\text{sgn}(e(k)) \cdot \text{sgn}(x_i(k))$, can only be $+1$, 0, or -1, successive n_i can only be different from your initial guess by an integer value. In other words we can only hope, at best, to cause each n_i to converge to the integer nearest n_i^o. If we want to allow n_i to move closer to n_i^o, we must use a smaller step size upon each pass through (2.2.13). Consider

$$n_i(k + 1) = n_i(k) + \mu_i[\text{sgn}(e(k)) \cdot \text{sgn}(x_i(k))], \qquad (2.2.20)$$

where all of the μ_i are positive. In (2.2.13), each μ_i is unity. To permit a closer ultimate fit to any real number we should choose a μ_i smaller than unity. For example, if the $\mu_i = 0.01$, we would hope to be able to ultimately improve the n_i such that they are accurate to within ± 0.01. It is easy to see that with a smaller μ_i it will take more iterations through (2.2.20) before we "converge." It is easiest to see this if we return, momentarily, to guessing only one number. If $n^o = 2$ but we guess $n(1) = 0$, then with $\mu = 1$ and $x(k) \neq 0$ for all k we reach $n = 2$ in two iterations, that is, $n(k) = 2$ for $k \geq 3$. If instead $\mu = 0.01$, then 200 iterations are required. Thus, we can state that the convergence rate is reduced as μ is reduced. However, with smaller μ the steady-state error is decreased, in general. Imagine that $n^o = 2.346$. If $\mu = 1$, then the sequence of $n(k)$ for $x(k) \neq 0$ is 0, 1, 2, 3, 2, 3, 2, 3, ... But with $\mu = 0.01$, $n(k)$ follows 0, 0.01, 0.02, ..., 2.34, 2.35, 2.34, ... This trade-off between initially rapid improvement in the parameter estimates and steady-state accuracy will repeatedly manifest itself in adaptive algorithms.

One way to reduce the steady-state coarseness of the "sign–sign" algorithm in (2.2.20), while retaining larger μ_i to speed initial convergence, is to somehow adjust the correction term magnitude so that it becomes smaller as we are closer to the correct answer. We again alter the "rules" of our number guessing game to permit this possibility. We now give you the actual values of $e(k)$ and $x_i(k)$ rather than just their signs. Using the smallness of $|e(k)|$ as an indicator of parameter estimate "closeness" to the desired parameters, we wish to reduce the adjustment of $n_i(k)$ when $|e(k)|$ is small. This suggests modifying (2.2.20) to the form of

$$n_i(k + 1) = n_i(k) + \mu_i[\text{sgn}(x_i(k))]e(k). \qquad (2.2.21)$$

As $e(k)$ decreases toward zero, the magnitude of the adjustment of $n_i(k)$ decreases to zero.

What (2.2.21) does not exploit is the information we have available in the magnitude of each x_i. Exploiting this will allow us to broaden the classes of $\{x_i\}$ sequences for which we can achieve successful (i.e., nondiverging) recursive number guessing strategies. Recall our objective in formulating (2.2.9), and subsequently (2.2.21), of adjusting each n_i so it changes y so as to reduce the magnitude of e. Thus, the n_i associated with the smaller x_i should receive less correction due to their less significant effect on y, and thus e, for the same absolute change in the parameter estimate. This suggests considering

$$n_i(k + 1) = n_i(k) + \mu_i x_i(k)e(k). \qquad (2.2.22)$$

Smaller $e(k)$ reduces the change in all n_i. The n_i receiving the least change, with all μ_i the same, is the one associated with the smallest x_i. We can actually prove that (2.2.22) has the property that the "distance" to the origin in parameter estimate error space is never increased. This property is analogous to that of (2.2.14) for (2.2.13). In the case of (2.2.22) we will find that

$$\sum_{i=1}^{N} \tilde{n}_i^2(k + 1)/\mu_i \leqslant \sum_{i=1}^{N} \tilde{n}_i^2(k)/\mu_i \qquad (2.2.23)$$

instead of (2.2.14) measures this nonincreasing distance for $N \geqslant 1$.

To directly verify (2.2.23) for (2.2.22) we begin by subtracting both sides of (2.2.22) from n_i^o to form

$$\tilde{n}_i(k + 1) = \tilde{n}_i(k) - \mu_i x_i(k)e(k). \qquad (2.2.24)$$

Using (2.2.24) we can rewrite the left side of (2.2.23) as

$$\sum_{i=1}^{N} \tilde{n}_i^2(k + 1)/\mu_i = \sum_{i=1}^{N} \{[\tilde{n}_i(k) - \mu_i x_i(k)e(k)]^2/\mu_i\}$$

$$= \sum_{i=1}^{N} \{\tilde{n}_i^2(k)/\mu_i - 2\tilde{n}_i(k)x_i(k)e(k) + \mu_i x_i^2(k)e^2(k)\}. \qquad (2.2.25)$$

Recall (2.2.16), written as $e(k) = \sum_{i=1}^{N} \tilde{n}_i(k)x_i(k)$, which allows (2.2.25) to be rewritten as

$$\sum_{i=1}^{N} \tilde{n}_i^2(k + 1)/\mu_i = \sum_{i=1}^{N} \{\tilde{n}_i^2(k)/\mu_i\} - e^2(k)\left[2 - \sum_{i=1}^{N} \mu_i x_i^2(k)\right]. \qquad (2.2.26)$$

With all of the μ_i positive, (2.2.23) is satisfied if the last term on the right side of (2.2.26), that is, $-e^2(k)[2 - \sum_{i=1}^{N} \mu_i x_i^2(k)]$, is nonpositive. This is true, due to the nonnegativity of $e^2(k)$, if

$$2 \geqslant \sum_{i=1}^{N} \mu_i x_i^2(k), \qquad \text{for all } k. \qquad (2.2.27)$$

If we knew that the $|x_i(k)|$ were bounded by some value, say β_i, for all k, then we could choose each μ_i to satisfy

$$2/(N\beta_i^2) > \mu_i \qquad (2.2.28)$$

and thereby verify the satisfaction of (2.2.23) by (2.2.22).

Conversely, if the μ_i were selected too large (or the x_i were larger than expected) such that (2.2.27) was dissatisfied for all k, then the inequality opposite to that in (2.2.23) would be established. Thus, the weighted summed squared parameter error would increase (or at least never decrease) with each iteration. The parameter estimates could progressively worsen; they could diverge. (Construct such an example, if you doubt this conclusion.)

To avoid the magnitude constraints on "satisfactory" $\{x_i\}$ for a particular μ_i modify (2.2.22) to

$$n_i(k + 1) = n_i(k) + \left[\mu_i \bigg/ \left(1 + \sum_i x_i^2(k) \right) \right] x_i(k) e(k). \qquad (2.2.29)$$

The algorithm of (2.2.29) adds a step-size normalization absent in (2.2.22). It can be proven that (2.2.29) satisfies (2.2.23) if all $0 < \mu_i < 1$, which is far less restrictive on μ_i selection (and related $\{x_i\}$ constraints) than (2.2.28) for large β_i. The disadvantage with (2.2.29) is the extra computation, including division, relative to (2.2.22).

2.3 ADAPTIVE FILTER ALGORITHM INTERPRETATION

A comparison of (2.2.13), (2.2.22), and (2.2.29) suggests a general form for parameter adaptation

$$\begin{bmatrix} \text{New parameter} \\ \text{estimate} \end{bmatrix} = \begin{bmatrix} \text{Old parameter} \\ \text{estimate} \end{bmatrix}$$
$$+ \begin{bmatrix} \text{Bounded} \\ \text{step size} \end{bmatrix} \cdot \begin{bmatrix} \text{Function} \\ \text{of input} \end{bmatrix} \cdot \begin{bmatrix} \text{Function of} \\ \text{error} \end{bmatrix}. \qquad (2.3.1)$$

In (2.2.13) the step size is bounded by unity to allow capture of all possible integer values for the parameters. The "input" is x_i and the function used is the sign function. The error e is also passed through the sign function. For (2.2.22), (2.2.27) bounds the step-size multipliers μ_i, and the functions of x_i and e are simply identity functions. In (2.2.29) the functions of x_i and e remain identity functions, while the step size multipliers are the time-varying $\mu_i/(1 + \sum_i x_i^2(k))$, which are bounded since $0 < \mu_i < 1$. We will discover that (2.3.1) is an apt description of the structure of all of the adaptive algorithms we encounter in this book. In fact, the principal distinctions between various algorithms come in the form of the step-size specification and the various functions applied to the input and the error for adaptation via (2.3.1). These distinctions provide more than a mere cataloging index. As we have seen in these number guessing games, and will observe with their translation to adaptive filter algorithms, the distinctions in the various elements of (2.3.1) are related to differences in performance and thus applicability.

We can now turn our number guessing game solutions into adaptive filter algorithms. Note that associating the coefficients w_i in (2.1.1) with the n_i in (2.2.11) and the input samples $x(k - i)$ with the $x_i(k)$ equates (2.2.11) and (2.1.1). We could now repeat the preceding heuristic development in this signal-processing setting with the appropriate redefinitions and show that (2.3.1) applies to adaptive algorithms for the filter parameterization of (2.1.1). Instead, in the following chapters we will develop strategies for adaptation of the w_i in a more direct association with signal-processing objectives. This approach will eventually bring us to an algorithm similar to (2.2.22), among others. However, based on our number guessing game observations, we can summarize four characteristics we should look for in the adaptive filter algorithms to be developed:

  (i) The generic form of (2.3.1) should emerge with the step size appropriately bounded to insure convergence.

  (ii) Distinctions due to different realizations of (2.3.1) can be related to differences in performance and applicability.

  (iii) The character of the input sequence $\{x(k)\}$ will determine whether or not (and how) the parameter estimates converge to their desired values.

  (iv) Step-size selection represents a compromise between convergence speed and steady-state coarseness (especially when the problem does not yield $e = 0$ for $\mathbf{W} = \mathbf{W}^o$). Appropriate compromise is crucial to the ability to adequately track time-varying $\mathbf{W}^o$, in other words, to successfully adapt.

3

An Analytical Framework for Developing Adaptive Algorithms

● *PRECIS In many cases the best adaptive filter impulse response can be determined by solving the appropriate multivariable optimization problem. Careful examination of the underlying optimization problem provides both the adaptive algorithms themselves and a formal structure for evaluating their performance.*

3.1 BACKGROUND AND DIRECTION

In Chapter 1 we saw that several practical problems could be solved if one had available a filter which could somehow be adjusted to produce just the right transfer function at the right time. In Chapter 2 we chose a fairly simple type of filter, a tapped-delay-line FIR filter, and heuristically developed techniques for updating the impulse response vector. Given the filter input and a measurement of how good or bad the resulting filter output is, a "guessing game" algorithm can be used in an attempt to improve (or at least not worsen) the estimate of the proper impulse response vector.

In this chapter we take a different, more formal approach. We still use a tapped-delay-line FIR filter and we still measure the error between our desired version of the filter output and the one we actually formed. The difference here is that we develop techniques for finding impulse-response vectors which minimize some average measurement of the error. As we shall see, this requires us to make more mathematical assumptions, but this increase in analytical formalism reveals methods for analyzing the performance of the adaptive algorithms we develop and leads to the development of additional, more complex algorithms with various performance improvements. The analytical structure and methods are introduced in this chapter, while the actual adaptive algorithms are presented in Chapter 4.

The mathematical notation used in this chapter will be basically the same as before. An input waveform $x(k)$ is applied to an FIR filter to produce an output waveform $y(k)$ according to the convolution sum

$$y(k) = \sum_{l=0}^{N-1} x(k-l)w_l \tag{3.1.1}$$

where $\{w_l\}$ is the filter impulse response. Since the filter is assumed to have a finite duration, causal impulse response, w_l equals zero for l less than zero and l greater than $N-1$. To facilitate our development it is useful to define an input data vector $\mathbf{X}(k)$ and the impulse response vector $\mathbf{W}$ according to the following expressions

$$\mathbf{X}(k) = [x(k)\, x(k-1)\, x(k-2) \cdots x(k-N+1)]^t \tag{3.1.2}$$

and

$$\mathbf{W}^t = [w_0\ w_1\ \cdots\ w_{N-1}]. \tag{3.1.3}$$

The data vector contains the contents of the delay line, that is, the current input sample and the $N-1$ most recent past inputs. The impulse response vector, also often called the coefficient or weight vector, contains the nonzero portion of the impulse response. This vector notation allows a certain compactness of expression. Equation (3.1.1), for example, can be written as

$$y(k) = \mathbf{W}^t\mathbf{X}(k). \tag{3.1.4}$$

The waveforms $x(k)$, $y(k)$, and $d(k)$, the template or "desired" waveform, are in general complex-valued. The weight vector $\mathbf{W}$ can also be complex, but for many problems it will be considered to be real-valued.

3.2 THE LEAST SQUARES PROBLEM

3.2.1 Basic Formulation

We begin by assuming that we have L samples of the input sequence, $x(0)$ through $x(L-1)$, where $L > N$. Reviewing the definition of $\mathbf{X}(k)$ in (3.1.2), this means we have the delay line vectors $\mathbf{X}(N-1)$ to $\mathbf{X}(L-1)$. Earlier values of $\mathbf{X}(k)$ such as $\mathbf{X}(0)$ are not defined because we do not know about $x(k)$ before $k=0$. Therefore, $y(k)$ is also defined only for k between $N-1$ and $L-1$.

We also assume that we are provided the reference signal $d(k)$ for k ranging from $k=0$ to $k=L-1$. Our objective is to find the best impulse response vector $\mathbf{W}$, which we call $\mathbf{W}_{ss}^o$. In this section we define "best" to mean the choice of **W** that makes the summed squared difference between $d(k)$ and $y(k)$, the filter output, as small as possible. In mathematical terms we want to find $\mathbf{W}_{ss}^o$ such that

$$J_{ss} = \sum_{k=N-1}^{L-1} |d(k) - y(k)|^2 \tag{3.2.1}$$

is minimized. The resulting vector $\mathbf{W}_{ss}^o$ is termed the optimum choice in the sense of "least squares." The variable J_{ss} is the sum of the squared errors and is called the "performance function." It is obviously a function of $d(k)$, and also $\mathbf{X}(k)$ and **W** since they determine $y(k)$. For a given sequence of vectors $\{\mathbf{X}(k)\}$ and scalars $\{d(k)\}$, J_{ss} is a function of **W** only, and therefore J_{ss} is a measure of how well **W** performs as a filter impulse response to produce an output $y(k)$ which matches the reference sequence $d(k)$. The choice of **W** that minimizes J_{ss} is that value that has the best performance. In this case J_{ss} is the sum of the squared error $e(k) = d(k) - y(k)$ over all values of k for which $\mathbf{X}(k)$ and hence $y(k)$ are defined. The best J_{ss} can attain is zero, which is attained if **W** can be chosen so that

$$\mathbf{W}^t\mathbf{X}(k) \triangleq y(k) = d(k) \qquad \text{for all valid } k, \tag{3.2.2}$$

that is, $N - 1 \leqslant k \leqslant L - 1$. If so, then each term of the sum is zero and so is the sum itself. There is no worst value of J_{ss} since J_{ss} can be made arbitrarily positive with the proper choice of **W**. Note that J_{ss} can not be made negative by any choice of **W**.

3.2.2 Reduction to the Normal Equations

To find the best value of **W** we first expand (3.2.1) as

$$J_{ss} = \sum_{k=N-1}^{L-1} [d(k) - y(k)][d^*(k) - y^*(k)]$$

$$= \sum_{k=N-1}^{L-1} |d(k)|^2 - \sum_{k=N-1}^{L-1} y(k)d^*(k) - \sum_{k=N-1}^{L-1} d(k)y^*(k) + \sum_{k=N-1}^{L-1} |y(k)|^2, \tag{3.2.3}$$

where the asterisk denotes conjugation of a complex variable or vector. Using the fact that

$$y(k) = \mathbf{X}^t(k)\mathbf{W} = \mathbf{W}^t\mathbf{X}(k)$$

the performance function can be written as

$$J_{ss} = \sum_{k=N-1}^{L-1} |d(k)|^2 - \sum_{k=N-1}^{L-1} \mathbf{W}^t\mathbf{X}(k)d^*(k)$$

$$- \sum_{k=N-1}^{L-1} d(k)\mathbf{X}^h(k)\mathbf{W}^* + \sum_{k=N-1}^{L-1} \mathbf{W}^t\mathbf{X}(k)\mathbf{X}^h(k)\mathbf{W}^*, \tag{3.2.4}$$

where h denotes the conjugate transpose of a vector or matrix. Since $\mathbf{W}$ is assumed here to be a constant vector and not a function of k,

$$J_{ss} = \sum_{k=N-1}^{L-1} |d(k)|^2 - \mathbf{W}^t \left\{ \sum_{k=N-1}^{L-1} \mathbf{X}(k)d^*(k) \right\}$$

$$- \left\{ \sum_{k=N-1}^{L-1} d(k)\mathbf{X}^h(k) \right\} \mathbf{W}^* + \mathbf{W}^t \left\{ \sum_{k=N-1}^{L-1} \mathbf{X}(k)\mathbf{X}^h(k) \right\} \mathbf{W}^*. \qquad (3.2.5)$$

To simplify J_{ss} further we can define the following terms:

$$D_{ss} \triangleq \sum_{k=N-1}^{L-1} |d(k)|^2 \qquad (3.2.6)$$

$$\mathbf{P}_{ss} \triangleq \sum_{k=N-1}^{L-1} \mathbf{X}(k)d^*(k) \qquad (3.2.7)$$

and

$$R_{ss} \triangleq \sum_{k=N-1}^{L-1} \mathbf{X}(k)\mathbf{X}^h(k). \qquad (3.2.8)$$

With these definitions J_{ss} can be written as

$$J_{ss} = D_{ss} - \mathbf{W}^t\mathbf{P}_{ss} - \mathbf{P}^h_{ss}\mathbf{W}^* + \mathbf{W}^t R_{ss}\mathbf{W}^*$$

$$= D_{ss} - \mathbf{W}^t\mathbf{P}_{ss} - (\mathbf{W}^t\mathbf{P}_{ss})^* + \mathbf{W}^t R_{ss}\mathbf{W}^*$$

$$= D_{ss} - 2Re(\mathbf{W}^t\mathbf{P}_{ss}) + \mathbf{W}^t R_{ss}\mathbf{W}^*. \qquad (3.2.9)$$

If both $x(k)$ and $d(k)$ are real-valued instead of complex, then $\mathbf{W}$, $\mathbf{X}(k)$, $\mathbf{P}_{ss}$, and R_{ss} are also real-valued, and (3.2.9) becomes

$$J_{ss} = D_{ss} - 2\mathbf{W}^t\mathbf{P}_{ss} + \mathbf{W}^t R_{ss}\mathbf{W}. \qquad (3.2.10)$$

We now face the problem of how to find the vector $\mathbf{W}$ that minimizes J_{ss}. We do this by using the result from vector calculus [Noble and Daniel, 1977] which states that $\mathbf{W}^o_{ss}$ is the value of $\mathbf{W}$ which minimizes J_{ss} if and only if two conditions are satisfied:

$$\nabla_W J_{ss}|_{\mathbf{W}=\mathbf{W}^o_{ss}} = \mathbf{0} \qquad (3.2.11)$$

and

$$H_W \text{ is positive definite,} \qquad (3.2.12)$$

where $\nabla_W J$ is the gradient of J_{ss} with respect to the elements of $\mathbf{W}$, and H_W is the

Hessian matrix of J_{ss} with respect to the elements of $\mathbf{W}$. Taking the gradient of J_{ss} is the same as forming a vector of the partial derivatives of J_{ss} with respect to the impulse response coefficients. Thus (3.2.11) can be written as

$$\left.\frac{\partial J_{ss}}{\partial w_j}\right|_{\mathbf{W}=\mathbf{W}^o_{ss}} = 0, \qquad 0 \leqslant j \leqslant N-1, \qquad (3.2.13)$$

that is, the first derivative of J with respect to each of the impulse response coefficients is zero when the derivative is evaluated at $\mathbf{W} = \mathbf{W}^o_{ss}$. The Hessian of J_{ss} is just the matrix of second derivatives. The i,jth element of the matrix is

$$h_{ij} = \frac{\partial^2 J_{ss}}{\partial w_i \partial w_j}, \qquad (3.2.14)$$

i.e., the second derivative with respect to element w_i and element w_j. The test described in (3.2.12) is satisfied if, for any nonzero vector $\mathbf{V}$,

$$\mathbf{V}^t H_W \mathbf{V} > 0. \qquad (3.2.15)$$

Any matrix which satisfies this condition is called "positive definite."

The intuitive meanings of conditions (3.2.11) and (3.2.12) are revealed by looking at the one-dimensional case where $\mathbf{W} = w$. When this is true, w^o is optimum when

$$\left.\frac{\partial J}{\partial w}\right|_{w=w^o} = 0 \qquad (3.2.16)$$

and

$$\left.\frac{\partial^2 J}{\partial w^2}\right|_{w=w^o} > 0. \qquad (3.2.17)$$

These are just the conditions used in functional analysis to find the local minimum of a curve. In particular,

● (3.2.16) states that the slope at the optimum point must equal zero, and

● (3.2.17) requires that the curvature at the optimum point must be positive, i.e. the function curves upward in both directions from the optimum point.

Conditions (3.2.11) and (3.2.12) are the N-dimensional versions of (3.2.16) and (3.2.17). Equation (3.2.11) requires the first derivative with respect to each weight to equal zero, setting all slopes to zero. Equation (3.2.12) ensures that the curvature is upward in all directions, which allows us to say that $\mathbf{W}^o_{ss}$ specifies a local minimum point for J_{ss}.

We can now employ the tests (3.2.11) and (3.2.12) to find necessary conditions for $\mathbf{W}_{ss}^{o}$ for the real data case. First, we evaluate the gradient of J_{ss}. Using (3.2.10), we find that

$$\nabla_W J_{ss} = \nabla_W D_{ss} - 2\nabla_W(\mathbf{W}^t\mathbf{P}_{ss}) + \nabla_W(\mathbf{W}^t R_{ss}\mathbf{W}). \qquad (3.2.18)$$

Since D_{ss} is a constant and not a function of $\mathbf{W}$, its gradient is zero. The gradient of $\mathbf{W}^t\mathbf{P}_{ss}$ can be found directly by evaluating each partial derivative

$$\frac{\partial}{\partial w_j}(\mathbf{W}^t\mathbf{P}_{ss}) = \frac{\partial}{\partial w_j}\left\{ \sum_{n=0}^{N-1} w_n p_n \right\} = p_j, \qquad 0 \leqslant j \leqslant N-1. \qquad (3.2.19)$$

Thus,

$$\nabla_W\{\mathbf{W}^t\mathbf{P}_{ss}\} = \mathbf{P}_{ss}. \qquad (3.2.20)$$

Using the same approach $\nabla_W\{\mathbf{W}^t R_{ss}\mathbf{W}\}$ can be evaluated as

$$\frac{\partial}{\partial w_l}\{\mathbf{W}^t R_{ss}\mathbf{W}\} = \sum_{i=0}^{N-1} \sum_{j=0}^{N-1} \frac{\partial}{\partial w_l}\{w_i r_{ij} w_j\}$$

$$= 2 \sum_{j=0}^{N-1} r_{lj} w_j, \qquad (3.2.21)$$

which implies that

$$\nabla_W\{\mathbf{W}^t R_{ss}\mathbf{W}\} = 2R_{ss}\mathbf{W}. \qquad (3.2.22)$$

Combining (3.2.20) and (3.2.22) with (3.2.18), the gradient of J becomes

$$\nabla_W J_{ss} = -2\mathbf{P}_{ss} + 2R_{ss}\mathbf{W}. \qquad (3.2.23)$$

Thus the first necessary condition for optimality, i.e., finding $\mathbf{W}_{ss}^{o}$ such that the gradient is zero, becomes

$$0 = -\mathbf{P}_{ss} + R_{ss}\mathbf{W}_{ss}^{o} \qquad (3.2.24)$$

or

$$R_{ss}\mathbf{W}_{ss}^{o} = \mathbf{P}_{ss}. \qquad (3.2.25)$$

The set of N linear simultaneous equations described by the matrix equation are called the "normal equations." Their satisfaction is a requirement for $\mathbf{W}_{ss}^{o}$ to be considered the optimal solution.

The second condition, described by (3.2.12), is that the matrix of second derivatives be positive definite. We evaluate this matrix by evaluating each term of H_W:

$$h_{ij} = \frac{\partial^2 J_{ss}}{\partial w_i \partial w_j} = \frac{\partial^2}{\partial w_i \partial w_j} \{D_{ss}\} - 2 \frac{\partial^2}{\partial w_i \partial w_j} \{\mathbf{W}^t \mathbf{P}_{ss}\} + \frac{\partial^2}{\partial w_i \partial w_j} \{\mathbf{W}^t \mathbf{R}_{ss} \mathbf{W}\}. \quad (3.2.26)$$

Since the first derivative of the constant D_{ss} is zero, so is the second derivative. Since the elements of $\mathbf{P}_{ss}$ are constants, it follows that

$$\frac{\partial^2}{\partial w_i \partial w_l} \left\{ \sum_{m=0}^{N-1} w_m p_m \right\} = \frac{\partial}{\partial w_i} \{p_l\} \equiv 0, \quad 0 \leqslant i \leqslant N-1. \quad (3.2.27)$$

Thus, the second term is also zero. The third term can be simplified as follows

$$\frac{\partial^2}{\partial w_i \partial w_j} \left\{ \sum_{l=0}^{N-1} \sum_{m=0}^{N-1} w_l r_{lm} w_m \right\} = \frac{\partial}{\partial w_i} \left\{ \sum_{l=0}^{N-1} \sum_{m=0}^{N-1} \frac{\partial}{\partial w_j} \{w_l r_{lm} w_m\} \right\}$$

$$= \frac{\partial}{\partial w_i} \left\{ 2 \sum_{m=0}^{N-1} r_{mj} w_m \right\}$$

$$= 2 r_{ij}, \quad 0 \leqslant i, j \leqslant N-1. \quad (3.2.28)$$

Since $h_{ij} = 2r_{ij}$, for all possible indices i and j, this means that the Hessian matrix, which describes the curvature of J at $\mathbf{W} = \mathbf{W}_{ss}^o$, is exactly

$$H_W = 2R_{ss}, \quad (3.2.29)$$

i.e., twice the matrix R_{ss}. Thus the two conditions for $\mathbf{W}^o$ to be optimum are

$$R_{ss} \mathbf{W}_{ss}^o = \mathbf{P}_{ss} \quad (3.2.30)$$

and

$$R_{ss} \text{ is positive definite.} \quad (3.2.31)$$

3.2.3 Direct Solution for the Optimal Vector $\mathbf{W}_{ss}^o$

If the matrix R_{ss} can be inverted, then the normal equations (3.2.30) can be used to find $\mathbf{W}_{ss}^o$, that is, if R_{ss}^{-1} exists, then

$$\mathbf{W}_{ss}^o = R_{ss}^{-1} \mathbf{P}_{ss}. \quad (3.2.32)$$

From vector calculus [Noble and Daniel, 1977], it is known that R_{ss}^{-1} exists if R_{ss} is positive definite. Thus, if R_{ss} is positive definite, then condition (3.2.31) is satisfied, meaning that the solution $\mathbf{W}_{ss}^o$ is unique and (3.2.30) can be used to produce $\mathbf{W}_{ss}^o$ according to (3.2.32).

From this analysis, we conclude that given real $\mathbf{X}(k)$ and $d(k)$, for $k = N - 1$ to $k = L - 1$, we can find the optimum least squares solution $\mathbf{W}_{ss}^o$ using the following steps:

(a) Use $x(k)$ to form $\mathbf{X}(k)$ and hence $R_{ss} = \sum_{k=N-1}^{L-1} \mathbf{X}(k)\mathbf{X}^t(k)$.

(b) Use $\mathbf{X}(k)$ and $d(k)$ to form $\mathbf{P}_{ss} = \sum_{k=N-1}^{L-1} \mathbf{X}(k)d(k)$.

(c) If R_{ss} is positive definite, use R_{ss} and $\mathbf{P}_{ss}$ to form $\mathbf{W}_{ss}^o$ using $\mathbf{W}_{ss}^o = R_{ss}^{-1}\mathbf{P}_{ss}$.

3.2.4 The Meaning of $\mathbf{P}_{ss}$ and R_{ss}

We turn now to trying to determine exactly what $\mathbf{P}_{ss}$ and R_{ss} are and why is it reasonable that they control the optimal value of the impulse-response vector. We start with R_{ss}.

Referring back to its definition in (3.2.8), we see that r_{ij}, the i, jth term of R_{ss}, is given by

$$r_{ij} = \sum_{k=N-1}^{L-1} \{\mathbf{X}(k)\mathbf{X}^h(k)\}_{i,j}$$

$$= \sum_{k=N-1}^{L-1} \{\mathbf{X}(k)\}_{\text{row } i} \cdot \{\mathbf{X}^h(k)\}_{\text{column } j}$$

$$= \sum_{k=N-1}^{L-1} x(k-i) \cdot x^*(k-j). \tag{3.2.33}$$

When $x(k)$ is a real-valued rather than a complex-valued process,

$$r_{ij} = \sum_{k=N-1}^{L-1} x(k-i) \cdot x(k-j). \tag{3.2.34}$$

On the main diagonal of R, i equals j. At element ii,

$$r_{ii} = \sum_{k=N-1}^{L-1} x^2(k-i), \tag{3.2.35}$$

which is the energy of waveform $x(k)$ measured over the interval from $k = N - 1 - i$ to $k = L - 1 - i$. Thus, the main diagonal elements of R_{ss} measure the energy in windowed portions of the waveform $x(k)$. In general, we expect these entries to be different since the instantaneous power of practical signals varies and so, therefore, does a windowed estimate of its energy.

The off-diagonal elements measure the "cross-energy" between $L-N+1$ points of the waveform $x(k-i)$ and the same number of points of the time-shifted waveform $x(k-j)$. In general, if $x(k-j)$ closely resembles the waveshape of $x(k-i)$, then the cross-energy r_{ij} is very close to the energy of $x(k-i)$ or $x(k-j)$. If so, $x(k-j)$ is said to be "highly correlated" with $x(k-i)$. If the shifted waveforms bear little resemblance to one another, then the cross-energy r_{ij} will be much smaller than r_{ii} or r_{jj} and the waveforms are said to be relatively uncorrelated for shifts i and j.

We see now that the matrix R_{ss} contains energy measurements over an $L-N+1$ point window for all possible shifted versions of $x(k)$. The term r_{ij} indicates the amount of correlation between shifted versions of $x(k)$ and itself. Because it measures the internal correlation properties of one signal sequence $\{x(k)\}$, R_{ss} is called the autocorrelation matrix for the sequence $\{x(k)\}$. Since the $\{r_{ij}\}$ are based on time averages, R_{ss} is sometimes referred to as the time-averaged autocorrelation matrix.

We progress in the same way with the vector $\mathbf{P}_{ss}$. From (3.2.7), and assuming the waveform $d(k)$ to be real, the ith element of $\mathbf{P}_{ss}$ is

$$p_i = \sum_{k=N-1}^{L-1} x(k-i)d(k), \qquad 0 \leqslant i \leqslant N-1. \tag{3.2.36}$$

From the previous discussion, it is clear that p_i is a measurement of the cross-energy of the reference waveform $d(k)$ and a shifted version of $x(k)$. The magnitude of p_i would be expected to be high when $x(k-i)$ and $d(k)$ closely resemble each other and to be small when they differ. As a result, $\mathbf{P}_{ss}$ is referred to as the time-averaged cross-correlation vector describing the correlation properties of $d(k)$ and $x(k)$ over a N-point choice of shifts.

Thus, we see that the optimal filter solution $\mathbf{W}_{ss}^o$ is just a function of the correlation properties of the input $x(k)$ and the cross-properties between $x(k)$ and the reference waveform $d(k)$. The implications of this will be revealed as we continue, but one is obvious here. Suppose that $x(k-i)$ and $d(k)$ have no substantial cross energy for shifts $i=0$ to $N-1$; if this is true then $p_i=0$, $0 \leqslant i \leqslant N-1$ and hence

$$\mathbf{P}_{ss} = \mathbf{0}. \tag{3.2.37}$$

Assuming R_{ss}^{-1} exists, we find that $\mathbf{W}^o$ is also zero, meaning that choosing the all-zero filter (that is, $\mathbf{W}=\mathbf{0}$), is the best filter. This is intuitively reasonable. The filter output $y(k)$ consists of scaled versions of $x(k-i)$, for shifts $i=0$ to $N-1$. The error $d(k)-y(k)$ is reduced when some or several past versions of $x(k)$ can be scaled and then subtracted off from $d(k)$. If no delayed version $x(k-i)$ resembles $d(k)$ (i.e., is correlated with $d(k)$), then none can be used to reduce the error. When this is the case, all the scaling terms w_i^o should be set to zero.

3.2.5 Examples

We now consider two simple examples.

3.2.5.1 Matching a Pure Delay

First, suppose we have an N-point FIR filter and desire to choose $\mathbf{W}$, the impulse response vector, to minimize the summed squared error. Suppose further that $x(k)$ and $d(k)$ are given as

$$x(k) = A\delta(k - k_o), \qquad N - 1 \leqslant k \leqslant L - 1, \qquad 0 \leqslant k_o \leqslant N - 1 \quad (3.2.38)$$

and

$$d(k) = B\delta(k - k_o - \Delta), \qquad N - 1 \leqslant k \leqslant L - 1, \qquad 0 \leqslant \Delta \leqslant N - 1, \qquad (3.2.39)$$

where $\delta(k)$ is the unit pulse function. To find $\mathbf{W}_{ss}^o$, we first determine the autocorrelation matrix R_{ss} and the cross-correlation vector $\mathbf{P}_{ss}$. From (3.2.8) and (3.2.4),

$$r_{ij} = \sum_{k=N-1}^{L-1} x(k - i) \cdot x(k - j), \qquad 0 \leqslant i, j \leqslant N - 1$$

$$= A^2 \sum_{k=N-1}^{L-1} \delta(k - k_o - i) \cdot \delta(k - k_o - j), \qquad 0 \leqslant i, j \leqslant N - 1$$

$$= A^2 \cdot \delta(i - j). \qquad (3.2.40)$$

All diagonal elements are equal and all off-diagonal elements are zero. Thus,

$$R_{ss} = A^2 \cdot I_N, \qquad (3.2.41)$$

where I_N is the $N \times N$ identity matrix. From (3.2.7)

$$p_i = \sum_{k=N-1}^{L-1} x(k - i) \cdot d(k)$$

$$= \sum_{k=N-1}^{L-1} A \cdot \delta(k - k_o - i) \cdot B\delta(k - k_o - \Delta)$$

$$= AB \sum_{k=N-1}^{L-1} \delta(k - k_o - i) \cdot \delta(k - k_o - \Delta). \qquad (3.2.42)$$

Just as with R_{ss}, it can be verified that the product of the two delta functions equals zero unless their two arguments are identical. Thus,

$$p_i = AB \cdot \delta(i - \Delta), \qquad 0 \leqslant i \leqslant N - 1$$

or

$$\mathbf{P}_{ss} = AB \cdot \mathbf{e}_{\Delta}, \tag{3.2.43}$$

where $\mathbf{e}_{\Delta}$ is an N-element column vector with all zeros except for unity in the Δth element. This $\mathbf{P}$ vector tells us that only $x(k - \Delta)$ is correlated with $d(k)$.

Equations (3.2.30) and (3.2.31) are used to find the optimal solution. As long as A is nonzero, then R_{ss} is positive definite, R_{ss} is therefore invertible, and a unique solution $\mathbf{W}_{ss}^{o}$ exists. We see that

$$\mathbf{W}_{ss}^{o} = R_{ss}^{-1} \mathbf{P}_{ss} = \frac{1}{A^2} I_N \cdot AB \cdot \mathbf{e}_{\Delta} = \frac{B}{A} \mathbf{e}_{\Delta}. \tag{3.2.44}$$

The optimal impulse response $\mathbf{W}_{ss}^{o}$ has only one nonzero coefficient; that is,

$$w_{\Delta}^{o} = \frac{B}{A}. \tag{3.2.45}$$

It can be confirmed that this choice of impulse response delays $x(k)$ by Δ time steps and scales it by B/A. Doing this makes the output $y(k)$ identical with $d(k)$ and makes the output error $d(k) - y(k)$ identically equal to zero. One can do no better. Making other filter coefficients differ from zero or making the Δth coefficient differ from B/A all increase the error.

3.2.5.2 Matching a Rotated Phasor

Now suppose that $x(k)$ is a complex-valued sinusoid described by

$$x(k) = A e^{j\omega_0 kT} \tag{3.2.46}$$

where ω_o is the radian frequency and T is the sampling interval. Suppose further that $d(k)$ is given by another complex waveform

$$d(k) = B e^{j\omega_0 T(k - \Delta)} \tag{3.2.47}$$

The FIR filter has N complex coefficients and we desire to find the optimal (possibly complex) impulse response $\mathbf{W}_{ss}^{o}$.

A comparison of this problem with the one in Section 3.2.5.1 shows a great similarity, and in fact it can be quickly confirmed that

$$\mathbf{W}^{o} = \frac{B}{A} \mathbf{e}_{\Delta} \tag{3.2.48}$$

results in a zero output error when used in this situation. We press ahead with the analytic solution in any case.

First, we find R_{ss}:

$$r_{lm} = \sum_{k=N-1}^{L-1} x(k-l) \cdot x^*(k-m)$$

$$= A^2 \sum_{k=N-1}^{L-1} e^{j\omega_0 T(k-l)} e^{-j\omega_0 T(k-m)}$$

$$= A^2 \sum_{k=N-1}^{L-1} e^{j\omega_0 T(m-l)}$$

$$= A^2(L-N+1) \, e^{j\omega_0 T(m-l)}. \tag{3.2.49}$$

Similarly, for $\mathbf{P}_{ss}$

$$p_l = \sum_{k=N-1}^{L-1} x(k-l) \cdot d^*(k)$$

$$= AB \sum_{k=N-1}^{L-1} e^{j\omega_0 T(k-l)} e^{-j\omega_0 T(k-\Delta)}$$

$$= AB \cdot (L-N+1) \cdot e^{j\omega_0 T(\Delta-l)}. \tag{3.2.50}$$

Given these expressions for R_{ss} and $\mathbf{P}_{ss}$, we set out to determine $\mathbf{W}_{ss}^o$. As before, the first step is to confirm that R_{ss} is positive definite and then to invert it. And here we run into trouble. Consider the almost trivial case where $N = 2$:

$$R_{ss} = A^2(L-1) \cdot \begin{bmatrix} 1 & e^{jT\omega_o} \\ e^{-jT\omega_o} & 1 \end{bmatrix}. \tag{3.2.51}$$

The determinant of the matrix is

$$1 \cdot 1 - e^{j\omega_o T} \cdot e^{-j\omega_o T} \equiv 0. \tag{3.2.52}$$

Therefore, R_{ss} does not have full rank, is not positive definite, and cannot be inverted. Thus, while (3.2.30) is still true, for $N = 2$ the other condition for a unique solution (3.2.31) is not. Does this mean there is no optimal solution? No, it means there are many. To see why, we now try again with $N = 1$.

If $N = 1$, then $R_{ss} = A^2 L$ and $\mathbf{P}_{ss} = A \cdot B \cdot L \cdot \exp j\omega_0 T\Delta$. Since R_{ss} is a positive scalar, it is certainly positive definite and invertible. The optimum and unique solution $\mathbf{W}_{ss}^o$ is then the scalar

$$w_{ss}^o = \frac{ABL e^{-j\omega_0 T\Delta}}{A^2 L}$$

$$= \frac{B}{A} e^{-j\omega_0 T\Delta}. \tag{3.2.53}$$

This can be confirmed to be a good solution since

$$y(k) = x(k) \cdot w_{ss}^o = A\, e^{j\omega_0 Tk} \cdot \frac{B}{A}\, e^{-j\omega_0 T\Delta}$$

$$= B\, e^{j\omega_0 T(k-\Delta)}$$

$$\equiv d(k). \tag{3.2.54}$$

We see that the optimal filter inserts a gain of B/A and a phase rotation of $-\omega_0 T\Delta$ radians. This phase rotation exactly compensates for the apparent delay in $d(k)$.

Why, then, is the second and higher-order filter solution not unique? The answer is this: There are too many degrees of freedom in the impulse response to uniquely specify its value. All the optimal filter need do is provide a gain of B/A and a phase shift of $-\omega_0 T\Delta$ radians at frequency ω_0. Any number of filters with $N = 2$ can do that, and all satisfy (3.2.30). Obviously, for $N = 3$ and above, multiple solutions are possible, including (3.2.44).

What we find here is that solutions are available, but there may not be just one for any given N. The source of this nonuniqueness is the character of the input. Clearly something is different about the input in (3.2.38) compared to the complex sinusoidal input in (3.2.46). The effect of the character of the input on the uniqueness of the solution was also noted in the number guessing games in Chapter 2. This issue and its impact on the performance of adaptive filters will be revisited in Section 3.4.

3.2.6 Two Solution Techniques

Suppose we have followed the procedure outlined earlier to solve for the optimal impulse response $\mathbf{W}_{ss}^o$. We collected L samples of $x(k)$, $L - N + 1$ samples of $d(k)$, computed R_{ss} and $\mathbf{P}_{ss}$, confirmed the invertibility of R_{ss}, and now desire to compute $\mathbf{W}_{ss}^o$. We examine two techniques here; *neither* is commonly used in practice, however both are theoretically sound and serve as benchmarks by which we can measure other methods.

3.2.6.1 Direct Inversion
The most straightforward approach to solving (3.2.32) is the direct inversion of R_{ss} followed by its multiplication with the vector $\mathbf{P}_{ss}$. Since R_{ss} is of dimension N by N, this inversion requires approximately N^3 multiplications, a roughly equal number of additions, and a smaller number of divisions, if done with the classical Gaussian elimination technique. Multiplication of R_{ss}^{-1} by $\mathbf{P}_{ss}$ requires an additional N^2 multiply-adds.

3.2.6.2 Iterative Approximation
Suppose now that we try to iteratively determine $\mathbf{W}_{ss}^o$ by using the following scheme. Given R_{ss} and $\mathbf{P}_{ss}$ we form an estimate $\mathbf{W}(l)$ and try to improve the

estimate according to the following rule:

$$\mathbf{W}(l+1) = (I - \mu R_{ss}) \cdot \mathbf{W}(l) + \mu \mathbf{P}_{ss}, \qquad \mathbf{W}(0) = \mathbf{0}, \qquad (3.2.55)$$

where μ is a small positive constant and l is allowed to increase by one after computation of each $\mathbf{W}(l+1)$ until $\mathbf{W}(l)$ approximates $\mathbf{W}_{ss}^o$ well enough.

Does this scheme work? We can show that

$$\mathbf{W}(1) = (1 - \mu R_{ss}) \cdot \mathbf{W}(0) + \mu \mathbf{P}_{ss} = \mu \mathbf{P}_{ss}$$

$$\mathbf{W}(2) = \mu(I - \mu R_{ss})\mathbf{P}_{ss} + \mu \mathbf{P}_{ss}$$

$$\mathbf{W}(k) = \mu \left[\sum_{m=0}^{k-1} (I - \mu R_{ss})^m \right] \mathbf{P}_{ss} \qquad (3.2.56)$$

Recall from the study of power series [Knopp, 1956] that if some constant α has magnitude less than 1, then

$$\sum_{m=0}^{k-1} |\alpha|^m = \frac{1 - |\alpha|^k}{1 - |\alpha|} \qquad (3.2.57)$$

and

$$\lim_{k \to \infty} \sum_{m=0}^{k-1} |\alpha|^m = \frac{1}{1 - |\alpha|}. \qquad (3.2.58)$$

In a similar manner, if the matrix $I - \mu R_{ss}$ is "less than 1" in the sense that multiplying any vector $\mathbf{V}$ by $I - \mu R_{ss}$ decreases its length, then it can be shown that

$$\mathbf{W}(k) = \mu\{[I - (I - \mu R_{ss})^k] \cdot [I - (I - \mu R_{ss})]^{-1}\}\mathbf{P}_{ss}$$

$$= \{I - (I - \mu R_{ss})^k\} R_{ss}^{-1} \mathbf{P}_{ss}. \qquad (3.2.59)$$

If, as assumed,

$$\lim_{k \to \infty} (I - \mu R_{ss})^k \to 0, \qquad (3.2.60)$$

then

$$\lim_{k \to \infty} \mathbf{W}(k) = R_{ss}^{-1} \mathbf{P}_{ss} = \mathbf{W}_{ss}^o. \qquad (3.2.61)$$

Thus, by making μ small enough to make $I - \mu R_{ss}$ "less than 1," we see that repeated iteration of (3.2.55) will force $\mathbf{W}(l)$ toward the optimal vector $\mathbf{W}_{ss}^o$.

The number of mathematical operations needed to determine $\mathbf{W}^o_{ss}$ in this way is more difficult to specify clearly than with the direct inversion method. Neglecting the calculations needed to initialize the iterative procedure, each iteration requires N^2 multiply-adds, where N is the filter length. What is harder to specify is how many iterations are required to make $\mathbf{W}(l)$ close enough to $\mathbf{W}^o_{ss}$. By iteration K, $K \cdot N^2$ multiply-adds have been done.

3.2.6.3 Computational Comparisons
At this point it is interesting to examine these two methods in terms of the computation required and in terms of their general philosophies.

First, we count computations. Both assume the availability of R_{ss} and $\mathbf{P}_{ss}$. Assuming that $x(k)$ and $d(k)$ are real-valued functions for the moment, it can be confirmed that each vector outer product $\mathbf{X}(k) \cdot \mathbf{X}^t(k)$ requires N^2 multiplications and therefore that R_{ss} requires

$$C_R = (L - N + 1) \cdot N^2 \text{ multiply-adds.} \tag{3.2.62}$$

Similarly, computing $\mathbf{P}_{ss}$ requires

$$C_P = (L - N + 1) \cdot N \text{ multiply-adds.} \tag{3.2.63}$$

Computing $\mathbf{W}^o_{ss}$ with the direct inversion method uses an additional

$$C_D = N^3 + N^2 \text{ operations,} \tag{3.2.64}$$

while the iterative matrix inversion technique uses

$$C_I = K \cdot N^2, \tag{3.2.65}$$

where K is the number of iterations considered to be sufficient to approximate $\mathbf{W}^o_{ss}$ well enough. Combining these terms, we see that

$$C_{\text{direct inversion}} = C_R + C_P + C_D = (L - N + 1) \cdot (N^2 + N) + N^3 + N^2 \tag{3.2.66}$$

and

$$C_{\text{iterative conversion}} = C_R + C_P + C_I = (L - N + 1) \cdot (N^2 + N) + K \cdot N^2. \tag{3.2.67}$$

In most practical problems, K must be significantly larger than N to attain adequate approximation of $\mathbf{W}(k)$ to $\mathbf{W}^o_{ss}$. On the surface, then, the direct method will almost always use fewer computations than the iterative techniques. Even here, though, several factors make the issue harder to decide:

(a) In most practical problems, the amount of data L is much longer than the filter length N. When this is true, C_R, the computation required to compute the correlation matrix R_{ss}, overwhelms C_P, C_D, and C_I. When this is true, the difference between C_D and C_I may be irrelevant compared to the total computation required.

(b) If R_{ss} does not have full rank, then R_{ss} cannot be inverted and no value of $\mathbf{W}^o_{ss}$ can be found by the direct method. As we shall see later, (3.2.55) can still be used in this case to find a solution to the normal equations. Such a solution is not necessarily unique but does minimize the summed squared performance function.

(c) The iterative technique is less sensitive to numerical problems such as roundoff errors.

In reality neither of these techniques is used very much for adaptive filters, since substantially more efficient techniques of computing $\mathbf{W}^o_{ss}$ have been developed, as we will show in Chapter 4. It is useful to go through the exercise, however, since the issues examined here will arise later. In particular, some algorithms will yield an exact solution for $\mathbf{W}^o_{ss}$ while others will approximate it. Some algorithms take a predictable number of computations to reach their solutions, while the computation time for others may depend on the vagaries of the data itself. Properties of the data may in fact preclude any solution at all for some techniques.

3.2.7 Consolidation

In Chapter 2 we developed an iterative algorithm which accepts a data vector $\mathbf{X}(k)$ and a reference sample $d(k)$ at each time k and uses it to improve (or at least not degrade) the weight vector $\mathbf{W}(k)$. In this chapter we have taken an alternative approach; we form an objective or performance function which describes how good a choice $\mathbf{W}$ is, and then we develop procedures for finding the best possible choice.

In particular, Section 3.2 has focused on the least squares problem where the performance function J_{ss} is the sum of the squared difference between the filter output $\mathbf{W}^t\mathbf{X}(k)$ and the reference waveform $d(k)$. We showed that this best choice of the impulse response vector $\mathbf{W}$ satisfies the "normal equations":

$$R_{ss}\mathbf{W}^o_{ss} = \mathbf{P}_{ss}, \tag{3.2.68}$$

where the time-averaged autocorrelation matrix R_{ss} and the time-averaged cross-correlation vector $\mathbf{P}_{ss}$ are computed directly from the input data vectors $\mathbf{X}(k)$ and the reference waveform $d(k)$. If the number of data vectors $\mathbf{X}(k)$ exceeds the filter order N, and if the input $x(k)$ has the right properties, then R_{ss} is positive definite and invertible. When this is true (as it usually is in practice), the solution for the best impulse response vector is unique and is given by

$$\mathbf{W}^o_{ss} = R_{ss}^{-1}\mathbf{P}_{ss}. \tag{3.2.69}$$

Given this expression we examined two techniques for computing $\mathbf{W}_{ss}^o$. The first was a brute force technique, computing R_{ss} and $\mathbf{P}_{ss}$ directly, inverting R_{ss} and then multiplying $\mathbf{P}_{ss}$. The second was a iterative technique based on repeating a simple matrix formula.

This effort shows that it is possible to state a performance function and then find a closed-form expression for the best weight vector. It also illustrates that even though the closed-form solution can be written, one might want to compute the expression by some technique other than the most obvious one.

We have bypassed some important aspects of the behavior of the optimum summed squared error solution, such as how the input process $x(k)$ affects the rank of R_{ss} and hence the uniqueness of $\mathbf{W}_{ss}^o$. We do address this in Section 3.4, but only after we develop a class of optimal filters based on a different averaging technique.

3.3 THE LEAST MEAN SQUARES PROBLEM

3.3.1 Formulation

The least squares technique described in the last section forms the basis for most practical adaptive filtering algorithms. Even so, learning about its properties is difficult simply because the input $x(k)$ can be any bounded sequence. Given L points of input $x(k)$ and the reference waveform $d(k)$, the exact optimum impulse-response vector can be found. Given another set of L points, however, there is no guarantee that the resulting optimum vector $\mathbf{W}_{ss}^o$ is at all close to, or even related to, the first solution. Intuitively, we expect the impulse response vector to be essentially constant for different sections of the same waveform if the "properties" of the waveform do not change. Unfortunately, it is difficult to translate these concepts into mathematical definitions for the least squares problem. To do so, we adopt a different analytical direction and develop another way of representing the signals and their averages.

Suppose now that the input sequence $x(k)$ is now defined over all time, instead of just from $k = 0$ to $k = L - 1$, and that it is a stochastic process. The desired or reference waveform $d(k)$ is also defined for all time and is assumed to be a stochastic process as well. We define the data vector $\mathbf{X}(k)$ and the filter output $y(k)$ as before, but now they exist for all time since $x(k)$ does. The error $e(k)$ is defined by

$$e(k) = d(k) - y(k), \tag{3.3.1}$$

and is also a stochastic process.

With these differences in the pertinent signals, we now define a new performance function called J_{ms}, the mean squared error, by the expression

$$J_{ms} = E\{e^2(k)\}, \tag{3.3.2}$$

where $E\{\cdot\}$ denotes the expectation operator. To properly interpret this expression and the ensuing analysis, we digress temporarily to discuss stochastic processes and some of the associated definitions.

3.3.2 A Brief Review of Stochastic Processes
[Melsa and Sage, 1973]

3.3.2.1 Working Definition of a Stochastic Process
The signals we dealt with in Section 3.2 were functions of only the time index k. In its simplest form the stochastic process $f(k, q)$ is a function of two variables, the time index k and the realization index q. As we have discussed, k ranges from $-\infty$ to ∞. The index q selects one of Q "realizations" of the signal. The number of realizations can in principle be finite or infinite. For any choice of index q, say q_0, there is exactly one deterministic function $f(k, q_0)$ which is defined for all time. Randomness is introduced into $f(k, q)$ by making q a random variable. We assume that q is chosen randomly and that a particular value of q is chosen with a probability of $p(q)$. As usual with probability functions,

$$0 \leqslant p(q = q_0) \leqslant 1, \qquad \text{for all } q \tag{3.3.3}$$

and

$$\sum_{\text{all } q} p(q) = 1. \tag{3.3.4}$$

This implies that if we randomly pick a realization of the stochastic process f, then the probability of it being $f(k, q_0)$ is given by $p(q = q_0)$. The set of all possible realizations of $f(k, q)$ is called the "ensemble" of functions, and there are Q members of this ensemble.

3.3.2.2 Ensemble Averages
The procedures developed in Section 3.2 assumed the existence of two finite-duration signals $x(k)$ and $d(k)$ and used those to find a single optimal impulse-response vector $\mathbf{W}^o_{ss}$. Our goal in this section is to extend this approach across other function pairs (x, d) within a particular class and to find the impulse response which is the best over all such pairs. To do this, we must develop some tools which allow us to perform averages within the class (or ensemble) of pairs (x, d).

Consider the function g which has as its argument some function of the realization $f(k, q)$. Given the ensemble of realizations over all Q of the set of indices q, we can compute the "ensemble average" of g, denoted $\bar{g}$, by

$$\bar{g}(k) = \sum_{i=1}^{Q} p(q = q_i) \cdot g(f(k, q_i)). \tag{3.3.5}$$

Here g is computed for all realizations of f and then all are weighted by their probability of occurrence and summed to produce the ensemble average $\bar{g}$ of g. When Q becomes infinite or uncountable, this expression gracefully turns into an infinite sum or an integral.

This ensemble average is also sometimes called the mean of g or the expected value of g, that is, $E\{g\}$. Since the probability weighting sums to unity, the sum in (3.3.5) will always exist if $g(\cdot)$ is bounded. We should also observe that $\bar{g}$ is in general a function of k since the averaging is done over the realization index only.

We now make some useful definitions.

Mean of $f(k, q)$
Suppose that

$$g(f(k, q)) \equiv f(k, q), \qquad (3.3.6)$$

that is, g is just f itself. The mean of g, hence f, is

$$\bar{f}(k) = \sum_{i=1}^{Q} p(q = q_i) \cdot f(k, q_i), \qquad (3.3.7)$$

simply a weighted average of $f(k, q)$ over all realizations. Notice that $\bar{f}$ is in fact a function of time.

Average Power of $f(k, q)$
Suppose now that

$$g(f(k, q)) \equiv |f(k, q)|^2. \qquad (3.3.8)$$

The ensemble average of g is

$$\bar{g} = \overline{f^2}(k) = \sum_{i=1}^{Q} p(q = q_i)|f(k, q_i)|^2 . \qquad (3.3.9)$$

Since $|f(k, q)|^2$ is the instantaneous power of realization q at time k, $\overline{f^2}(k)$ is the (ensemble) average power of f at time k.

Autocorrelation of $f(k, q)$
Define $g(k, q)$ by the expression

$$g(k, q) = f(k + \Delta, q) \cdot f^*(k, q), \qquad (3.3.10)$$

the product of a time-shifted version of f and an unshifted version of its complex conjugate. The ensemble average of this function g is function g is

$$\bar{g} \equiv r(k, \Delta) = \sum_{i=1}^{Q} p(q = q_i) \cdot f(k + \Delta, q_i) \cdot f^*(k, q_i). \qquad (3.3.11)$$

This average is called the ensemble average autocorrelation function of f and indicates on the average how much f resembles (i.e., is "correlated" with) a time-shifted version of itself. We note that for zero shift, $r(k, \Delta)$ is just the average power

$$r(k, 0) = \overline{f^2}(k).$$

Cross-Correlation of $f_1(k, q)$ and $f_2(k, q)$

Suppose now that we have two ensembles, $f_1(k, q)$ and $f_2(k, q)$, both with the same number of members Q and both using the same realization index q. Clearly the mean, average power, and autocorrelation function for either f_1 or f_2 can be computed as described above. Here we desire to measure the degree of correlation between realizations of the two stochastic processes. We do this by defining g as

$$g(k, \Delta, q) = f_1(k + \Delta, q) \cdot f_2^*(k, q). \qquad (3.3.12)$$

The average of $\bar{g}$ across the two ensembles is called the cross-correlation function of f_1 and f_2, and is defined by

$$c(k, \Delta) \equiv \bar{g}(k, \Delta)$$

$$= \sum_{i=1}^{Q} p(q = q_i) \cdot f_1(k + \Delta, q) \cdot f_2^*(k, q). \qquad (3.3.13)$$

In general, we expect $c(k, \Delta)$ to be near zero if $f_1(k + \Delta, q)$ and $f_2^*(k, q)$ do not resemble each other on the (ensemble) average and to be nonzero if they do.

3.3.2.3 An Example

Suppose we define an ensemble of functions by the expression

$$f(k, q) = A \cos(\omega k T + 2\pi q/Q), \qquad 1 \leqslant q \leqslant Q, \qquad (3.3.14)$$

where the amplitude A, the radian frequency ω, and the sampling interval T are constant. The index realization q affects only the "starting phase" of the cosine, i.e., the phase of the cosine's argument when k equals zero. Suppose further that the probability of occurrence of any specific value of q, say q_0, is given by

$$p(q = q_0) = \frac{1}{Q}, \qquad (3.3.15)$$

for all choices of q_0 between 1 and Q. Thus, any value of q is equally likely. Following (3.3.7), the mean value of $f(k, q)$ is

$$\bar{f}(k) = \sum_{i=1}^{Q} A \cos(\omega kT + 2\pi q_i/Q) \cdot p(q = q_i)$$

$$= \frac{A}{Q} \cdot \sum_{i=1}^{Q} \cos(\omega kT + 2\pi q_i/Q)$$

$$= \frac{A}{Q} \cos \omega kT \cdot \sum_{i=1}^{Q} \cos 2\pi q_i/Q - \frac{A}{Q} \sin \omega kT \cdot \sum_{i=1}^{Q} \sin 2\pi q_i/Q,$$

$$= 0, \tag{3.3.16}$$

since both sums equal zero.

Similarly, we determine the average power by using (3.3.9):

$$\overline{f^2}(k) = A^2 \sum_{i=1}^{Q} p(q = q_i) \cdot |\cos(\omega kT + 2\pi q_i/Q)|^2$$

$$= \frac{A^2}{Q} \cdot \sum_{i=1}^{Q} \left(\frac{1}{2} + \frac{1}{2} \cos 4 \pi q_i/Q \right)$$

$$= \frac{A^2}{Q} \left\{ \frac{Q}{2} + \frac{1}{2} \cdot \sum_{i=1}^{Q} \cos 4\pi q_i/Q \right\}$$

$$= \frac{A^2}{2}. \tag{3.3.17}$$

As in the calculation of the mean, the fact that phases 180° apart are equally likely causes the sum in the next to last line of (3.3.17) to equal zero. A different probability distribution $p(q)$ might ruin this symmetry. Because of the uniform $p(q)$, however, neither the mean nor the average power are functions of the time index k.

The autocorrelation function of $f(k, q)$ is determined using (3.3.11) and various sine/cosine identities:

$$r(k, \Delta) = A^2 \sum_{i=1}^{Q} p(q = q_i) \cdot \cos(\omega(k + \Delta)T + 2\pi q_i/Q) \cdot \cos(\omega kT + 2\pi q_i/Q)$$

$$= \frac{A^2}{Q} \sum_{i=1}^{Q} \{ \tfrac{1}{2} \cos[\omega(k + \Delta)T + 2\pi q_i/Q - \omega kT - 2\pi q_i/Q]$$

$$+ \tfrac{1}{2} \cos[\omega k + \Delta)T + 2\pi q_i/Q + \omega kT + 2\pi q_i/Q] \}$$

$$= \frac{A^2}{2Q} \cdot \sum_{i=1}^{Q} \{ \cos(\omega \Delta T) + \cos(2\omega kT + \Delta \omega T + 4\pi q_i/Q) \}$$

$$\equiv \frac{A^2}{2} \cdot \cos \omega \Delta T. \tag{3.3.18}$$

Again, the portion of the final sum containing q_i disappears because a full rotation of the cosine sums to zero. Two other points:

(a) Since $r(k, \Delta)$ turns out not to be a function of k, it may be written as $r(\Delta)$, i.e., a function only of the time *difference* between the two waveforms.
(b) Note again that at a zero shift ($\Delta = 0$), the autocorrelation of $f(k, q)$ equals its average power, that is, $r(k, 0) = \overline{f^2(k)}$.

To provide an example of the cross-correlation function, as well as of the complex arithmetic version of these formulas, define $f_1(k, q)$ and $f_2(k, q)$ by the following:

$$f_1(k, q) = A e^{j(\omega k T + 2\pi q/Q)} \tag{3.3.19}$$

and

$$f_2(k, q) = B e^{j(\omega k T + \theta + 2\pi q/Q)}, \tag{3.3.20}$$

where A, B, ω, T, and θ are constants. The Q realizations of both f_1 and f_2 are indexed by q and each is assumed to be equally likely with probability $1/Q$. The cross-correlation function $c(k, \Delta)$ is given by

$$c(k, \Delta) = \sum_{i=1}^{Q} p(q = q_i) \cdot f_1(k + \Delta, q) \cdot f_2^*(k, q)$$

$$= \frac{AB}{Q} \cdot \sum_{i=1}^{Q} e^{j(\omega T(k + \Delta) + 2\pi q_i/Q)} \cdot e^{-j(\omega k T + 2\pi q_i/Q + \theta)}$$

$$= AB \, e^{j(\omega \Delta T - \theta)}. \tag{3.3.21}$$

Again, $c(k, \Delta)$ is not a function of time, but is directly affected by the phase offset θ which appears in f_2.

This example has been considered assuming that the realization was chosen from a finite set of Q functions. In fact, the reader should verify that these examples are easily carried over to the countably infinite and uncountably infinite cases. For example, suppose $f(k, \theta)$ is complex sinusoid of the form

$$f(k, \theta) = A \, e^{j(\omega k T + \theta)}, \tag{3.3.22}$$

where θ is chosen from the continuous interval $(-\pi, \pi]$ with a uniform probability density of

$$p_\theta(\theta = \theta_0) = \frac{1}{2\pi}. \tag{3.3.23}$$

If so,

(a) $\bar{f}(k, \theta) = \int_{-\pi}^{\pi} p_\theta(\theta = \theta_0) \cdot A \ e^{j(\omega Tk + \theta)} d\theta = 0$ (3.3.24)

(b) $\overline{f^2}(k) = \overline{f^2} = A^2$ (3.3.25)

(c) $r(k, \Delta) = r(\Delta) = A^2 \ e^{j\omega T\Delta}.$ (3.3.26)

3.3.2.4 Stationarity
A stochastic process is strictly stationary if $f(k)$ and $f(k + \Delta)$ have the same statistics, that is, the mean power, variance, and all other statistics must be invariant with time [Melsa and Sage, 1973]. For a stochastic process to be strictly stationary the probability density and distribution functions of all orders must be invariant under any time translation.

A wide-sense stationary (WSS) process is one for which the first- and second-order statistics (mean and correlation) are time-invariant. This is a much relaxed version of stationarity but is adequate for most of the purposes of this book. We note that from (3.3.24)–(3.3.26) the stochastic process (3.3.22) can be proclaimed WSS since $\bar{f}, \overline{f^2}$, and $r(\Delta)$ are all time-invariant.

3.3.2.5 The Concept of White Noise
The concept of white noise is widely used for both analysis and as a practical benchmark for testing systems and algorithms. For reasons to be immediately seen, it is also sometimes called uncorrelated noise.

Suppose we construct a discrete-time stochastic process by allowing each sample $x(k)$ to be a random variable with zero mean. The probability distribution of each of these variables is not too important, but we do assume that each random variable is statistically independent of all others. Suppose α and β are two such variables and they are assigned to $x(k_1)$ and $x(k_2)$. Since α and β are statistically independent, their joint probability distribution is just the product of their individual ones, that is

$$p_{\alpha\beta}(\alpha = \alpha_0, \beta = \beta_0) = p_\alpha(\alpha = \alpha_0) \cdot p_\beta(\beta = \beta_0).$$ (3.3.27)

We now examine the statistics of $x(k)$.

(a) Mean:
$$\bar{x}(k_1) = \bar{\alpha} = \int_\alpha p_\alpha(\alpha = \alpha_0) \cdot \alpha d\alpha = 0,$$ (3.3.28)

since all variables were assumed to have zero mean.

(b) Variance:
$$\overline{x^2}(k_1) = \overline{\alpha^2} = \int_\alpha p_\alpha(\alpha = \alpha_0) \cdot (\alpha_0 - \bar{\alpha})^2 d\alpha,$$ (3.3.29)

where the exact value depends on the distribution p_α.

(c) Correlation between $x(k_1)$ and $x(k_2)$:

$$\overline{x(k_1) \cdot x(k_2)} = \int_{\alpha, \beta} p_{\alpha, \beta}(\alpha = \alpha_0, \beta = \beta_0) \cdot \alpha_0 \cdot \beta_0 d\alpha d\beta$$

$$= \int_{\alpha} p_{\alpha}(\alpha = \alpha_0) \cdot \alpha_0 d\alpha \cdot \int_{\beta} p_{\beta}(\beta = \beta_0) \cdot \beta_0 d\beta$$

$$= \bar{\alpha} \cdot \bar{\beta} = 0 \cdot 0 = 0 \qquad (3.3.30)$$

unless $k_1 = k_2$, and, if so,

$$\overline{x(k_1) \cdot x(k_2)} = \overline{\alpha^2}. \qquad (3.3.31)$$

From (c) we observe that since the process is zero mean and since each sample of $x(k)$ is independent of all others, then the autocorrelation function has the simple form of

$$r(k, \Delta) = \overline{x^2(k)} \cdot \delta(\Delta), \qquad (3.3.32)$$

that is, at time k, the autocorrelation function is a pulse function, zero for all Δ other than zero.

We note that this process is not stationary since the power and the autocorrelation function depend on exactly which random variable is used for each value of $x(k)$. It is common to make the white noise process better behaved by assuming that *all* values of $x(k)$ are chosen from the same probability distribution, that is, that they are independent and identically distributed (i.i.d.). If so, the power is the same for each value of k, and

$$r(k, \Delta) = r(\Delta) = \overline{x^2} \cdot \delta(\Delta). \qquad (3.3.33)$$

Now the white noise process is also wide-sense stationary. It is also common, but not necessary, to assume that the random variables are chosen from a particular type of probability distribution, such as Gaussian or Poisson. For the remainder of Section 3.3 we assume that the samples of the stochastic process are i.i.d., but we need make no assumption about the distribution itself.

A process such as $x(k)$ is termed a "white noise" process. To see this we use the Wiener–Khinchine relationship, which states that the power spectrum of a WSS stochastic process is given by the Fourier transform of the process's autocorrelation function [Melsa and Sage, 1973], i.e.,

$$S(\omega) \triangleq \sum_{\Delta = -\infty}^{\infty} r(\Delta) e^{-j\omega \Delta T}, \qquad -\frac{\pi}{T} \leqslant \omega < \frac{\pi}{T}. \qquad (3.3.34)$$

Using the autocorrelation function specified in (3.3.33), the power spectrum of $x(k)$ is

$$S_x(\omega) = \sum_{\Delta = -\infty}^{\infty} \overline{x^2} \cdot \delta(\Delta) e^{-j\omega\Delta T} = \overline{x^2}, \qquad \text{for all } \omega. \qquad (3.3.35)$$

Note that the process has a uniform power density of $\overline{x^2}$ at all frequencies. Since all frequencies are equally represented, like white light, the process is termed "white." Any correlation between samples will "color" the process and destroy the spectral uniformity.

3.3.3 Development of the Normal Equations

We return now to (3.3.2) and set out to find an impulse response vector $\mathbf{W}_{ms}^o$ which minimizes this new objective function J_{ms}, the expected value of the squared error. As we shall see, much of the analysis parallels that of Section 3.2, except that ensemble averaging instead of time summing is employed. We begin by assuming that $x(k)$ and $d(k)$ are stochastic processes. As the development proceeds, we add a few other conditions.

Using the definition of $e(k)$ in (3.3.1), we first expand J_{ms} in (3.3.2) as

$$J_{ms} = E\{|e(k)|^2\} = E\{|d(k) - y(k)|^2\}$$
$$= E\{|d(k)|^2\} - E\{d(k)y^*(k)\} - E\{d^*(k)y(k)\} + E\{|y(k)|^2\}. \quad (3.3.36)$$

Since $y(k) \triangleq \mathbf{X}^t(k)\mathbf{W}$, J_{ms} becomes

$$J_{ms} = E\{|d(k)|^2\} - 2\mathbf{W}^t Re(E\{d^*(k)\mathbf{X}(k)\}) + \mathbf{W}^t E\{\mathbf{X}(k)\mathbf{X}^h(k)\}\mathbf{W}^*. \qquad (3.3.37)$$

For convenience we make the following definitions:

(a) The expected power of $d(k)$:

$$D_{ms} \triangleq E\{|d(k)|^2\}. \qquad (3.3.38)$$

(b) The ensemble autocorrelation matrix of $x(k)$

$$R_{ms} \triangleq E\{\mathbf{X}(k)\mathbf{X}^h(k)\}. \qquad (3.3.39)$$

(c) The ensemble average cross-correlation vector

$$\mathbf{P}_{ms} \triangleq E\{d^*(k)\mathbf{X}(k)\}. \qquad (3.3.40)$$

To make D, P, and R time-invariant we can assume that $d(k)$ and $x(k)$ are at least wide-sense stationary. Then, for real-valued processes, J_{ms} can be written compactly as

$$J_{ms} = D_{ms} - 2\mathbf{W}^t \mathbf{P}_{ms} + \mathbf{W}^t R_{ms} \mathbf{W}. \qquad (3.3.41)$$

Note the similarity between this expression and J_{ss} in (3.2.10).

To find that choice of $\mathbf{W}$ which minimizes J_{ms} we follow the same steps as those used to minimize J_{ss}. We find the gradient of J_{ms} with respect to $\mathbf{W}$ and find the value of $\mathbf{W}$ which sets it to zero. This leads to the normal equations,

$$R_{ms} \mathbf{W}_{ms}^\circ = \mathbf{P}_{ms}. \qquad (3.3.42)$$

We determine the uniqueness of this solution by evaluating the Hessian matrix of J_{ms} and checking it for positive definiteness. Exactly the same analysis which led to (3.2.29) shows us that

$$H_w = 2R_{ms}; \qquad (3.3.43)$$

the Hessian matrix is just twice the autocorrelation matrix. If this matrix has full rank, then the solution is unique, R_{ms} is invertible, and

$$\mathbf{W}_{ms}^\circ = R_{ms}^{-1} \mathbf{P}_{ms}. \qquad (3.3.44)$$

We obtain this result promptly since, once P and R are defined, the solution for $\mathbf{W}_{ms}^\circ$ comes directly from vector analysis of the quadratic form of the performance function. We now look at the meaning of $\mathbf{P}_{ms}$ and R_{ms}, as a start toward understanding the relationship between J_{ss} in (3.2.1) and J_{ms} in (3.3.2).

3.3.4 The Ensemble Average Auto- and Cross-Correlation Functions

We turn first to the autocorrelation matrix R_{ms}. The i,jth term of R_{ms} is given by

$$r_{ij} = E\{x(k - i) \cdot x(k - j)\}. \qquad (3.3.45)$$

Since we have assumed $x(k)$ to be WSS, the statistic r_{ij} should be invariant to a time shift. Thus,

$$r_{ij} = E\{x(k + (j - i)) \cdot x(k)\}. \qquad (3.3.46)$$

For real-valued $x(k)$, this is just the definition of $r(\Delta)$ from (3.3.11). Therefore,

$$r_{ij} = r(j - i) \qquad (3.3.47)$$

and the matrix R_{ms} is built up completely with values of $r(\Delta)$. Since $0 \leqslant i$, $j \leqslant N - 1$, we can note a few things about R_{ms}:

(a) It is Toeplitz or "banded", i.e., terms in the main diagonal or the same off-diagonal have the same value of $j - i$ and hence the same value $r(j - i)$.

(b) For real-valued $x(k)$,

$$r(j - i) = r(i - j). \tag{3.3.48}$$

Thus R_{ms} is symmetric and symmetrically banded. If $x(k)$ is complex-valued, then R_{ms} becomes conjugate symmetric (sometimes called "Hermitian").

(c) The average power appears N times on the main diagonal.

(d) The highest value of shift Δ in $r(\Delta)$ used to build R_{ms} is $\pm(N - 1)$, since $\Delta = j - i$ and both i and j are limited to $N - 1$. Thus only a $2N - 1$ point window of the whole autocorrelation function is used to build R_{ms}.

The cross-correlation vector $\mathbf{P}_{ms}$ is given by

$$p_i = E\{d(k) \cdot x(k - i)\}. \tag{3.3.49}$$

Again, following the cross-correlation definition in (3.3.13) and invoking the assumed stationarity of d and x,

$$
\begin{aligned}
p_i &= E\{d(k + i) \cdot x(k)\} \\
&= c(i), \qquad 0 \leqslant i \leqslant N - 1,
\end{aligned}
\tag{3.3.50}
$$

where $c(i)$ is just the cross-correlation function of $d(k)$ and $x(k)$. The cross-correlation vector is built directly from an N-point window of the cross-correlation function.

3.3.5 Examples

To show how (3.3.44) can be used to develop the optimal filter in the mean square sense, we shall do two examples both closely related to the examples in Section 3.2.6. The key difference is that the stochastic framework is used here.

3.3.5.1 Matching a Pure Delay with White Noise Excitation

Suppose we have an N-point FIR filter and desire to choose $\mathbf{W}$, the impulse response vector, to minimize the ensemble average of the squared error, that is, J_{ms}. Suppose further that $x(k)$ and $d(k)$ are given as

$$x(k) = An(k), \qquad -\infty < k < \infty, \tag{3.3.51}$$

where $n(k)$ is a zero mean white process with average power σ^2, and

$$d(k) = Bn(k - \tau), \qquad 0 \leqslant \tau \leqslant N - 1, \tag{3.3.52}$$

that is, $d(k)$ is a scaled and delayed version of the filter input $x(k)$.

To determine R_{ms} and $\mathbf{P}_{ms}$, we must first find $r(\Delta)$ and $c(\Delta)$. Since $n(k)$ has been defined as white, with power σ^2 and zero mean, the autocorrelation function for $x(k)$ is

$$r(\Delta) = A^2\sigma^2\delta(\Delta). \tag{3.3.53}$$

Since

$$r_{ij} = r(j - i),$$

the autocorrelation matrix R_{ms} has $A^2\sigma^2$ on the main diagonal (where $j - i = 0$) and zero on all other diagonals since $j - i \neq 0$ there. Therefore,

$$R_{ms} = \sigma^2 A^2 I_N, \tag{3.3.54}$$

where I_N is the N-dimensional identity matrix.

We find $c(\Delta)$ as follows:

$$c(\Delta) \triangleq E\{d(k) \cdot x(k - \Delta)\}$$

$$= E\{Bn(k - \tau) \cdot A \cdot n(k - \Delta)\}$$

$$= A \cdot B \cdot E\{n(k - \tau + \Delta) \cdot n(k)\}$$

$$= A \cdot B \cdot \sigma^2\delta(\Delta - \tau). \tag{3.3.55}$$

Thus $c(\Delta)$ equals $AB \cdot \sigma^2$ when $\Delta = \tau$ but is zero otherwise. The vector $\mathbf{P}_{ms}$ is then

$$\mathbf{P}_{ms} = A \cdot B \cdot \sigma^2\mathbf{e}_\tau, \tag{3.3.56}$$

where $\mathbf{e}_\tau$ has zeros in all locations except for the τth entry, which is unity.

The optimal vector $\mathbf{W}^o_{ms}$ is found using (3.3.44). Since A^2 and σ^2 are positive, R_{ms} is invertible and

$$\mathbf{W}^o_{ms} = R_{ms}^{-1}\mathbf{P}_{ms} = (A^2\sigma^2 I_N)^{-1} AB \cdot \sigma^2\mathbf{e}_\tau = \frac{B}{A}\mathbf{e}_\tau. \tag{3.3.57}$$

It can be confirmed that the impulse response vector $\mathbf{e}_\tau$ simply delays the filter input by τ samples. Thus, the optimal filter scales the input by B/A and delays it by τ samples.

3.3.5.2 Matching a Rotating Phasor

Now suppose that $x(k)$ is a complex-valued sinusoid with random starting phase,

$$x(k) = A e^{j(\omega Tk + \theta)}, \tag{3.3.58}$$

where ω and T are constants and θ is a random variable chosen from a uniform distribution over $(-\pi, \pi]$. Suppose further that $d(k)$ is given by another complex waveform

$$d(k) = B e^{j(\omega T(k - \tau) + \theta)}. \tag{3.3.59}$$

The autocorrelation function $r(\Delta)$ can be determined by reference to (3.3.26) as

$$r(\Delta) = A^2 e^{j\omega T\Delta} \tag{3.3.60}$$

and $c(\Delta)$ can easily be found to be

$$c(\Delta) = AB e^{j\omega T(\Delta - \tau)}. \tag{3.3.61}$$

At this point, the correlation functions, plus R and $\mathbf{P}$, are identical to the time-average correlation functions used in Section 3.2.5.2, except for a constant scaling factor. As a result, the same solution holds. For $N = 1$, the solution is unique,

$$w^0 = \frac{B}{A} e^{-j\omega T\tau}, \tag{3.3.62}$$

but for high-order filters there are an infinite number of choices for $\mathbf{W}^o_{ms}$ which minimize J_{ms}.

3.3.5.3 Filtering a Noisy Sinusoid

We now consider a problem which represents a simplified form of many practical situations. We assume that the filter input $x(k)$ is the sum of two components, one a complex valued, zero-mean white noise process $An(k)$ and the other a complex sinusoid $B \exp j(\omega_0 kT + \theta)$. Since $n(k)$ is white, we know that each time sample is statistically independent of all others, but here we also assume that θ, the random starting phase of the sinusoid, is also independent of all samples of $n(k)$. As in Section 3.3.5.2, we further assume that the starting phase θ is uniformly distributed over the interval $(-\pi, \pi]$. For this example the desired signal $d(k)$ is given by

$$d(k) = S \exp (\omega_0(k - \tau)T + \theta), \tag{3.3.63}$$

which is a delayed and scaled version of the sinusoidal component present at the input. The amplitudes A, B, and S are constants and the variance of $n(k)$, its power, is given by σ^2. The filter input is a complex sinusoid plus broadband noise, and the desired output is the sinusoid only. Intuitively, then, it seems that the optimal filter should have a bandpass response which selects the sinusoid and suppresses the noise. To see if this is true, we will solve for the optimal impulse response vector W°.

Since θ and $n(k)$ are statistically independent, the autocorrelation function of $x(k)$, called $r_x(\Delta)$, is just the sum of the autocorrelation functions of $n(k)$ and the sinusoid. Using (3.3.53) and (3.3.60), we can write

$$r_x(\Delta) = A^2\sigma^2\delta(\Delta) + B^2 \exp j\omega_0\Delta T, \tag{3.3.64}$$

and the elements of the autocorrelation matrix as

$$r_{il} = r_x(l - i) = A^2\sigma^2\delta(l - i) + B^2 \exp j\omega_0 T(l - i). \tag{3.3.65}$$

The results of the two previous sections can also be used to determine the cross-correlation function $c(\Delta)$. By definition,

$$c_{xd}(\Delta) = E\{x(k)d(k + \Delta)\} = E\{d(k)x(k - \Delta)\}$$

$$= E\{d(k)\cdot[A \cdot n(k - \Delta) + B \exp\{j\omega_0(k - \Delta)T + \theta\}]\}$$

$$= c_{nd}(\Delta) + c_{sd}(\Delta), \tag{3.3.66}$$

which is the sum of the cross-correlation functions of $d(k)$ with the two components of $x(k)$. But θ has been assumed to be independent of $n(k)$, thereby making c_{nd} equal to zero for any value of Δ. Thus, reusing (3.3.61),

$$c_{xd}(\Delta) = c_{sd}(\Delta) = BS \exp j\omega_0(\Delta - \tau)T. \tag{3.3.67}$$

Now that we have $r_x(\Delta)$ and $c_{xd}(\Delta)$ we could simply plug values into R_{ms} and P_{ms} and solve, but first we will find convenient forms for R and P. Suppose we define the complex N-vector $\Gamma(\omega_0)$ as

$$\Gamma(\omega_0) = [1 \quad e^{j\omega_0 T} \cdots e^{j\omega_0(N-1)T}]^t. \tag{3.3.68}$$

If we evaluate the i, l-th term of $\Gamma(\omega_0)\Gamma^h(\omega_0)$, we find it to be

$$\{\Gamma(\omega_0)\Gamma^h(\omega_0)\}_{il} = e^{j\omega_0 iT}e^{-j\omega_0 lT}$$

$$= e^{j\omega_0(i-l)T}. \tag{3.3.69}$$

Comparing this with the second term of (3.3.65), we see that R_{ms}, the autocorrelation matrix of $x(k)$, can be written as the sum of a scaled identity matrix and an outer product of two vectors

$$R_{ms} = A^2\sigma^2 I_N + B^2\Gamma(\omega_0)\Gamma^h(\omega_0). \tag{3.3.70}$$

The cross-correlation vector can also be expressed in terms of $\Gamma(\omega_0)$, since

$$p_i = c_x(\Delta - i) = BS \exp j\omega_0(i - \Delta)T = SBe^{-j\omega_0\Delta T}e^{j\omega_0 iT}$$

or

$$\mathbf{P} = SB \cdot e^{-j\omega_0\tau T} \cdot [1 \quad e^{j\omega_0 T} \quad \dots \quad e^{j\omega_0(N-1)T}]^t$$

$$= SB \cdot e^{j\omega_0\tau T} \cdot \Gamma(\omega_0). \tag{3.3.71}$$

Now that we have R_{ms} and $\mathbf{P}_{ms}$ we can solve for $\mathbf{W}^o_{ms}$. As we shall show in a later section, an autocorrelation matrix of this form always has full rank if σ^2, the power of the noise process, is not zero. It is easily confirmed that the inverse of R is given by

$$R_{ms}^{-1} = \frac{1}{A^2\sigma^2}\left\{I_N - \frac{B^2}{A^2\sigma^2 + B^2 N}\,\Gamma(\omega_0)\Gamma^h(\omega_0)\right\}.$$

Multiplying by $\mathbf{P}$ yields

$$\mathbf{W}^o = \frac{1}{A^2\sigma^2}\left\{I_N - \frac{NB^2}{A^2\sigma^2 + B^2 N}\right\} \cdot \{SB \cdot e^{-j\omega_0\tau T}\Gamma(\omega_0)\}$$

$$= \frac{SB}{A^2\sigma^2 + B^2 N}\cdot e^{-j\omega_0\tau T}\cdot\Gamma(\omega_0); \tag{3.3.72}$$

thus the optimal coefficient vector $\mathbf{W}^o_{ms}$ is a scaled version of the vector $\Gamma(\omega_0)$. The implications of this result become clearer when we determine the frequency response of this filter,

$$W(\omega) = \sum_{i=0}^{N-1} w_i^o e^{-j\omega iT}$$

$$= \frac{SB}{A^2\sigma^2 + B^2 N}\, e^{-j\omega_0\tau T}\left\{\sum_{i=0}^{N-1} e^{j\omega_0 iT}e^{-j\omega iT}\right\}$$

$$= \begin{cases} \dfrac{SB}{A^2\sigma^2 + B^2 N}\cdot e^{-j\omega_0\tau T}\left\{\dfrac{1 - \exp jN(\omega_0 - \omega)T}{1 - \exp j(\omega_0 - \omega)T}\right\}. & \omega \neq \omega_0 \\[4mm] \dfrac{SNB}{A^2\sigma^2 + B^2 N}\cdot e^{-j\omega_0\tau T}, & \omega = \omega_0. \end{cases} \tag{3.3.73}$$

A plot of this frequency response is shown in Figure 3.1, where $\omega_0 = 0$ rad/sec, $N = 25$, and the DC gain of the filter is unity. The optimal filter clearly has a

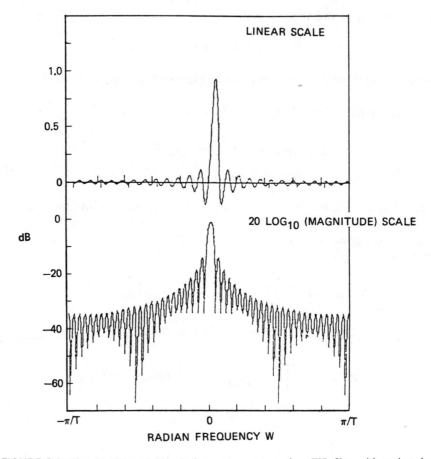

FIGURE 3.1. Linear and log magnitude frequency response of an FIR filter with an impulse response length N of 25.

bandpass response just as intuition predicted. The maximum gain is attained at the frequency of the complex sinusoid, and the noise is suppressed over the remainder of the band.

What may be unexpected, however, is the gain of the filter. Reviewing the problem again, we see that a filter gain of S/B is required to scale the sinusoidal input component so that it precisely cancels template waveform $d(k)$. In light of that, we ask when $W(\omega_0)$ can attain this value. Rewriting $W(\omega_0)$, we see that

$$|W(\omega_0)| = \frac{S}{B} \left\{ \frac{B^2 N}{A^2 \sigma^2 + B^2 N} \right\} \triangleq \frac{S}{B} \left\{ \frac{N\rho}{1 + N\rho} \right\} \qquad (3.3.74)$$

where $\rho = B^2/A^2\sigma^2 =$ the signal-to-noise ratio. Thus if ρ, the ratio of the tone to noise power at the input, is very high, then the gain approaches S/B asymptotically. If ρ is very low compared to $1/N$, then the gain falls. If ρ equals zero, then so do the gain and the optimal vector $\mathbf{W}^o$. The analytical explanation for this behavior will be developed in the next section, but an intuitive one can be offered already. The error signal consists of two components, the uncancelled portion of the sinusoid and any noise which is able to pass through the filter. If the tone-to-noise ratio is very high, then there is only a small amount of noise to pass through the filter, so the overall error power is minimized by cancelling the tone well. When ρ is very low, however, the filter input is mostly noise, and minimizing the error power is done best by "turning off" the filter, that is, reducing its gain to zero as ρ goes to zero. For moderate levels of tone and noise amplitude, the optimal solution is a compromise between the two extremes.

3.3.6 Consolidation

In this section we phrased a new problem. The objective was still to find the filter impulse response that minimizes the error $e(k)$, but averaging over an ensemble of possible waveforms was used instead of summing or averaging over a finite time segment of one waveform.

We find that much is the same about the summed squared error problem and the mean squared error problem. In both cases the average squared error can be written in terms of the auto- and cross-correlation functions of the signals $x(k)$ and $d(k)$. The only difference comes in the way these correlation functions are defined. Given those functions, hence R and $\mathbf{P}$, vector calculus techniques are used to find the impulse response vector(s) that minimize the solution itself, and even methods of computing the solution are identical, once given R and $\mathbf{P}$.

What, then, is different, and why should we pursue the analysis of the mean square solution? Consider again the sum-of-squares criterion of Section 3.2. If the data window from $k = 1$ to $k = L$ is used instead of the window from $k = 0$ to $L - 1$, there is no guarantee that the two associated optimal impulse response vectors will be close to each other at all. This comes simply from the fact that we do not have any good way of quantifying how similar two deterministic waveforms are. The stochastic framework allows us to do this, however. For example, section 3.3.5.3 tells us the optimal weight vector over a broad class of wideband noise signals and sinusoids of arbitrary frequency and uncertain phase.

The paradox then is this: In the real world we have only one realization and therefore can really only consider the summed squared solution. It is virtually impossible to generalize about this solution. As a result we create the analytical artifice of a stochastic process, one realization of which is our observed sequence. With these stochastic process definitions, however, we can predict the optimal solution over a broad class of possible realizations, including the one actually observed. Broadly speaking, we prove theorems about the mean square problem but in practice actually solve the sum-of-squares problem.

3.4 PROPERTIES OF THE SOLUTION

3.4.1 Evaluation of the Performance Function

In this section we use the solutions developed earlier for the optimal impulse response to find new expressions for the performance function J. In all cases the intent is to add some intuitive insight into what J means and how the optimal solution $\mathbf{W}^\circ$ depends on the characteristics of $x(k)$ and $d(k)$.

We know from Sections 3.2 and 3.3 that both the summed square and mean square performance functions can be written as

$$J = D - 2\mathbf{W}^t\mathbf{P} + \mathbf{W}^t R \mathbf{W} \qquad (3.4.1)$$

where D, $\mathbf{P}$, and R are computed with the appropriate averaging technique. For convenience here we also assume $x(k)$ and $d(k)$ to be real-valued functions. Both summed and mean square problems yield an optimal solution which satisfies the normal equations

$$R\mathbf{W}^\circ = \mathbf{P}, \qquad (3.4.2)$$

whether or not $\mathbf{W}^\circ$ is unique. We can then evaluate $J_{\min}$, the minimum value of J, as the value of J when $\mathbf{W}$ satisfies the normal equations:

$$
\begin{aligned}
J_{\min} &= D - 2\mathbf{W}^{\circ t}\mathbf{P} + \mathbf{W}^{\circ t} R \mathbf{W}^\circ \\
&= D - 2\mathbf{W}^{\circ t}\mathbf{P} + \mathbf{W}^{\circ t}\mathbf{P} \\
&= D - \mathbf{W}^{\circ t}\mathbf{P}.
\end{aligned}
\qquad (3.4.3)
$$

Thus the minimum value of J depends on the power or energy of $d(k)$, called D, on the optimal weight vector $\mathbf{W}^\circ$ and on the cross-correlation between $\mathbf{X}(k)$ and $d(k)$, called $\mathbf{P}$.

Given this minimum value, it is interesting to know how large a price in performance is paid by not using the optimal vector. In particular, let $\mathbf{W}$, the actual weight vector, be defined by

$$\mathbf{W} \triangleq \mathbf{W}^\circ + \mathbf{V}, \qquad (3.4.4)$$

where $\mathbf{V}$ is an N-dimensional vector representing the difference between $\mathbf{W}$ and $\mathbf{W}^\circ$. Substituting this value of $\mathbf{W}$ into (3.4.1), we find that

$$
\begin{aligned}
J &= D - 2(\mathbf{W}^{\circ t} + \mathbf{V}^t)\mathbf{P} + \{\mathbf{W}^{\circ t} + \mathbf{V}^t\}R\{\mathbf{W}^\circ + \mathbf{V}\} \\
&= D - 2\mathbf{W}^{\circ t}\mathbf{P} + \mathbf{W}^{\circ t}R\mathbf{W}^\circ - 2\mathbf{V}^t\mathbf{P} + \mathbf{W}^{\circ t}R\mathbf{V} + \mathbf{V}^t R\mathbf{W}^\circ + \mathbf{V}^t R\mathbf{V}.
\end{aligned}
\qquad (3.4.5)
$$

We recognize the first three terms to be $J_{\min}$. Using the normal equations again and the fact that R is a symmetric matrix for real signals, we can write J as

$$J = J_{min} - 2\mathbf{V}^t\mathbf{P} + \mathbf{P}^t\mathbf{V} + \mathbf{V}^t\mathbf{P} + \mathbf{V}^t R\mathbf{V},$$

$$= J_{min} + \mathbf{V}^t R\mathbf{V} \tag{3.4.6}$$

or ΔJ, the excess average squared error, as

$$\Delta J \triangleq J - J_{min} = \mathbf{V}^t R\mathbf{V}. \tag{3.4.7}$$

This is a key result and forms the basis for much intuition about optimal squared error filters. It states that any excess squared error is a quadratic function of the difference between the actual weight vector $\mathbf{W}$ and the optimal one $\mathbf{W}^o$. We have assumed earlier, and will shortly prove, that R is positive semidefinite, meaning that

$$\mathbf{V}^t R\mathbf{V} \geqslant 0 \tag{3.4.8}$$

for any nonzero vector $\mathbf{V}$ (that is, $\mathbf{V}^t\mathbf{V} \neq 0$). Since $\mathbf{V}^t R\mathbf{V}$ is never negative, J_{min} is in fact the minimum attainable value. For most well-behaved problems $\mathbf{V}^t R\mathbf{V}$ is always positive for nonzero $\mathbf{V}$, meaning that any difference between $\mathbf{W}$ and $\mathbf{W}^o$, hence nonzero $\mathbf{V}$, will result in J being greater than the minimum possible value. We notice that ΔJ, the penalty for poor choice of $\mathbf{W}$, depends only on the input process $x(k)$ and not on the desired template waveform $d(k)$. Thus once again the character of the filter input affects the nature of the solution.

The quadratic penalty for nonoptimal choices of $\mathbf{W}$ has many implications, including the idea of iteratively choosing $\mathbf{W}$ in such a way as to reduce J each time, thus ultimately reaching j_{min} and $\mathbf{W}^o$. We return to the expression for ΔJ first, however, and try to determine under what circumstances a nonzero difference vector $\mathbf{V}$ leads to no performance penalty.

3.4.2 The Positivity of R

If the correlation matrix R is positive definite, i.e., (3.4.8) is satisfied with a strict inequality, then several related facts are implied:

(a) $\Delta J > 0$ for any $\mathbf{V} \neq 0$; that is, any difference between $\mathbf{W}$ and $\mathbf{W}^o$ leads to a performance penalty;

(b) R has full rank;

(c) R is invertible;

(d) The normal equations have a unique solution; and

(e) $\mathbf{W}^o = R^{-1}\mathbf{P}$.

One might ask when R is in fact positive definite. To answer this inquiry we return to the definition of R as

$$R = \text{Avrg}\{\mathbf{X}(k)\mathbf{X}^t(k)\}, \tag{3.4.9}$$

where $\text{Avrg}\{\cdot\}$ denotes some form of averaging. The quadratic form $\mathbf{V}^t R \mathbf{V}$ then becomes

$$\mathbf{V}^t R \mathbf{V} = \mathbf{V}^t[\text{Avrg}\{\mathbf{X}(k)\mathbf{X}^t(k)\}]\cdot\mathbf{V}$$

$$= \text{Avrg}\{\mathbf{V}^t\mathbf{X}(k)\mathbf{X}^t(k)\mathbf{V}\}. \tag{3.4.10}$$

Including constant $\mathbf{V}^t$ and $\mathbf{V}$ inside the averaging operator is possible for any linear averaging technique. Now suppose we write $\mathbf{V}^t\mathbf{X}(k)$ as

$$\mathbf{V}^t\mathbf{X}(k) = s(k). \tag{3.4.11}$$

Notice that $s(k)$ is just the output of an N-tap FIR filter whose impulse-response vector is $\mathbf{V}$ and whose input is $x(k)$. Thus $s(k)$ is the output of a fictitious "difference filter." If $\mathbf{V}$ equals zero, then $s(k)$ also equals zero for any input $x(k)$, and hence for any state vector $\mathbf{X}(k)$. Using this output $s(k)$, we see that, since $\mathbf{V}^t\mathbf{X}(k) = s(k) = \mathbf{X}^t(k)\mathbf{V}$, the quadratic form is just

$$\mathbf{V}^t R \mathbf{V} = \text{Avrg}\{s^2(k)\}. \tag{3.4.12}$$

But $s^2(k)$ is never negative, and therefore no time or ensemble average of it will be. Therefore, we have proven that $\mathbf{V}^t R \mathbf{V}$ is never less than zero.

Under what conditions can this quadratic term equal zero? Looking back it is clear that $\text{Avrg}\{s^2(k)\}$ will equal zero only if $s(k)$ equals zero identically for every term in the average. For a time average, it means $\text{Avrg}\{s^2(k)\}$ equals zero for all appropriate k; for the ensemble average it means that $\text{Avrg}\{s^2(k)\}$ equals zero for every realization (or strictly speaking, for all except a set of measure zero). In other words, $\mathbf{V}^t R \mathbf{V}$ can equal zero if there is some value of $\mathbf{V}$ for which $\mathbf{V}^t\mathbf{X}(k) = s(k) = 0$ for all possible choices of $\mathbf{X}(k)$, over an ensemble or over time. If, for all possible $x(k)$, there is some choice of $\mathbf{V}$, call it $\mathbf{V}_o$, which makes $s(k) \equiv 0$, then R is not positive definite. If there is no such $\mathbf{V}_o$, then R is called strictly positive definite, any nonzero difference $\mathbf{V}$ is sensed by ΔJ, and a unique solution exists for $\mathbf{W}^o$.

What is the intuitive meaning of this? Suppose there is some vector $\mathbf{V}_o$ which makes the difference filter output $s(k)$ equal to zero. If so, then any amount of this difference vector can be added to $\mathbf{W}^o$ without increasing the output $y(k)$, without impacting the error $e(k)$, and therefore without influencing the performance function J. Thus the performance function J is blind to a difference between $\mathbf{W}^o$ and $\mathbf{W}^o + \alpha\cdot\mathbf{V}_o$ where α is any scalar. When this is true, the optimal solution is definitely not unique. If we want filters with unique solutions, we want to avoid situations where there exists some $\mathbf{V}_o$ which is unobservable in the output. A general way of analyzing this is presented in the next section, but for now it is useful to show one example of where this nonuniqueness can happen.

Suppose $x(k)$, the filter input, is a sinusoid given by

$$x(k) = B \cos(2\pi nk/N + \theta), \qquad 0 < n < N/2, \qquad (3.4.13)$$

where N is the filter length, n is an integer, and θ is an unknown (or random) phase. We can show that if $\mathbf{V}_o$ is given by $\mathbf{V}_o = [1 \ 1 \ \cdots \ 1]^t$, then $\mathbf{X}^t(k)\mathbf{V}_o = 0$. Why is this nonzero $\mathbf{V}_o$ allowed with no output error $s(k)$ from (3.4.11)? One answer is that the frequency response of the filter given by $\mathbf{V}_o$ has transmission zeros at all nonzero multiples of $f = 1/N$, including n/N and $-n/N$. Thus $\mathbf{V}_o$ has spectral nulls at exactly the frequency of the input, making the output zero in spite of any change in the magnitude of $\mathbf{V}_o$.

As it happens, there are many other choices of $\mathbf{V}_o$ for which $s(k) \equiv 0$, but all of them share the property that they have zero gain at the frequency of the input sinusoid. A more complete analytical answer is developed in the next section, but as a practical matter, the nonpositivity of R almost always stems from some form of spectral imbalance of the input signal. If the input signal does not have any energy in some band of frequencies, then the filter's solution may not be unique in that region. Even if lack of uniqueness does not bother the designer, the associated fact that R is not invertible may be a cause for concern, depending on how $\mathbf{W}^o$ is to be determined.

3.4.3 Examples Revisited

3.4.3.1 Matching a Pure Delay with Noise Excitation
Continuing from Section 3.3.5.1, we know that

$$R_{ms} = \sigma^2 A^2 I_N, \qquad \mathbf{P}_{ms} = AB\sigma^2 \mathbf{e}_\tau, \quad \text{and} \quad \mathbf{W}_{ms}^o = \frac{B}{A} \mathbf{e}_\tau. \qquad (3.4.14)$$

We find that J_{min} is given by

$$J_{min} = D - \mathbf{P}^t \mathbf{W}^o = B^2 \sigma^2 - AB\sigma^2 \mathbf{e}_\tau^t \cdot \frac{B}{A} \mathbf{e}_\tau.$$

$$= B^2 \sigma^2 - B^2 \sigma^2 \equiv 0, \qquad (3.4.15)$$

thus there is no residual error. The excess error ΔJ is given by

$$\Delta J = J - J_{min} = \mathbf{V}^t R \mathbf{V} = \sigma^2 A^2 \mathbf{V}^t \mathbf{V} = \sigma^2 A^2 |\mathbf{V}|^2. \qquad (3.4.16)$$

Thus every component of the difference vector can cause excess error.

3.4.3.2 Filtering a Noisy Sinusoid
Again continuing from Section 3.3.5.3, we have

$$R_{ms} = A^2 \sigma^2 I_N + B^2 \Gamma(\omega_0) \Gamma^h(\omega_o) \qquad (3.4.17)$$

and

$$\mathbf{P}_{ms} = SBe^{-j\omega_0 t T}\mathbf{\Gamma}(\omega_0) \tag{3.4.18}$$

The optimal solution yields a minimum performance penalty of

$$J_{\min} = D - \mathbf{P}^h\mathbf{W}^\circ = B^2 - BSe^{j\omega_0 t T}\mathbf{\Gamma}(\omega_0)\frac{B}{S}\left(\frac{\rho}{1+N\rho}\right)e^{-j\omega_0 t T}\mathbf{\Gamma}(\omega_0)$$

$$= B^2 - B^2\left(\frac{N\rho}{1+N\rho}\right) = B^2\left(\frac{1}{1+N\rho}\right), \quad \text{where } \rho = \frac{B^2}{A^2\sigma^2}. \tag{3.4.19}$$

If ρ, the SNR, is very high, then $J_{\min} \to 0$. If the SNR is very low, then $\mathbf{W}^\circ \to 0$ and $J_{\min} = B^2$, the power of the template signal alone.

The excess performance penalty is given by

$$\Delta J = \mathbf{V}^t R\mathbf{V} = A^2\sigma^2\|\mathbf{V}\|^2 + B^2\{\mathbf{V}^t\mathbf{\Gamma}(\omega_0)\}^2. \tag{3.4.20}$$

The first term on the right of (3.4.20) is always positive if $\mathbf{V}$ has any nonzero component and σ^2 is not zero. However the size of the second term depends on the frequency response of the filter described by $\mathbf{V}$ at frequency ω_0.

3.4.4 Decompositions Based on the Eigensystem of *R*

Each of the examples in Section 3.4.3 has in common the fact that the associated autocorrelation matrix *R* could be presented in some simplified form. In every case the simplicity sprang from the assumed form of the input $x(k)$. With a more general (and realistic) input signal, *R* cannot be so tidily decomposed. Even so, some such decomposition does exist and it can be exploited to understand both the convergence and transient behavior of important adaptive filtering algorithms.

To provide the background tools for *R* decomposition and exploitation we quickly review the linear algebra concepts of eigenvalues, eigenvectors, and the modal matrix. This modal matrix permits a rotation of the coordinate system used to describe *R*, **W**, and **P**. In this rotated coordinate system both the optimal filter impulse response and revealing properties of the adaptive algorithms can be written in a simplified manner.

Suppose we have an *N*-by-*N* matrix *A* and some *N*-element vector **V**. The vector **V** has a certain length, called the norm $\|\mathbf{V}\|$, and points in some specific direction in *N*-space. This direction is that established by associating each entry in **V** with a different coordinate location and drawing a vector from the origin to this point. The multiplication of **V** by the matrix *A* generally has the effect of changing (e.g., rotating) the direction of **V** and changing its length. For some matrices there are special vectors, called eigenvectors, which have the special property that they are not rotated when multiplied by A. When this is true, multiplying V by A results in a vector pointed in the same direction as V but perhaps of a different length. We write this condition as

$$AV = \lambda V, \tag{3.4.21}$$

where V is an eigenvector and λ is a scalar, called the eigenvalue, which accounts for a change in length introduced by multiplication with A. As far as the vector V is concerned, multiplication by A is accomplished equally well by a simple scalar multiplication with λ. There can be as many as N unique eigenvector–eigenvalue combinations for a given N-by-N matrix A, and these combinations are important to us because, as we shall see later, we can use the eigenvectors to decompose A.

We turn our attention now to the autocorrelation matrix R. We have already shown that R has some special properties. In particular,

(a) R is symmetric for real input data, and conjugate symmetric in the case of complex input data; that is, $r_{ij} = r_{ji}^*$, $1 \leqslant i, j \leqslant N$.

(b) R is positive semidefinite; that is, $V^h R V \geqslant 0$, if $V^h V \neq 0$.

Because of these properties, the eigenvalues and eigenvectors of R have special properties as well [Noble and Daniel, 1977], including:

(a) R has N linearly independent eigenvectors. We will call them U_1 through U_N. Since their length is arbitrary we define them to be unit length vectors; that is,

$$U_i^h U_i = \|U_i\| = 1, \qquad \text{for } 1 \leqslant i \leqslant N. \tag{3.4.22}$$

(b) The N eigenvectors of R are orthogonal; that is,

$$U_i^h U_j = 0, \qquad \text{for } 1 \leqslant i, j \leqslant N \quad \text{and} \quad i \neq j. \tag{3.4.23}$$

(c) Since the eigenvectors are orthogonal and normalized to unit length, we call them orthonormal and write their inner products using the Kronecker delta δ_{ij},

$$U_i^h U_j = \delta_{ij} = \begin{cases} 1, & i = j \\ 0, & i \neq j. \end{cases} \tag{3.4.24}$$

(d) If the input $x(k)$ is real-valued, then N real eigenvectors can be found. If $x(k)$ is complex, then the eigenvectors will in general be complex-valued.

We exploit the special nature of the "eigensystem" of R by using it to decompose the various expressions that depend on R. The key to this decomposition is the formation of the so-called modal matrix Q, composed of the eigenvectors of R; that is,

$$Q = [\mathbf{U}_1 \ldots \mathbf{U}_N]. \tag{3.4.25}$$

Since the vectors are orthonormal, the product $Q^h Q$ is simply

$$Q^h Q = I_N, \tag{3.4.26}$$

i.e., the N-by-N identity matrix, implying that $Q^{-1} = Q^h$. It is also easy to prove that

$$Q^h R Q = \Lambda \quad \text{and} \quad Q \Lambda Q^h = R, \tag{3.4.27}$$

where Λ is a N-by-N diagonal matrix whose diagonal elements are the eigenvalues of R. This relationship between R and Λ is called a similarity transformation. The matrix Q is called the modal matrix since, as we shall see later, it can be used to decompose various important equations into their uncoupled "modes."

Given Q, suppose we now define a coordination transformation of the impulse response vector $\mathbf{W}$ as

$$\mathbf{W} = Q\mathbf{W}' \quad \text{or} \quad Q^h \mathbf{W} = \mathbf{W}', \tag{3.4.28}$$

where $\mathbf{W}'$ is the transformed weight vector. This transformation changes the direction but not the length of the weight vector. This is proven by

$$\|\mathbf{W}\|^2 \triangleq \mathbf{W}^h \mathbf{W} = \mathbf{W}'^h Q^h Q \mathbf{W}' = \mathbf{W}'^h I_N \mathbf{W}' = \mathbf{W}'^h \mathbf{W}' \triangleq \|\mathbf{W}'\|^2. \tag{3.4.29}$$

We will use the transformation in (3.4.28) often. As an example of its usefulness we will first look at what it tells us about the normal equations. From (3.2.30) and (3.3.42), the optimal solution to either of the average squared error problems in Sections 3.2 and 3.3 is given by the matrix normal equation

$$R\mathbf{W}^o = \mathbf{P}. \tag{3.4.30}$$

If we substitute $Q \Lambda Q^h$ for R from (3.4.27) and recognize $Q^h \mathbf{W}^o$ to be $\mathbf{W}^{o\prime}$ from (3.4.28), we can write

$$Q \Lambda \mathbf{W}^{o\prime} = \mathbf{P}. \tag{3.4.31}$$

We define the transformed version of the cross-correlation vector as

$$\mathbf{P}' = Q^h \mathbf{P}. \tag{3.4.32}$$

Premultiplication of both sides of (3.4.31) by Q^h yields

$$\Lambda \mathbf{W}^{o\prime} = \mathbf{P}^\prime. \tag{3.4.33}$$

What makes this new form so useful is that Λ is a diagonal matrix. The whole set of N equations can be written as

$$\lambda_i w_i^{o\prime} = p_i^\prime, \qquad 1 \leqslant i \leqslant N, \tag{3.4.34}$$

where $w_i^\prime$ and $p_i^\prime$ are the ith scalar entries of $\mathbf{W}^{o\prime}$ and $\mathbf{P}^\prime$, respectively.

Each uncoupled weight $w_i^\prime$ can be written in terms of its own eigenvalue λ_i and its uncoupled cross correlation term $p_i^\prime$. If $\lambda_i \neq 0$, then

$$w_i^{o\prime} = \frac{p_i^\prime}{\lambda_i}, \qquad 1 \leqslant i \leqslant N, \tag{3.4.35}$$

and, if λ_i does equal zero, then the value of $w_i^\prime$ is indeterminate. It is this zero eigenvalue case which results in a lack of uniqueness in the solution for the optimal weight vector.

Other useful matrix formulas developed in Section 3.4 can also be decoupled into scalar equations. The minimum average squared error becomes

$$J_{\min} = D - \mathbf{P}^h \mathbf{W}^o$$

$$= D - (\mathbf{P}^{\prime h} Q^h)(Q \mathbf{W}^{o\prime}) = D - \mathbf{P}^{\prime h} \mathbf{W}^{o\prime}$$

$$= D - \sum_{i=1}^{N} p_i^\prime w_i^{o\prime} = D - \sum_{i=1}^{N} |p_i^\prime|^2 / \lambda_i. \tag{3.4.36}$$

If we define the uncoupled difference vector $\mathbf{V}^\prime$ by the expression

$$V \triangleq \mathbf{W} - \mathbf{W}^o = Q\mathbf{V}^\prime, \tag{3.4.37}$$

we can write the excess average squared error as

$$\Delta J = J - J_{\min} = \mathbf{V}^h R \mathbf{V}$$

$$= \mathbf{V}^{\prime h} \Lambda \mathbf{V}^\prime = \sum_{i=1}^{N} \lambda_i |v_i^\prime|^2. \tag{3.4.38}$$

Thus the excess squared error is quadratic with each uncoupled difference term $v_i^\prime$, and the eigenvalue determines the degree of penalty. A high value of λ_i means that a small change in $v_i^\prime$ makes a big difference in ΔJ. No change in ΔJ is observed for any change in the uncoupled weight if $\lambda_i = 0$.

3.5 SUMMARY AND PERSPECTIVE

This chapter has presented a structure for developing adaptive filtering algorithms based on the idea of optimizing some function of the data waveforms provided to the filter. While the formal structure discussed here (and the algorithms seen in Chapter 4) are the basis of most adaptive filters in current use, it is useful to remember that this chapter has considered a very specific and quite limited problem. We have assumed that the filter has a discrete-time finite impulse response and is implemented with a tapped-delay-line (transversal) structure. The filter's performance is judged based on some template or reference waveform also provided to the algorithm.

Given these choices for the filter structure and the manner in which the quality of the filter output would be judged, we then turned to the third aspect of adaptive filter design: the development of the methods to be used to choose the filter impulse response. In Chapter 2 we did this by developing an intuitively reasonable set of rules for updating an estimate of the desired impulse response in such a way as to improve the match between the filter output and the reference or template waveform. In this chapter a more rigorous approach was used. We defined a performance function based on an average of the squared output error, that is, the difference between the filter output and the reference, and then developed methodical procedures that allow us to determine the impulse response which provides the best performance.

In addition to the optimal impulse responses themselves, we also developed a set of tools in this chapter which will allow us to study some of the properties of the algorithms developed in the next chapter. We did this by putting the filter input and reference waveforms in a stochastic framework and then defining a new performance function based on the ensemble average of the squared error. Since this ensemble average reflects the behavior of the sum-squared error over all possible signal waveforms, analysis of the properties of this performance function allows us to judge the general or average behavior of adaptive algorithms. Using this framework, we characterized the performance function itself, examined the effects of input signals that do not fully "excite" the filter, and will be able to estimate the convergence rates of several important adaptive algorithms. These tools will be used in the next chapter to both suggest new algorithms and to analyze them.

4

Algorithms for Adapting FIR Filters

●*PRECIS   Once the adaptive filtering problem is expressed as a multivariable minimum average squared error optimization problem, two basic types of aaptive algorithms emerge: simple but somewhat inefficient techniques which search for the optimum solution, and more elegant, but complicated, techniques that maintain optimality with each new input sample. sample.*

4.1 INTRODUCTION

We now turn our attention to developing procedures, or algorithms, for finding the optimal weight vector $\mathbf{W}^o$. In principle this is simple. Given a record of data $\{x(k), d(k)\}$, one computes R_{ss} and $\mathbf{P}_{ss}$, and, inversion permitted, determines $\mathbf{W}^o_{ss}$ according to the equation $\mathbf{W}^o_{ss} = R_{ss}^{-1} \mathbf{P}_{ss}$. Even so, we often choose to compute $\mathbf{W}^o$ in other ways. Some reasons for this include the following:

(a) R_{ss} may not be invertible.
(b) Even if R_{ss} is theoretically invertible, the numerical precision required to invert R_{ss} properly is beyond the capabilities of the hardware or computer to be used in implementing the filter.
(c) There may be more efficient ways to calculate $\mathbf{W}^o$ than the direct path that has been discussed.

In this chapter we examine two different methodologies for developing adaptive filtering algorithms. The first class is based on so-called search techniques, which have the advantage of being simple to implement but at the price of some

73

inaccuracy in the final estimate. In Section 4.3, we examine techniques which compute $\mathbf{W}^o$ exactly but at the expense of more complicated algorithms.

4.2 SEARCH TECHNIQUES

Our goal is to find $\mathbf{W}^o$, that particular choice of the weight vector for which J, the performance function, is minimized. One relatively straightforward way of finding $\mathbf{W}^o$ is to search over the function J to find its minimum. This might be done by computing J for all possible values of $\mathbf{W}$ and then picking the smallest, or it might be done by randomly picking values of $\mathbf{W}$, computing J, and seeing if it is smaller than those seen already [Widrow and McCool, 1976]. A better approach yet is to develop an orderly search procedure which leads one methodically from a starting point to the value of $\mathbf{W}$ which minimizes J.

4.2.1 The Gradient Search Approach

From Section 3.4, we recall several facts which can be used to develop search procedures. If we presume that R_{ss} has full rank, they include:

(a) The optimal solution $\mathbf{W}^o$ is a unique choice of $\mathbf{W}$;
(b) Any difference between the actual weight vector $\mathbf{W}$ and the optimal one $\mathbf{W}^o$ leads to an increase in the performance function; that is, $\Delta J = J - J_{min} = \mathbf{V}^t R \mathbf{V}$, where $\mathbf{V} = \mathbf{W}^o - \mathbf{W}$; and
(c) $\Delta J > 0$ for $\mathbf{V} \neq 0$.

Suppose now that we iteratively estimate $\mathbf{W}^o$; that is, we start with some initial choice of $\mathbf{W}$, called $\mathbf{W}(0)$, and then choose a new value $\mathbf{W}(1)$ in some way which leads us closer to the optimal value. The points made above suggest how to do this. Unless $\mathbf{W}(0)$ equals $\mathbf{W}^o$, the value of J at $\mathbf{W}(0)$ is greater than J_{min}. Suppose we choose $\mathbf{W}(1)$ in such a way that ΔJ and hence J is reduced by some amount. If ΔJ is still greater than zero, an improvement is made but $\mathbf{W}(1)$ is not $\mathbf{W}^o$. Suppose we take additional steps, $\mathbf{W}(2)$, $\mathbf{W}(3)$, etc. and each time are able to reduce ΔJ by some amount. If this can be done every time and if the input excitation is adequate, then ultimately ΔJ will approach zero and $\mathbf{W}(l)$ will approach $\mathbf{W}^o$. The fact that $\mathbf{W}^o$ is unique and minimizes J guarantees that if a search procedure finds the minimum of the performance function, then it has also found $\mathbf{W}^o$.

How, then, should we move from $\mathbf{W}(l)$ to $\mathbf{W}(l + 1)$? One good technique is to use the gradient function to select the direction. The basic idea is shown in the one-dimensional example in Figure 4.1. The performance function J has a clear minimum, at $w = w^o$, and has a quadratic shape. At $w(l)$, ΔJ exceeds zero, indicating that $w(l) \neq w^o$. If we desired to improve $w(l)$, we want to move in the direction toward w^o. Without direct solution we do not know where w^o is, but, by evaluating the derivative of J at $w(l)$, we can get a clue. By evaluating the derivative we determine the first-order change in J caused by a small change in

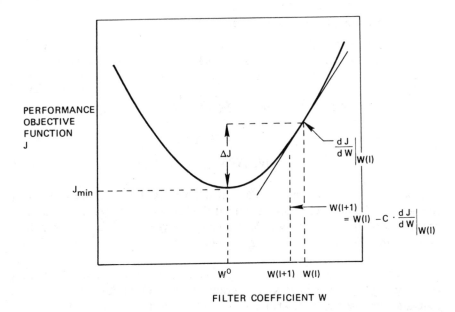

FIGURE 4.1. Gradient descent on one-dimensional projection of quadratic performance function.

w. Suppose, as shown, that dJ/dw is positive. Therefore, increasing w will make J increase. Our goal, however, is to decrease J, and this may be done by taking a step in the negative direction. This rule can be reduced to an equation as follows:

$$w(l + 1) = w(l) - c \cdot \frac{dJ}{dw}\bigg|_{w(l)}, \qquad (4.2.1)$$

where c is a small positive constant. Thus, whenever the derivative is positive, $w(l)$ is decreased to yield $w(l + 1)$, and if dJ/dw is negative, $w(l)$ is increased. Reference again to Figure 4.1 shows that repeated application of this rule will cause $w(l)$ to take steps down the parabola and to finally come to rest at the minimum point of J. This is exactly the intended behavior, and at that point $w(l)$ will equal w^o.

This derivative-based search is extended to the N-dimensional adaptive filter by using the gradient function, which is the vector of first derivatives of J with respect to the coefficients of the impulse response vector. This updating rule then becomes

$$\mathbf{W}(l + 1) = \mathbf{W}(l) - c \cdot \nabla_W J|_{W = W(l)}, \qquad l \geqslant 0, \qquad (4.2.2)$$

where c, as before, is a small positive constant. For each estimate $\mathbf{W}(l)$, the gradient is computed and a small step proportional to that gradient is taken, leading to a small reduction in the performance function J. Continuation of this procedure would, in theory, reduce J until it attained its minimum value.

4.2.2 Approximation of the Gradient

The gradient of J with respect to the impulse response vector $\mathbf{W}$ can be estimated directly from the input data $\{x, d\}$. Suppose we compute the gradient of the instantaneous squared error, that is,

$$\mathbf{G}(k) = \nabla_W[e^2(k)] = 2e(k)\cdot\nabla_W e(k)$$

$$= 2e(k)\nabla_W\{d(k) - \mathbf{W}^t\mathbf{X}(k)\}$$

$$= -2e(k)\nabla_W\{\mathbf{W}^t\mathbf{X}(k)\} = -2e(k)\mathbf{X}(k). \tag{4.2.3}$$

The last two steps rely on the fact that neither $d(k)$ nor $\mathbf{X}(k)$ are affected by changes in $\mathbf{W}$. This vector $\mathbf{G}(k)$ is termed the instantaneous gradient since it is based only on the value of e and $\mathbf{X}$ at time k. The vector $\mathbf{G}(k)$ can be ensemble-averaged to yield the gradient of J_{ms} or time-averaged to produce the gradient of J_{ss}.

We can use this instantaneous gradient to form an adaptive filtering algorithm which approximates a true gradient search of the desired performance function. Suppose we use $\mathbf{G}(k)$ in place of the true gradient in (4.2.2). The result is the Least Mean Squares (LMS) algorithm popularized by Widrow [Widrow and Hoff, 1960; Widrow et al., 1976]:

$$\mathbf{W}(l + 1) = \mathbf{W}(l) - c\mathbf{G}(l),$$

$$= \mathbf{W}(l) + \mu e(l)\mathbf{X}(l), \qquad l \geqslant 0, \tag{4.2.4}$$

where μ is a small positive constant.

Usually, but not always, the coefficient vector estimate $\mathbf{W}(l)$ is updated for every input sample. When this is true, $l = k$ and the complete LMS algorithm is written as

LMS:

$$y(k) = \mathbf{W}^t(k)\mathbf{X}(k) \qquad \text{filter output} \tag{4.2.5}$$

$$e(k) = d(k) - y(k) \qquad \text{error formation} \tag{4.2.6}$$

$$\mathbf{W}(k + 1) = \mathbf{W}(k) + \mu e(k)\mathbf{X}(k) \qquad \text{weight vector update} \tag{4.2.7}$$

This is a straightforward algorithm and has served as the cornerstone for most of the adaptive filtering field. Referring back to Chapter 2, we find that this LMS algorithm is of exactly the same form as those heuristically developed there. The new parameter estimate is based on the old one plus a term which is the product of a bounded step size (μ), a function of the input state $[\mathbf{X}(k)]$, and a function of the error $[e(k)]$. Even though the form is exactly the same, we developed it here in a very different way from the number guessing game logic of

Chapter 2. In particular, to arrive at the LMS algorithm in (4.2.5)–(4.2.7), we did the following:

(a) We developed an analytical performance function based on averaging the squared error;
(b) We developed a gradient-descent search procedure for finding that value of weight vector which minimizes the performance function; and
(c) We developed an approximation for the required gradient function which can be computed directly from the data.

The fact that the algorithms are the same provides a certain confidence. What is new, however, is that the analytical structure built up in Chapter 3 can now be used to predict the performance of this adaptive algorithm.

4.2.3 The Average Convergence Properties of LMS

The LMS algorithm stated in (4.2.5)–(4.2.7) can be analyzed in a number of different ways to determine where and how fast it converges. We choose here to model the LMS adaptation as if it had actually used the true gradient function instead of the approximate one. While not perfectly accurate, it provides many of the answers we seek. Moreover, in the limit of small adaptation constant μ, the approximation is quite good [Widrow et al., 1976].

From Sections 3.2, 3.3 and 3.4, we know that the gradient of a squared error function is given by (3.2.23) as

$$\nabla_W J = -2P + 2RW. \tag{4.2.8}$$

Evaluating $\nabla_W J$ at $W = W(l)$ yields

$$\nabla_W J|_{W(l)} = -2P + 2RW(l). \tag{4.2.9}$$

Combining this expression with (4.2.2), produces the expression

$$W(l + 1) = W(l) - \mu(-P + RW(l))$$
$$= (I - \mu R) \cdot W(l) + \mu P, \qquad l \geqslant 0. \tag{4.2.10}$$

This recursion expression might be recalled from Section 3.2.6.2, where it was shown that as $l \to \infty$, $W(l)$ converges to the optimum weight vector W°. To gain some additional insight, however, we will use the eigenvector decomposition developed in Section 3.4 to examine this recursion. We start by defining $W'(l)$ by the expression

$$W(l) \triangleq Q W'(l), \tag{4.2.11}$$

where Q is the N-by-N matrix formed by the eigenvectors of R. Similarly, we defined (in Section 3.4.3)

$$R = Q\Lambda Q^h, \qquad \Lambda = Q^h R Q, \quad \text{and} \quad \mathbf{P}' = Q^h \mathbf{P}, \tag{4.2.12}$$

where Λ is the diagonal matrix of eigenvalues of R. Applying these expressions to (4.2.10) leads to the expression

$$\mathbf{W}'(l + 1) = (I - \mu\Lambda) \cdot \mathbf{W}'(l) + \mu\mathbf{P}', \qquad l \geqslant 0. \tag{4.2.13}$$

While it looks very much like (4.2.10), this expression is much simpler to analyze since Λ is a diagonal matrix. Each of the N equations in the matrix recursion is now uncoupled and can be written separately as

$$w_i'(l + 1) = (1 - \mu\lambda_i) \cdot w_i'(l) + \mu p_i', \qquad 0 \leqslant i \leqslant N - 1. \tag{4.2.14}$$

In this simple scalar form we can determine several interesting aspects of the algorithm's behavior.

Convergence Points

The scalar equation can be solved to yield a closed form expression for $w_i'(l)$. It becomes

$$w_i'(l) = \left\{ \mu \sum_{n=0}^{l-1} (1 - \mu\lambda_i)^n \cdot p_i' \right\} + (1 - \mu\lambda_i)^l w_i'(0). \tag{4.2.15}$$

If we choose μ to be small enough so that

$$|1 - \mu\lambda_i| < 1, \tag{4.2.16}$$

then using (3.2.57),

$$\sum_{n=0}^{l-1} (1 - \mu\lambda_i)^n = \frac{1 - (1 - \mu\lambda_i)^l}{1 - (1 - \mu\lambda_i)} \tag{4.2.17}$$

and

$$\lim_{l \to \infty} (1 - \mu\lambda_i)^l = 0 \quad \text{and} \quad \lim_{l \to \infty} \frac{1 - (1 - \mu\lambda_i)^l}{1 - (1 - \mu\lambda_i)} = \frac{1}{\mu\lambda_i}. \tag{4.2.18}$$

In this case, $w_i'(l)$ converges to

$$\lim_{l \to \infty} w_i'(l) = \frac{1}{\mu\lambda_i} \cdot \mu p_i' = \frac{p_i'}{\lambda_i} \equiv w_i'^o. \tag{4.2.19}$$

Reference to (3.4.35) confirms that p'_i/λ_i equals the ith uncoupled optimal weight. Thus, each uncoupled coefficient converges to the proper point as $l \to \infty$.

Bounds on the Adaptive Constant μ

To obtain a closed form solution for $w'_i(l)$ we assumed that $|1 - \mu\lambda_i|$ was less than 1. In fact, we have to choose μ so that this condition is attained for all N uncoupled weights. This leads to the following condition on the selection of μ:

$$|1 - \mu\lambda_i| < 1 \quad \text{for} \quad 0 \leqslant i \leqslant N - 1 \quad \text{or} \quad 0 < \mu < \frac{2}{\lambda_{\text{max}}}, \qquad (4.2.20)$$

where λ_{max} is the maximum of the N eigenvalues λ_i of R. As a practical matter, it is found in most applications of the LMS algorithm that μ should be chosen to be substantially smaller than the upper bound. It is not unusual to use a value of μ two orders of magnitude smaller than μ_{max} to obtain smooth convergence.

Adaptive Time Constants

Since the uncoupled weights $w'_i(l)$ obey a first-order recursion expression, it is possible to describe their behavior in terms of a time constant, that is, the time required for any transient to decay to $1/e \,(\cong 37\%)$ of its initial value. We can determine this time duration by looking again at (4.2.15), assuming p'_i to equal zero, and asking how many iterations are required for $w'_i(l)$ to decay from $w'_i(0)$ to $w'_i(0)/e$. Suppose this number is τ_i. It can be found by taking the natural logarithm of both sides of the expression

$$\ln\left\{\frac{w'_i(0)}{e}\right\} = \ln\{(1 - \mu\lambda_i)^{\tau_i}w'_i(0)\} \qquad (4.2.21)$$

or

$$-1 = \tau_i \cdot \ln(1 - \mu\lambda_i). \qquad (4.2.22)$$

When x is small compared to unity, $\ln(1 + x) \approx x$. If μ is small enough so that $0 < \mu\lambda_i \ll 1$ (which corresponds to the "small μ" assumption made above), then

$$-1 \approx \tau_i \cdot (-\mu\lambda_i) \qquad (4.2.23)$$

and the time constant of the ith uncoupled coefficient is

$$\tau_i \approx \frac{1}{\mu\lambda_i}. \qquad (4.2.24)$$

Note that the time constants τ_i are different, in general, for various i since the eigenvalues λ_i are. Note also that τ_i depends only on the nature of the input signal $x(k)$ as manifested in the autocorrelation matrix R, and not on the desired signal $d(k)$.

Algorithm Convergence Time

The implication of (4.2.24) is that each uncoupled adaptive mode has its own time constant which is determined by the overall adaptation constant μ and the eigenvalue λ_i associated with that mode. Since the actual observed weight vector $\mathbf{W}(l)$ is a linear function of these uncoupled weights, it is clear that $\mathbf{W}(l)$ does not converge until all of the uncoupled weights do, and therefore that $\mathbf{W}(l)$ converges no faster than the slowest mode. We therefore define the time constant of $\mathbf{W}(l)$, called τ, by its worst case bound

$$\tau = \max_i \left\{ \frac{1}{\mu \lambda_i} \right\} = \frac{1}{\mu \cdot \min_i \{\lambda_i\}} \triangleq \frac{1}{\mu \lambda_{\min}}. \qquad (4.2.25)$$

Thus the mode with the smallest eigenvalue determines the asymptotic convergence time of the algorithm.

The dependence of LMS's convergence time on the eigenstructure of the R matrix can be further demonstrated. Suppose we define the normalized step size α by the expression

$$\alpha = \frac{\mu \lambda_{\max}}{2} \quad \text{or, equivalently,} \quad \mu = \frac{2\alpha}{\lambda_{\max}}. \qquad (4.2.26)$$

From (4.2.20) we know that for convergence α must be bounded by $0 < \alpha < 1$. Substituting (4.2.26) into (4.2.25) we see that the convergence time τ is given by

$$\tau = \frac{1}{2\alpha} \cdot \frac{\lambda_{\max}}{\lambda_{\min}}. \qquad (4.2.27)$$

For any given value of the normalized adaptation constant α, the ratio of maximum to minimum eigenvalues (sometimes called the "eigenvalue disparity" or "condition number of R") determines the speed of convergence of $\mathbf{W}(l)$.

4.2.4 The Effects of a Singular Autocorrelation Matrix

So far our analysis has assumed that the input process $x(k)$ is well enough behaved so that the autocorrelation matrix R is nonsingular. This in turn implies that $\mathbf{W}^o$ is unique and that a search procedure can be garanteed to find the optimum value of the impulse response vector. Suppose now that R is singular. To examine the effects of this, we revisit some of the points made in the previous section.

When R is singular it can be shown that at least one of its eigenvalues is zero.

Further, it can be shown that if $\lambda_i = 0$, then so does the associated uncoupled cross-correlation term p'_i. The uncoupled weight recursion (4.2.14) becomes

$$w'_i(l + 1) = (1 - \mu\lambda_i) \cdot w'_i(l) + \mu p'_i$$

$$= (1 - 0) \cdot w'_i(l) + 0 = w'_i(l).$$

In other words, the associated uncoupled coefficient is undriven and undamped. It does not decay nor does it grow. It remains unchanged at its initial value. All uncoupled weights with nonzero eigenvalues will converge as before.

Since $\lambda_i = 0$, initial conditions in $w'_i(l)$ cannot decay and the time constant is infinite. This in turn implies that the weight vector $\mathbf{W}(l)$ never converges. While strictly speaking this is true, it does not necessarily mean that the algorithm will not produce a useful solution. This apparent paradox can be resolved as follows.

Suppose there is one zero eigenvalue in R and that $\mathbf{U}$ is the associated eigenvector. This implies that

$$R \cdot \{\gamma\mathbf{U}\} = \gamma R\mathbf{U} = 0 \tag{4.2.28}$$

for any scalar γ. Because of this property, $\mathbf{U}$ is said to be in the "null space" of R. Now consider the problem of finding $\mathbf{W}^\circ$ so that the "normal equations" are satisfied; that is,

$$R\mathbf{W}^\circ = \mathbf{P}. \tag{4.2.29}$$

Suppose that instead of $\mathbf{W}^\circ$ we substitute $\mathbf{W}^\circ + \gamma\mathbf{U}$. Surprisingly, the normal equations are still satisfied; i.e.,

$$R(\mathbf{W}^\circ + \gamma\mathbf{U}) = R\mathbf{W}^\circ + \gamma R\mathbf{U} = \mathbf{P} + 0 = \mathbf{P}. \tag{4.2.30}$$

Thus if R is singular, then any vector or combination of vectors from the null space of R can be added to $\mathbf{W}^\circ$ without disturbing the satisfaction of the normal equations. This has two key implications. The first, already obvious, is that $\mathbf{W}^\circ$ is not unique. The second, however, is that the convergence of the "null space modes," i.e., those for which $\lambda_i = 0$, is irrelevant if the goal is only to find a, as opposed to *the*, solution of the normal equations. When null space modes are present, the bound on convergence time of the algorithm should be based on the smallest nonzero eigenvalue. In this case the expression for τ should be

$$\tau = \frac{1}{2\alpha} \frac{\lambda_{max}}{\lambda^*_{min}}, \qquad \lambda^*_{min} = \min_i \{\lambda_i\} \qquad \text{such that } \lambda_i \neq 0. \tag{4.2.31}$$

In applications where convergence of the weight vector to a unique $\mathbf{W}^\circ$ is the goal, as opposed to minimizing J, the infinite time constant correctly reflects the fact that this algorithm (or any other) will fail if R is singular.

4.2.5 Related Approximate Gradient Search Algorithms

The LMS algorithm described in Section 4.2.2 has served as the cornerstone for most of the adaptive filtering algorithms actually in use today. Its simplicity makes it relatively straightforward to implement, while its close relationship to the theory of optimum filtering has appeal in terms of its performance. We now examine a set of adaptive algorithms which are all direct variants of LMS.

4.2.5.1 The Complex LMS Algorithm [Widrow et al., 1975a]
When the input sequence $x(k)$, the output sequence $y(k)$, and the desired sequence $d(k)$ are all complex-valued, then a complex version of the LMS algorithm must be used. It is developed by assuming the coefficient vector to be complex as well and then determining the gradient of $|e(k)|^2$ with respect to that complex vector. Incorporating this instantaneous gradient into the gradient search formulation produces the complex LMS algorithm.

Complex LMS:

$$y(k) = \mathbf{X}^t(k)\mathbf{W}(k)$$

$$e(k) = d(k) - y(k)$$

$$\mathbf{W}(k + 1) = \mathbf{W}(k) + \mu e(k)\mathbf{X}^*(k). \tag{4.2.32}$$

The asterisk denotes complex conjugation.

4.2.5.2 Normalized LMS [Albert and Gardner, 1967]
All forms of the LMS algorithm examined so far use the adaptation constant μ, a small constant which determines, among other things, the speed of convergence of the algorithm. Various issues associated with the choice of μ will be discussed in Chapter 7, but one practical problem often confronted is that of finding some way to ensure that μ does not become large enough to cause the algorithm to become unstable. The theoretical limit to stability is provided in (4.2.20); i.e., the largest possible value of μ is determined by the largest eigenvalue of R. This result is of limited practical use since R is usually not available, and even if it were, computing its eigenvalues is an undesirable chore. A more reasonable approach is to find some bounds for the largest of the eigenvalues. To this end, it can be shown that

$$\text{Avrg}\{\mathbf{X}^t(k)\mathbf{X}(k)\} = \sum_{i=1}^{N} \lambda_i, \tag{4.2.33}$$

that is, the average value of the dot product of the data vector with itself equals the sum of the eigenvalues of R. Since all $\lambda_i \geq 0$,

$$\text{Avrg}\{\mathbf{X}^t(k)\mathbf{X}(k)\} \geq \lambda_{\max}, \tag{4.2.34}$$

meaning that the average value of the inner product is an upper bound to λ_{max}. This suggests defining μ as

$$\mu(k) = \frac{\alpha}{\mathbf{X}^t(k)\mathbf{X}(k)}, \tag{4.2.35}$$

where α is the normalized step size chosen to be between 0 and 2. It can be seen that on the average $\mu(k)\lambda_{max}$ abides by (4.2.20). Another insightful interpretation of (4.2.35), this one based on projection concepts, is presented in [Goodwin and Sin, 1984].

Using this form for μ results in the following.

Normalized LMS:

$$y(k) = \mathbf{W}^t(k)\mathbf{X}(k)$$

$$e(k) = d(k) - y(k)$$

$$\mathbf{W}(k + 1) = \mathbf{W}(k) + \frac{\alpha e(k)\mathbf{X}(k)}{\gamma + \mathbf{X}^t(k)\mathbf{X}(k)}. \tag{4.2.36}$$

The term α is the new "normalized" adaptation constant, while γ is a small positive term included to ensure that the update term does not become excessively large when $\mathbf{X}^t(k)\mathbf{X}(k)$ temporarily becomes small.

It first appears that the inclusion of $\mathbf{X}^t(k)\mathbf{X}(k)$ in the denominator increases the computation requirement by another N multiplications and additions, but this can be avoided if N extra storage locations are available. At time k, $\mathbf{X}^t(k)\mathbf{X}(k)$ is given by

$$\mathbf{X}^t(k)\mathbf{X}(k) = \sum_{i=0}^{N-1} x^2(k - i). \tag{4.2.37}$$

In principle the term $\mathbf{X}^t(k + 1)\mathbf{X}(k + 1)$ can be computed by adding in $x^2(k + 1)$ and subtracting off $x^2(k - N + 1)$. By storing the intermediate values of $x^2(\cdot)$, the computation required to update the inner product is reduced to a squaring, an addition, and a subtraction. A division operation is still needed, however.

4.2.5.3 Normalized LMS with Recursive Power Estimation

A closer examination of (4.2.35) shows the data vector inner product to be an N-point arithmetic average of the input signal power. Therefore, the normalized LMS algorithm gains its stability by normalizing the weight vector update with an estimate of the signal power. Another way of obtaining this normalization is by computing a recursive estimate of the signal power, called $\pi(k)$. This algorithm has the following form.

Normalized LMS with Recursive Power Estimation:

$$y(k) = \mathbf{X}^t(k)\mathbf{W}(k)$$

$$e(k) = d(k) - y(k)$$

$$\pi(k + 1) = (1 - \beta) \cdot \pi(k) + N\beta x^2(k)$$

$$\mathbf{W}(k + 1) = \mathbf{W}(k) + \frac{\alpha e(k)\mathbf{X}(k)}{\gamma + \pi(k)}. \tag{4.2.38}$$

Instead of N locations of extra storage to compute $\mathbf{X}^t(k)\mathbf{X}(k)$ this algorithm uses one location. The positive constant β is used to control the averaging time constant. If β is chosen according to the rule

$$\beta = 1 - \frac{1}{N}, \tag{4.2.39}$$

then the exponential power averaging has a time constant of N samples. When β is chosen in this fashion, (4.2.38) exhibits roughly the same performance as the normalized LMS algorithm in (4.2.36).

4.2.5.4 Accelerated Algorithms

The approximate gradient search algorithms examined so far are often called "steepest-descent" algorithms since, by following the gradient, each adaptation step moves down the gradient slope, not necessarily directly toward the minimum of the performance function. By changing the direction of the adaptation it is theoretically possible to force the adaptation steps to lead more directly to the minimum of the performance function and thereby converge faster. Many of these algorithms take the form

$$y(k) = \mathbf{X}^t(k)\mathbf{W}(k)$$

$$e(k) = d(k) - y(k)$$

$$\mathbf{W}(k + 1) = \mathbf{W}(k) + \mu e(k)C\mathbf{X}(k). \tag{4.2.40}$$

The matrix C is chosen based on some a priori knowledge to improve the convergence performance. In particular, if C can be chosen to approximate R^{-1}, then the convergence time of the algorithm can be reduced substantially in cases of high eigenvalue disparity, that is, $\lambda_{max} \gg \lambda_{min}$. As a practical matter, these algorithms have limited utility, however. The reasons for this include:

(a) The proper choice of the accelerating matrix C is dependent on R and hence on the properties of the data $x(k)$. Choosing C without knowing the characteristics of the data can be counterproductive.

(b) The matrix C is in general an N-by-N matrix, and computing $C\mathbf{X}(k)$ requires N^2 multiplications and $N(N - 1)$ additions. This overwhelms the computational requirements for LMS itself (about $2N$ multiplications), and this in itself usually precludes the use of accelerated algorithms.

4.2.5.5 Griffiths's Algorithm [Griffiths, 1967]

All of the algorithms described so far have assumed that the reference waveform $d(k)$ is available. While this assumption is valid in some cases there are practical situations where adaptive filtering would be useful but no reference waveform can be provided (see Chapter 6 for examples). The absence of the reference implies that the error $e(k)$ cannot be formed and therefore that algorithms such as LMS cannot be used directly. Griffiths developed an agorithm which can be used in applications in which $d(k)$ is not available, but the correlation function between $d(k)$ and the data vector $\mathbf{X}(k)$ is. First we rewrite the LMS update (4.2.7) as

$$\begin{aligned} \mathbf{W}(k+1) &= \mathbf{W}(k) + \mu e(k)\mathbf{X}(k) \\ &= \mathbf{W}(k) - \mu y(k)\mathbf{X}(k) + \mu d(k)\mathbf{X}(k). \end{aligned} \tag{4.2.41}$$

The average behavior of this update equation can be evaluated by determining the ensemble average of both sides of the equation

$$E\{\mathbf{W}(k+1)\} = E\{\mathbf{W}(k)\} - \mu \cdot E\{y(k)\mathbf{X}(k)\} + \mu \cdot E\{d(k)\mathbf{X}(k)\}. \tag{4.2.42}$$

Using (3.3.40), (4.2.42) can be rewritten as

$$E\{\mathbf{W}(k+1)\} = E\{\mathbf{W}(k)\} - \mu E\{y(k)\mathbf{X}(k)\} + \mu \mathbf{P}_{ms}. \tag{4.2.43}$$

Based on this expression, Griffiths suggested the algorithm variously known as Griffiths's algorithm, the **P**-vector algorithm, and the steering vector algorithm.

Griffiths's algorithm:

$$y(k) = \mathbf{X}^t(k)\mathbf{W}(k) \tag{4.2.44}$$

$$\mathbf{W}(k+1) = \mathbf{W}(k) - \mu y(k)\mathbf{X}(k) + \mu \mathbf{P}_{ms}. \tag{4..45}$$

The basic idea of this approach is to substitute the expected average behavior of $d(k)\mathbf{X}(k)$ for the vector itself and thus avoid needing $d(k)$.

In general, $\mathbf{P}_{ms}$ is not known a priori, and more powerful techniques, such as those in Chapter 6, are required to circumvent the need for $d(k)$. However, Griffiths's algorithm does have practical application. An example is the case where the input $x(k)$ consists of a signal with known correlation characteristics and interference which is statistically uncorrelated with the signal. Suppose $x(k)$ is given by

$$x(k) = s(k) + i(k), \tag{4.2.46}$$

where $s(k)$ and $i(k)$ are the signal of interest and the interference, respectively. If $s(k)$ were available, we might employ it as the template waveform $d(k)$ to adapt an LMS-based adaptive filter. If so, the cross-correlation vector $\mathbf{P}_{ms}$ is given by

$$E\{d(k)\mathbf{X}(k)\} = E\{s(k)\mathbf{S}(k)\} + E\{s(k)\mathbf{I}(k)\}$$
$$= \mathbf{P}_s + \mathbf{0} = \mathbf{P}_s, \tag{4.2.47}$$

where $s(k) = d(k)$, $\mathbf{S}(k)$ and $\mathbf{I}(k)$ are the vectors of current and $N - 1$ past values of $s(k)$ and $i(k)$, respectively, and $\mathbf{P}_s$ is the cross-correlation of $s(k)$ with delayed versions of itself. The use of $\mathbf{P}_s$ for $\mathbf{P}_{ms}$ in Griffiths's algorithm allows the filter to adapt to suppress the interference and pass the desired signal with the same average effectiveness as an LMS-based adaptive filter would have.

This same basic algorithm was suggested by [Kaczmarz, 1937] to iteratively invert matrices and forms the basis for the algebraic reconstruction technique (ART) used for computerized axial tomography (CAT) scanners. It was developed in the signal processing context [Griffiths, 1967] to adapt antenna arrays to reject interference.

4.2.6 Simplified and Modified Versions of the LMS Algorithm

The previous section examined several algorithms which extended the basic gradient search approach along some analytically motivated line. This section examines several more, but in this case the motivation for each is some practical consideration, such as simplicity in implementation or robustness in operation. The resulting algorithms no longer necessarily exhibit gradient-descent behavior on J.

4.2.6.1 Sign-error LMS [Gersho, 1984]

Looking again at the real arithmetic version of the LMS, it can be seen that it requires almost exactly $2N$ real multiplications and additions to compute $y(k)$ and the update $\mathbf{W}(k)$ for each iteration, where N is the length of the filter impulse response. While this amount of computation is lower than required for many more complicated adaptive algorithms (e.g., see Section 4.3) even this level of computational cost has motivated efforts to simplify the algorithm and reduce the number of multiplications required. One such simplified algorithm employs only the sign of the error.

Sign-error LMS:

$$y(k) = \mathbf{X}^t(k)\mathbf{W}(k)$$

$$\bar{e}(k) \triangleq \text{sgn}\{d(k) - y(k)\} = \begin{cases} 1 & \text{if} \quad d(k) - y(k) \geqslant 0 \\ 0 & \text{if} \quad d(k) - y(k) = 0 \\ -1 & \text{if} \quad d(k) - y(k) < 0 \end{cases}$$

$$\mathbf{W}(k + 1) = \mathbf{W}(k) + \mu\bar{e}(k)\mathbf{X}(k). \tag{4.2.48}$$

Computation of each output $y(k)$ still requires N multiplications and $N-1$ additions but the weight update equation is much simpler. If μ is chosen to be a power of two, then each term of the update $w_i(k)$ can be done by shifting $x(k-i)$ to accomplish the μ multiplication, followed by adding or subtracting it with $w_i(k)$ to obtain $w_i(k+1)$. Thus, the update is done with N shifts and adds rather than N multiplications and adds. Depending on the relative execution times for multiplications and additions, this algorithm can provide a significant speed improvement over the normal LMS algorithm. This reduction in computation comes at the expense of performance, however. The sign-error algorithm effectively uses a noisy estimate of the instantaneous gradient as its basis for searching for the minimum of the squared error performance function. The noisy estimate is reflected into noisy estimates of the optimal weight vector. Obtaining the same steady-state estimate quality as LMS generally requires using a smaller value of the adaptation constant, thus slowing down the adaptation of the algorithm. This effect is examined more in Section 4.2.7.

An alternative method can be used to reduce the number of multiplications required for each update of the LMS algorithm. Instead of using the sign of the error $e(k)$, one can use the sign of the elements of $\mathbf{X}(k)$. This algorithm is described by the following equations.

Sign-data LMS:

$$y(k) = \mathbf{X}^t(k)\mathbf{W}(k)$$

$$e(k) = d(k) - y(k)$$

$$w_i(k+1) = w_i(k) + \mu e(k) \cdot \text{sgn}\{x(k-i)\}, \qquad 0 \leqslant i \leqslant N-1. \quad (4.2.49)$$

Just as with the sign-error algorithm, this technique does not change the number of multiply-adds required for the filtering operation, but does reduce the computation needed for the updating to just N shifts and additions. Neither technique mimics the LMS algorithm closely. The sign-data algorithm actually modifies the direction of the update vector of LMS, while the sign-error algorithm only scales it differently. Note that the sign-error and sign-data use can be combined to generate a sign-sign algorithm, where

$$w_i(k+1) = w_i(k) + \mu \cdot \text{sgn}[e(k)] \cdot \text{sgn}[x(k-i)]. \qquad (4.2.50)$$

The popularity of this sign–sign algorithm is based on the fact that no multiplication at all is required for the updating equation. Even though this method uses a very "noisy" gradient estimate, its implementation ease has led to its frequent use, for example in the CCITT standard for ADPCM transmission. Even so, the algorithm is not as well behaved as the full LMS algorithm and can even be shown to diverge in situations where LMS would not [Dasgupta and Johnson, 1986].

4.2.6.2 Coefficient Leakage

In Section 4.2.4 it was shown that the convergence of approximate gradient search algorithms such as LMS can be problematical when the input process $x(k)$ does not fully excite all the modes of the adaptive algorithm. When the autocorrelation matrix associated with the input process has one or more zero eigenvalues, then the associated modes of the adaptive algorithm are undriven and undamped. This can lead to several undesirable situations:

(a) The adaptive filter will not converge to a unique solution;

(b) The uncoupled nodes do not converge; and, worst,

(c) The uncoupled nodes can be driven by second-order terms, such as those resulting from an imperfect implementation of the filter and/or algorithm. It is not uncommon in this third situation for the uncoupled coefficients to grow without bound until hardware overflow or underflow occurs, with the usual catastrophic results.

To deal with problems such as this several solutions have been proposed, including the injection of a small amount of white noise into the filter input to "quiet" the coefficients [Zahm, 1973]. A related, and preferred, method is that of using coefficient leakage. Using LMS as an example, the "leaky" algorithm can be written as

$$\mathbf{W}(k+1) = (1 - \mu\gamma) \cdot \mathbf{W}(k) - \mu\hat{\mathbf{V}}(k)$$

$$= (1 - \mu\gamma) \cdot \mathbf{W}(k) + \mu e(k)\mathbf{X}(k). \qquad (4.2.51)$$

As before, the adaptation constant μ is a small positive value, as is the leakage coefficient γ. Obviously, with conventional LMS, $\gamma = 0$.

Since μ and γ are both small compared to 1 and positive in sign, the factor $1 - \mu\gamma$ is slightly less than 1. The first-order effect of this term can be seen by assuming for the moment that the gradient estimate $e(k)\mathbf{X}(k)$ equals zero. If so,

$$\mathbf{W}(k+1) = (1 - \mu\gamma)\mathbf{W}(k)$$

$$\mathbf{W}(k+m) = (1 - \mu\gamma)^m\mathbf{W}(k) \qquad (4.2.52)$$

and

$$\lim_{m \to \infty} \mathbf{W}(k+m) = \mathbf{0}, \qquad (4.2.53)$$

since $1 - \mu\gamma$ is slightly less than unity. Thus, in the absence of the driving term $e(k)\mathbf{X}(k)$, the filter coefficients $\mathbf{W}(k)$ tend to decay or "leak" to zero.

The effect of this leakiness can be evaluated directly by using the modal decomposition of the coefficient update equation developed in Section 4.2.3.

With the introduction of coefficient leakage the uncoupled weight vector $\mathbf{W}'(k)$ is given by

$$\mathbf{W}'(k + 1) = (1 - \mu\gamma) \cdot \mathbf{W}'(k) - \mu(-\mathbf{P}' + \Lambda\mathbf{W}'(k))$$

$$= (I - \mu\gamma I - \mu\Lambda)\mathbf{W}'(k) + \mu\mathbf{P}'$$

$$= \{I - \mu(\gamma I + \Lambda)\} \cdot \mathbf{W}'(k) + \mu\mathbf{P}'. \qquad (4.2.54)$$

From this equation, and analysis of the type used in Section 4.2.3, we draw the following conclusions:

(a) Leakage has the effect of modifying the correlation matrix of the input process. In particular,

$$R_{\text{new}} = \gamma I + R_{\text{old}} \qquad (4.2.55)$$

$$\Lambda_{\text{new}} = \gamma I + \Lambda_{\text{old}} \qquad (4.2.56)$$

and

$$\lambda_{i,\text{new}} = \gamma + \lambda_{i,\text{old}}, \qquad 0 \leqslant i \leqslant N - 1. \qquad (4.2.57)$$

(b) If $\gamma > 0$, then all eigenvalues of the update process are positive even if some of the input eigenvalues equal zero.

(c) The positivity of all eigenvalues implies a unique solution and bounded time constants for all uncoupled weights.

(d) Since all time constants are bounded, the overall algorithm will always converge with a time constant of less than or equal to

$$\tau_{\text{max}} = \frac{1}{\mu\lambda_{\text{new,min}}} \leqslant \frac{1}{\mu\gamma}. \qquad (4.2.58)$$

Thus the introduction of coefficient leakage leads to convergence to a unique solution with a finite time constant. The price of leakage, however, is some added complexity in the updating equation and the introduction of some bias into the convergent solution. This bias is introduced since the weight vector $\mathbf{W}$ converges to the solution

$$\lim_{k \to \infty} \mathbf{W}(k) = (R + \gamma I)^{-1}\mathbf{P}, \qquad (4.2.59)$$

which is not in general the solution of

$$RW = \mathbf{P}. \qquad (4.2.60)$$

As a practical matter, designers use the amount of leakage which just suffices to mask the offending noise source or second-order effect. The choice of γ thus represents a compromise between biasing the convergence weight vector away from the optimum and "quieting" undamped filter modes. Even though it must be applied carefully, coefficient leakage is an important part of algorithms which must operate with bandlimited signals. It has been employed successfully in fractionally spaced adaptive equalizers for telephone data modems [Gitlin et al., 1982], in ADPCM (see Chapters 5 and 7), and in adaptive interference suppressors.

4.2.7 Gradient Noise, Weight Noise, and the Concept of Misadjustment

Central to the concept of the gradient search methods is the idea of iteratively stepping down the squared-error performance function until the minimum point of the function is reached. At that point the gradient equals zero and therefore no change is made to the weight vector estimate $\mathbf{W}(k)$. Since the minimum has been so attained, the weight vector would then equal the optimal weight vector. As soon as we choose to employ an approximation of the gradient, instead of the gradient itself, then various imperfections in this procedure begin to appear. In this section we briefly examine their cause and effects.

The cause of these problems will be termed here "gradient noise," the vector difference between the theoretical gradient and the instantaneous estimate employed in its place. If we term this vector $\mathbf{N}(k)$, it is given by

$$\mathbf{N}(k) = \hat{\nabla} - \nabla_W J. \qquad (4.2.61)$$

In the particular case of the LMS algorithm it is given by

$$\mathbf{N}(k) = e(k)\mathbf{X}(k) - (R\,\mathbf{W}(k) - \mathbf{P}). \qquad (4.2.62)$$

The effect of gradient noise is most obvious when adaptation has progressed far enough that $\mathbf{W}(k)$ is very close to the optimal value $\mathbf{W}^o$. At this point the true gradient is very nearly equal to zero, and therefore any extra component in the gradient estimate tends to push the weight estimate away from the optimal choice. Because of this gradient noise the weight estimate $\mathbf{W}(k)$ obtains a noise component, again most obvious near convergence. The weights of approximate gradient algorithms are often said to "rattle" near convergence. The magnitude of this weight noise and its effect on the output $y(k)$ depend on many signal- and application-related factors, but in almost every case it depends directly on the adaptation constant μ. In general, a reduction in μ reduces the weight noise since it reduces the amount the weights can be adapted in response to a noise component in the gradient estimate.

Since the filter output $y(k)$ depends on $\mathbf{W}(k)$, one can expect $y(k)$ to contain a small extraneous component $s(k)$ that is attributable to weight noise. In particular, if $\mathbf{W}(k)$ is written as

$$\mathbf{W}(k) = \mathbf{W}^{\circ} + \mathbf{V}(k), \qquad (4.2.63)$$

then the filter output can be written as

$$y(k) = \mathbf{X}^t(k)\mathbf{W}^{\circ} + \mathbf{X}^t(k)\mathbf{V}(k) \triangleq y^{\circ}(k) + s(k), \qquad (4.2.64)$$

where $y^{\circ}(k)$ is the output expected if the optimal filter were used. The term $s(k)$ results directly from the weight noise, and its presence degrades the quality of the adaptive filter's output. Its effect is quantified by defining as misadjustment the increase in the average squared error induced by $s(k)$. The exact theoretical form of the misadjustment depends on the specific algorithm employed and input signal characteristics, but it is usually proportional to both μ and the filter order N. For an example, see [Widrow et al., 1976]. Decreasing μ decreases the weight noise, while decreasing N reduces the number of noisy weights acting on the data vector $\mathbf{X}(k)$. It is common to make comparisons of algorithm convergence rates while holding some measure of steady-state performance constant. Misadjustment is often used as that measure.

4.3 RECURSIVE SOLUTION TECHNIQUES

4.3.1 The Motivation for Recursive Algorithms

Section 4.2 showed how adaptive algorithms can be developed by employing the concept of a search. The available data samples are used in an algorithm which attempts to move the current estimate of the impulse response vector to the optimum value. The algorithms which result from this approach have the advantage of being quite simple to implement but carry with them the disadvantages that they can be slow to approach the optimal weight vector and, once close to it, will usually "rattle around" the optimal vector rather than actually converge to it due to the effects of approximations made in the estimate of the performance function gradient.

To overcome those difficulties we examine another approach in this section. Here we develop algorithms which use the input data $\{x, d\}$ in such a way as to ensure optimality at each step. If this can be done, then clearly the result of the algorithm for the last data point is the overall optimal weight vector.

We now make this more precise. Suppose that we redefine the sum squared performance function J_{ss} by the expression

$$J_k = \sum_{l=N-1}^{k-1} |y(l) - d(l)|^2, \qquad N-1 \leqslant k \leqslant L-1. \qquad (4.3.1)$$

This form of J simply reflects how many samples have been used so far. Clearly J_L uses all the available data from $k = N - 1$ to $k = L - 1$ and has exactly the same form as j_{ss} in (3.2.1). Suppose we define $\mathbf{W}_k^o$ as the impulse response vector which minimizes J_k. By this definition $\mathbf{W}_L^o$ equals $\mathbf{W}_{ss}^o$, the optimal impulse vector over all the data.

The motivation for developing "recursive-in-time" algorithms can be seen as follows. Suppose $x(l)$ and $d(l)$ have been received for time up through $k-1$ and that $\mathbf{W}_k^o$ has been computed. Now suppose that $x(k)$ and $d(k)$ are received, allowing us to form

$$J_{k+1} = \sum_{l=N-1}^{k} |y(l) - d(l)|^2 = J_k + |y(k) - d(k)|^2. \tag{4.3.2}$$

We desire to find some procedure by which $\mathbf{W}_k^o$ can be updated to produce $\mathbf{W}_{k+1}^o$, the new optimal vector. If we can develop such a procedure, then we can build up the optimal weight vector step by step until the final pair of data points $x(L-1)$ and $d(L-1)$ are received. With these points $\mathbf{W}_L^o$ can be computed, which, by definition, is the global optimum vector $\mathbf{W}_{ss}^o$.

4.3.2 Recursive Least Squares (RLS)

4.3.2.1 The Update Formula
The simplest approach to updating $\mathbf{W}_k^o$ is the following procedure:

(a) Update R_{ss} via

$$R_{ss,k+1} = R_{ss,k} + \mathbf{X}(k)\mathbf{X}^t(k). \tag{4.3.3}$$

(b) Update $\mathbf{P}_{ss}$ via

$$\mathbf{P}_{ss,k+1} = \mathbf{P}_{ss,k} + d(k)\mathbf{X}(k). \tag{4.3.4}$$

(c) Invert $R_{ss,k+1}$.
(d) Compute $\mathbf{W}_{k+1}^o$ via

$$\mathbf{W}_{k+1}^o = R_{ss,k+1}^{-1}\mathbf{P}_{ss,k+1}. \tag{4.3.5}$$

The autocorrelation matrix and the cross-correlation vectors are updated and then used to compute $\mathbf{W}_{k+1}^o$. While direct, this technique is computationally wasteful. Approximately $N^3 + 2N^2 + N$ multiplications are required at each update, where N is the impulse response length, and of that N^3 is required for the matrix inversion if done with the classical Gaussian elimination technique.

In an effort to reduce the computational requirement for this algorithm we focus first on this inversion. We notice that Gaussian elimination makes no use whatsoever of the special form of $R_{ss,k}$ or of the special form of the update from $R_{ss,k}$ to $R_{ss,k+1}$. We now set out to take advantage of it. We do so by employing the well-known matrix inversion lemma [Kailath, 1980], also sometimes called the "ABCD lemma,"

$$(A + BCD)^{-1} = A^{-1} - A^{-1}B(DA^{-1}B + C^{-1})^{-1}DA^{-1}. \quad (4.3.6)$$

We use this lemma by making the following associations:

$$A = R_k$$
$$B = \mathbf{X}(k)$$
$$C = 1$$

and

$$D = \mathbf{X}'(k). \quad (4.3.7)$$

With these associations, R_{k+1} can be represented as

$$R_{k+1} = R_k + \mathbf{X}(k)\mathbf{X}'(k) = A + BCD \quad (4.3.8)$$

and R_{k+1}^{-1} is given by

$$R_{k+1}^{-1} = R_k^{-1} - \frac{R_k^{-1}\mathbf{X}(k)\mathbf{X}'(k)R_k^{-1}}{1 + \mathbf{X}'(k)R_k^{-1}\mathbf{X}(k)}. \quad (4.3.9)$$

Thus, given R_k^{-1} and a new input $x(k)$, hence $\mathbf{X}(k)$, we can compute R_{k+1}^{-1} directly. We never compute R_{k+1}, nor do we invert it directly.

The optimal weight vector $\mathbf{W}_{k+1}^o$ is given by

$$\mathbf{W}_{k+1}^o = R_{k+1}^{-1}\mathbf{P}_{k+1}, \quad (4.3.10)$$

which can be obtained by combining (4.3.9) with (4.3.4)

$$\mathbf{W}_{k+1}^o = \left\{ R_k^{-1} - \frac{R_k^{-1}\mathbf{X}(k)\mathbf{X}'(k)R_k^{-1}}{1 + \mathbf{X}'(k)R_k^{-1}\mathbf{X}(k)} \right\} \cdot \{\mathbf{P}_k + d(k)\mathbf{X}(k)\}$$

$$= R_k^{-1}\mathbf{P}_k - \frac{R_k^{-1}\mathbf{X}(k)\mathbf{X}'(k)R_k^{-1}\mathbf{P}_k}{1 + \mathbf{X}'(k)R_k^{-1}\mathbf{X}(k)}$$

$$+ d(k)R_k^{-1}\mathbf{X}(k) - \frac{d(k) \cdot R_k^{-1}\mathbf{X}(k)\mathbf{X}'(k)R_k^{-1}\mathbf{X}(k)}{1 + \mathbf{X}'(k)R_k^{-1}\mathbf{X}(k)}. \quad (4.3.11)$$

To simplify this result we make the following associations and definitions.

The kth optimal weight vector:

$$R_k^{-1}\mathbf{P}_k = \mathbf{W}_k^o, \quad (4.3.12)$$

The filtered information vector:

$$\mathbf{Z}_k \triangleq R_k^{-1}\mathbf{X}(k). \tag{4.3.13}$$

The a priori output:

$$y_0(k) \triangleq \mathbf{X}^t(k)\mathbf{W}_k^o. \tag{4.3.14}$$

The normalized input power:

$$y_0(k) \triangleq \mathbf{X}^t(k)\mathbf{W}_k^o. \tag{4.3.14}$$

With these expressions, the optimal weight vector $\mathbf{W}_{k+1}^o$ becomes

$$\mathbf{W}_{k+1}^o = \mathbf{W}_k^o - \frac{\mathbf{Z}_k\mathbf{X}^t(k)\mathbf{W}_k^o}{1 + \mathbf{X}^t(k)\mathbf{Z}_k} + d(k)\mathbf{Z}_k - \frac{d(k)\mathbf{Z}_k\mathbf{X}^t(k)\mathbf{Z}_k}{1 + \mathbf{X}^t(k)\mathbf{Z}_k}$$

$$= \mathbf{W}_k^o - \frac{\mathbf{Z}_k y_0(k)}{1 + q} + d(k)\mathbf{Z}_k - \frac{d(k)q\mathbf{Z}_k}{1 + q}$$

$$= \mathbf{W}_k^o - \frac{\mathbf{Z}_k y_0(k)}{1 + q} + \frac{d(k)\mathbf{Z}_k}{1 + q}$$

$$= \mathbf{W}_k^o + \frac{\{d(k) - y_0(k)\} \cdot \mathbf{Z}_k}{1 + q}. \tag{4.3.16}$$

Equations (4.3.9) and (4.3.12)–(4.3.16) together comprise the "recursive least squares" (RLS) algorithm.

4.3.2.2 Interpretation of the Update Equations

We now examine (4.3.16) and interpret it in terms of the quantities named in (4.3.12)–(4.3.15). The update equation for $\mathbf{W}_{k+1}^o$ is based on starting with $\mathbf{W}_k^o$, and then adding a correction term which depends on $x(k)$ and $d(k)$. This dependence comes in three ways. The first factor is the "a priori" error,

$$e_0(k) = d(k) - y_0(k), \tag{4.3.17}$$

which is the difference between the new template sample $d(k)$ and the "a priori" output $y_0(k)$. The name for y_0 is a result of the timing of (4.3.14). When $x(k)$ is received, the new data vector $\mathbf{X}(k)$ is formed. The output $y_0(k)$ is formed by using the best filter available at time $k - 1$, called $\mathbf{W}_k^o$, but with the new vector of data $\mathbf{X}(k)$. Therefore, $y_0(k)$ is the output estimate before $x(k)$ is used to update $\mathbf{W}$; that is, it is a prediction of the optimal filter output given $\mathbf{X}(k)$ but with the old impulse response $\mathbf{W}_k$. Since $y_0(k)$ is computed before $\mathbf{W}_{k+1}^o$ is formed, y_0 is called the a priori output, and $e_0(k)$ is called the a priori error. It is also sometimes called the a priori "prediction error." The dependence of the update on $e_0(k)$ is also clear. If the old filter $\mathbf{W}_k^o$ allows $d(k)$ to be perfectly predicted, that is, $y_0(k) = d(k)$, then no update is needed. With $e_0 = 0$, then update is added on. If they differ, that is, $e_0 \neq 0$, a correction is made.

The term $\mathbf{Z}_k$ is called the filtered information vector, because R_k^{-1} acts to influence or "filter" the direction and length of the data (information) vector. The importance of this vector will emerge later.

The term q is a measure of the input signal power, just as $\mathbf{X}^t(k)\mathbf{X}(k)$ would be, but with a normalization introduced by R_k^{-1}. The effect of this normalization is to make q have an average value of N rather than being proportional to the actual signal level. Instantaneously, however, q combines with $\mathbf{Z}_k$ and e_0 to produce exactly the optimal weight update. Since R_k is nonnegative definite, $1 + q$ always equals or exceeds 1.

4.3.2.3 Related Gradient Descent Algorithms

The adaptive updating algorithm shown in (4.3.9) and (4.3.13)–(4.3.16) was developed based on the concept of updating the optimal vector $\mathbf{W}_k^o$ by just the right amount to generate the optimal vector $\mathbf{W}_{k+1}^o$. This differs from the approach used in Section 4.2 based on searching a performance surface. Even so, the form of (4.3.16) closely resembles some of the gradient descent algorithms cataloged in Section 4.2.5. To gain some perspective, we briefly examine two such algorithms.

Accelerated LMS

The accelerated LMS update equation can be written as

$$\mathbf{W}(k + 1) = \mathbf{W}(k) + \mu e(k) C \mathbf{X}(k), \tag{4.3.18}$$

where μ is the adaptation constant and C is a matrix chosen to improve the algorithm rate of convergence by acting on the information vector $\mathbf{X}(k)$ to change its direction or its length. Using (4.3.16) and (4.3.18), we can write the recursive least squares (RLS) update as

$$\mathbf{W}_{k+1}^o = \mathbf{W}_k^o + \frac{1}{1 + q} \cdot e_0(k) \cdot R_k^{-1}\mathbf{X}(k). \tag{4.3.19}$$

By making the following equivalences,

$$\mu = \frac{1}{1 + q} = \frac{1}{1 + \mathbf{X}^t(k)R_k^{-1}\mathbf{X}(k)} \tag{4.3.20}$$

and

$$C = R_k^{-1}, \tag{4.3.21}$$

the accelerated LMS algorithm becomes the RLS algorithm, which is in some sense the ultimate in algorithm convergence acceleration. The RLS algorithm defines just the right step size and the right direction to retain optimality over each time sample.

Normalized LMS

The normalized LMS algorithm uses a time-varying adaptation constant of the form

$$\mu(k) = \frac{\alpha}{\gamma + \mathbf{X}^t(k)\mathbf{X}(k)}, \qquad 0 < \alpha < 2. \qquad (4.3.22)$$

This form was developed using a simple bound for the maximum eigenvalue of R_k. The term γ was introduced heuristically just to avoid the possibility of a zero denominator. Clearly the RLS derivation shows that γ is a theoretically important part of the update. It also shows that the time-varying gain concept is closer to the optimal updating technique than conventional LMS with its fixed gain.

4.3.2.4 The RLS Algorithm

With the analytical derivation developed in Section 4.3.2.1 we can now write down a step-by-step procedure for updating $\mathbf{W}_k^o$. This set of steps is efficient in the sense that no unneeded variable is computed and that no needed variable is computed twice. We do, however, need assurance that R_k^{-1} exists. The procedure then goes as follows:

   (i)   Accept new samples $x(k)$, $d(k)$.

  (ii)   Form $\mathbf{X}(k)$ by shifting $x(k)$ into the information vector.

 (iii)   Compute the a priori output $y_0(k)$:

$$y_0(k) = \mathbf{W}_k^{ot}\mathbf{X}(k). \qquad (4.3.23)$$

 (iv)   Compute the "a priori" error $e_0(k)$:

$$e_0(k) = d(k) - y_0(k). \qquad (4.3.24)$$

  (v)   Compute the filtered information vector $\mathbf{Z}_k$:

$$\mathbf{Z}_k = R_k^{-1}\mathbf{X}(k). \qquad (4.3.25)$$

 (vi)   Compute the normalized error power q:

$$q = \mathbf{X}^t(k)\mathbf{Z}_k. \qquad (4.3.26)$$

(vii) Compute the gain constant v:

$$v = \frac{1}{1 + q}. \tag{4.3.27}$$

(viii) Compute the normalized filtered information vector $\tilde{\mathbf{Z}}_k$:

$$\tilde{\mathbf{Z}}_k = v \cdot \mathbf{Z}_k. \tag{4.3.28}$$

(ix) Update the optimal weight vector $\mathbf{W}_k^o$ to $\mathbf{W}_{k+1}^o$:

$$\mathbf{W}_{k+1}^o = \mathbf{W}_k^o + e_0(k)\tilde{\mathbf{Z}}_k. \tag{4.3.29}$$

(x) Update the inverse correlation matrix R_k^{-1} to R_{k+1}^{-1}:

$$R_{k+1}^{-1} = R_k^{-1} - \tilde{\mathbf{Z}}_k \mathbf{Z}_k^t. \tag{4.3.30}$$

This procedure assumes that R_k^{-1} exists at the initial time in the recursion. As a result, two initialization procedures are commonly used. The first is to build up R_k and $\mathbf{P}_k$ according to (4.3.3) and (4.3.4) until R has full rank, i.e., at least N input vectors $\mathbf{X}(k)$ are acquired. At this point R_k^{-1} is computed directly and then $\mathbf{W}_k$. Given these, the recursion can proceed as described above indefinitely or until $k = L - 1$. The advantage of this first technique is that optimality is preserved at each step. The major price paid is that about N^3 computations are required once to perform that initial inversion. A second, much simpler approach is also commonly used. In this case R_{N-1}^{-1} is initialized as

$$\hat{R}_{N-1}^{-1} = \eta I_N, \tag{4.3.31}$$

where η is a large positive constant and I_N is the N-by-N identity matrix. Since R_{N-1}^{-1} almost certainly will not equal ηI_N, this inaccuracy will influence the final estimate of R_k and hence $\mathbf{W}_k$. As a practical matter, however, η can usually be made large enough to avoid significant impact on $\mathbf{W}_{L-1}^o$ while still making R_{N-1} invertible. Because of the simplicity and the low computational cost, the second approach is the one most commonly used. It becomes even more theoretically justifiable when used with the exponentially weighted RLS algorithm to be discussed in Section 4.3.2.6.

4.3.2.5 The Computational Cost for the RLS Algorithm
As a prelude to developing even more efficient adaptive algorithms, we first should determine how much computation is required to execute the RLS algorithm laid out in the previous section.

We find that the ten steps in the procedure can be grouped by their computational complexity:

(a) Order 1: Steps (iv) and (vii) require only a few simple operations, such as a subtraction or an addition and division. These will be termed order 1 and denoted $O(1)$, since the amount of computation required is not related to the filter order.

(b) Order N: Steps (iii), (vi), (viii), and (ix) each require a vector dot product, a scalar–vector product, or a vector scale and sum operation. Each of these requires N multiplications and up to N additions for each iteration of the algorithm. The actual number of multiplications required for these steps is $4N$, but we refer to them as order N, or $O(N)$, since the computation requirement is proportional to N, the length of the filter impulse response.

(c) Order N^2: Step (v), a matrix-vector product, and step (x), the vector outer product, both require N^2 multiplications and approximately N^2 additions. These are termed $O(N^2)$ procedures.

The total number of computations needed to execute the RLS algorithm for each input sample pair $\{x(k), d(k)\}$ is

$$2N^2 + 4N \text{ multiplications,} \tag{4.3.32}$$

an approximately equal number of additions, and one division. Since this amount of computation is required for each sample pair, the total requirement of multiplications over the $N - 1$ to $L - 1$ sample interval is

$$C_{RLS} = (L - N + 1)\cdot 2N^2 + (L - N + 1)\cdot 4N. \tag{4.3.33}$$

It is interesting to compare this to the number of multiplications needed to do a brute-force calculation of $\mathbf{W}^o_{ss}$ after all the data are received, instead of updating with each new sample. In (3.2.66) we found this brute-force method's computational requirements to be

$$C_{ss} = (L - N + 1)\cdot N^2 + (L - N + 1)\cdot N + N^3 + N. \tag{4.3.34}$$

The N^3 term in C_{ss} comes from the inversion of R_{ss}, a requirement avoided by the recursive updating scheme used with the RLS algorithm. Depending on how large N is with respect to $L - N + 1$, this inversion can be significant $(L - N + 1 < N)$ or insignificant $(L \gg N)$. We see that the RLS algorithm is more computationally expensive in the $O(N)$ and $O(N^2)$ terms. For example, the RLS algorithm requires $(L - N + 1)\cdot 2N^2$ multiplications compared to $(L - N + 1)\cdot N^2$ for the brute force technique. Similarly, the respective $O(N)$ terms are $4N\cdot(L - N + 1)$ and $(L - N + 2)\cdot N$. Thus, avoiding the inversion with its $O(N^3)$ computational requirement leads to a growth in the $O(N^2)$ and $O(N)$ cost.

When the length of the data record L is substantially larger than the filter length N, the cost of the one-time inversion is overwhelmed by the $O(N^2)$ terms, and in this case the brute force method is computationally cheaper than the more elegant RLS procedure. Since this situation $(1 < N \ll L)$ is typical of adaptive filtering problems, one might question why the RLS method should be explored at all. There are, in fact, several reasons for exploring and using RLS techniques:

(a) RLS can be numerically better behaved than the direct inversion of R_{ss};
(b) RLS provides an optimal weight vector estimate at every sample time, while the direct method produces a weight vector estimate only at the end of the data sequence; and
(c) This recursive formulation leads the way to even lower-cost techniques.

Each of these aspects will be discussed in later sections.

4.3.2.6 The RLS Algorithm with Exponential Data Weighting

We now modify the definitions of R_k and $\mathbf{P}_k$ slightly as

$$R_k = \sum_{l=N-1}^{k-1} \rho^{k-1-l} \mathbf{X}(l)\mathbf{X}^t(l) \tag{4.3.35}$$

and

$$\mathbf{P}_k = \sum_{l=N-1}^{k-1} \rho^{k-1-l} \mathbf{X}(l)d(l), \tag{4.3.36}$$

where ρ, the averaging or "forgetting" factor, is a positive constant. It is usually chosen to be slightly less than 1, thereby diminishing the contribution of the "older" data. The exponential weighting emphasizes the most recently received data and has a time constant, or effective averaging period, of $1/(1 - \rho)$. When $\rho = 1$, the expressions for R_k and $\mathbf{P}_k$ are identical to those introduced in Section 4.3.2.1, and the averaging interval is infinite, implying that all data are averaged with equal weighting.

The principal motivation for introducing this type of weighting stems from the problem of data nonstationarity, that is, the situation where the input data $\{x, d\}$ changes its "character" within the record of L data points. As the examples revisited in Chapter 8 will show, this is a common practical situation and in fact is one of the principal reasons for employing an adaptive filter. By choosing an averaging interval shorter than the time interval over which the data changes character, the adaptive updating algorithm will "track" the changes.

As we show, exponential weighting can be introduced into the RLS procedure quite easily. We start by noting that R_k and $\mathbf{P}_k$ can be recursively defined by the expressions:

$$R_{k+1} = \sum_{l=N-1}^{k} \rho^{k-l} \mathbf{X}(l)\mathbf{X}^t(l) = \mathbf{X}(k)\mathbf{X}^t(k) + \sum_{l=N-1}^{k-1} \rho^{k-l} \mathbf{X}(l)\mathbf{X}^t(l)$$

$$= \rho \left\{ R_k + \frac{1}{\rho} \mathbf{X}(k)\mathbf{X}^t(k) \right\} \tag{4.3.37}$$

and

$$\mathbf{P}_{k+1} = \sum_{l=N-1}^{k} \rho^{k-l} \mathbf{X}(l)d(l)$$

$$= \left\{ \sum_{l=N-1}^{k-1} \rho^{k-l} \mathbf{X}(l)d(l) \right\} + \mathbf{X}(k)d(k)$$

$$= \rho \left\{ \mathbf{P}_k + \frac{1}{\rho} d(k)\mathbf{X}(k) \right\}. \tag{4.3.38}$$

The optimal weight vector at time step k is given by

$$\mathbf{W}_{k+1}^o = R_{k+1}^{-1}\mathbf{P}_{k+1} = \left\{ R_k + \frac{1}{\rho} \mathbf{X}(k)\mathbf{X}^t(k) \right\}^{-1} \cdot \left\{ \mathbf{P}_k + \frac{1}{\rho} d(k)\mathbf{X}(k) \right\}. \tag{4.3.39}$$

At this point the general approach used in Section 4.3.2.1 can be employed. The only significant difference is that C in (4.3.7) equals $1/\rho$ instead of 1. In fact, it can be easily shown that the weight vector estimate is updated according to the expression

$$\mathbf{W}_{k+1}^o = \mathbf{W}_k^o + \left\{ \frac{d(k) - y_0(k)}{\rho + q} \right\} \mathbf{Z}_k. \tag{4.3.40}$$

This leads to trivial differences in the procedure described in Section 4.3.2.4. In addition to (4.3.40), the only other modification is that needed to properly update R_k^{-1} as in (4.3.30),

$$R_{k+1}^{-1} = \frac{1}{\rho} \left\{ R_k^{-1} - \frac{\mathbf{Z}(k)\mathbf{Z}^t(k)}{\rho + q} \right\}. \tag{4.3.41}$$

Note that (a) the usual RLS algorithm is attained when $\rho = 1$, (b) no change in the amount of computation is required by introducing the exponential averaging, and (c) the algorithm is almost always initialized with $R_{N-1}^{-1} = \eta I$, since the effect of this erroneous initial condition will be "forgotten" as time evolves.

4.3.3 "Fast" RLS Algorithms

4.3.3.1 Approach

The basic motivation for developing the recursive algorithms examined in the previous section was to reduce the amount of computation required at each time step. We accomplished this by exploiting a special property of R_k to obtain a computationally less expensive procedure for computing the inverse. In particular we noticed that R_k becomes R_{k+1} with the addition of a vector outer product. We then used the "ABCD lemma" in (4.3.6) to produce a recursive procedure for obtaining R_{k+1}^{-1} from R_k^{-1}, reducing the computational load from N^3 to $O(N^2)$ per iteration. We now desire to reduce the computational cost still more, if possible to the level that only $O(N)$ operations are required at each time step. Algorithms which attain this goal have been termed "fast" algorithms, and, as we shall see, they obtain the desired improvement in efficiency by exploiting another special property of the adaptive filtering problem.

We start by making two apparently unrelated observations. The first is that the vector $\mathbf{X}(k)$ evolves to $\mathbf{X}(k + 1)$ with the incorporation of only one new input sample $x(k + 1)$. Of the N samples in the "delay line" of the filter, the oldest is dropped out, a new one $x(k + 1)$ is incorporated, and $N - 1$ remain in the delay line. This property is not exploited with the RLS algorithm, and in fact RLS will work with any choices of $\mathbf{X}$ data input vectors. This is useful, for example, in the case of adaptive sensor arrays, where this "shifting" property does not hold. One might expect the computational efficiency to improve if this shifting property of $\mathbf{X}$ could be exploited.

The second observation is that if $\mathbf{Z}_k$ could be simply propagated to $\mathbf{Z}_{k+1}$ in the RLS procedure, then the updating of R_k^{-1} and its multiplication by $\mathbf{X}(k)$ would not be necessary. This is very important since these two steps are the ones which require N^2 multiplications each. If they could be avoided, then the computational load would drop to the desired level of $O(N)$.

4.3.3.2 A Recursive Formula for $\bar{\mathbf{Z}}_k$

In the light of these observations, we will seek a procedure which allows us to exploit the shift properties of $\mathbf{X}(k)$ to produce an $O(N)$ updating scheme for the filtered information vector. We start by defining the slightly modified filtered information vector $\bar{\mathbf{Z}}_k$. An expression for this vector can be developed by solving for $\mathbf{V}_k = \mathbf{W}_{k+1}^o - \mathbf{W}_k^o$, the difference between the optimum weight vectors at times $k + 1$ and k.

$$R_{k+1}\mathbf{W}_{k+1}^o = R_{k+1}(\mathbf{W}_k^o + \mathbf{V}_k) = \mathbf{P}_{k+1}$$

$$= R_{k+1}\mathbf{W}_k^o + R_{k+1}\mathbf{V}_k = \mathbf{P}_{k+1}. \tag{4.3.42}$$

Substituting the recursive definitions of R_k and $\mathbf{P}_k$, we find that

$$\{R_k + \mathbf{X}(k)\mathbf{X}^t(k)\}\mathbf{W}_k^o + R_{k+1}\mathbf{V}_k = \mathbf{P}_k + d(k)\mathbf{X}(k). \tag{4.3.43}$$

Recall the following relationships:

$$R_k \mathbf{W}_k^o = \mathbf{P}_k \qquad (4.3.44)$$

$$y_0(k) \triangleq \mathbf{X}^t(k)\mathbf{W}_k^o. \qquad (4.3.45)$$

Using these, (4.3.43) becomes

$$R_{k+1} \mathbf{V}_k = \{d(k) - y_0(k)\} \mathbf{X}(k), \qquad (4.3.46)$$

and, since

$$e_0(k) = d(k) - y_0(k), \qquad (4.3.47)$$

we have

$$\mathbf{V}_k = e_0(k) \cdot R_{k+1}^{-1} \mathbf{X}(k). \qquad (4.3.48)$$

The vector $\mathbf{V}_k$ is the update to $\mathbf{W}_k^o$ which preserves optimality in the step from sample $k - 1$ to k. This update depends directly on the prediction error $e_0(k)$ and the vector $\bar{\mathbf{Z}}_k$ given by

$$\bar{\mathbf{Z}}_k = R_{k+1}^{-1} \mathbf{X}(k) = \left\{ \sum_{l=N-1}^{k} \mathbf{X}(l)\mathbf{X}^t(l) \right\}^{-1} \mathbf{X}(k). \qquad (4.3.49)$$

We see now that $\bar{\mathbf{Z}}_k$ depends on R_{k+1}^{-1} rather than R_k^{-1} as $\mathbf{Z}_k$ does. It is this vector which we endeavor to recursively update.

We note that $\bar{\mathbf{Z}}_k$ depends solely on the input sequence $x(k)$. The computationally efficient techniques for updating this vector hinge on the concept of prediction; that is, a block of N adjacent data points is used to predict the data samples immediately before and after the block. The resulting prediction errors are used to improve the predictors and, implicitly, to develop a model for the data. This model can then be used to update some function of that data, namely, $\bar{\mathbf{Z}}_k$.

One procedure for computing $\bar{\mathbf{Z}}_k$ is laid out below. It uses the auxiliary vectors $\mathbf{A}(k)$ and $\mathbf{B}(k)$. For reasons to be seen, $\mathbf{A}$ is known as the forward predictor and $\mathbf{B}$ is called the backward predictor.

(i) Compute the "a priori" forward prediction error ε_0:

$$\varepsilon_0(k + 1) = x(k + 1) + \mathbf{A}^t(k)\mathbf{X}(k). \qquad (4.3.50)$$

(ii) Update the forward prediction vector $\mathbf{A}$:

$$\mathbf{A}(k + 1) = \mathbf{A}(k) - \bar{\mathbf{Z}}_k \varepsilon_0(k + 1). \qquad (4.3.51)$$

(iii)   Compute the "a posteriori" forward prediction error ε:

$$\varepsilon(k + 1) = x(k + 1) + \mathbf{A}^t(k + 1)\mathbf{X}(k). \qquad (4.3.52)$$

(iv)   Compute prediction crosspower Σ:

$$\Sigma(k + 1) = \Sigma(k) + \varepsilon(k + 1)\varepsilon_0(k + 1). \qquad (4.3.53)$$

(v)   Form the augmented vector $\mathbf{F}$:

$$\mathbf{F} \triangleq \begin{bmatrix} \varepsilon(k + 1)/\Sigma(k + 1) \\ \bar{Z}(k) + \mathbf{A}(k + 1)\varepsilon(k + 1)/\Sigma(k + 1) \end{bmatrix} \quad \begin{array}{l} \text{1 element} \\ N \text{ elements} \end{array}. \qquad (4.3.54)$$

(vi)   Partition $\mathbf{F}$:

$$\mathbf{F} = \begin{bmatrix} M(k + 1) \\ \mu(k + 1) \end{bmatrix} \quad \begin{array}{l} N \text{ elements} \\ \text{1 element} \end{array} \qquad (4.3.55)$$

(vii)   Compute the "a priori" backward prediction error η_0:

$$\eta_0(k + 1) = x(k - N + 1) + \mathbf{B}^t(k)\mathbf{X}(k + 1). \qquad (4.3.56)$$

(viii)   Update the backward prediction vector $\mathbf{B}$:

$$\mathbf{B}(k + 1) = [\mathbf{B}(k) - M(k + 1)\eta_0(k + 1)]/[1 - \mu(k + 1)\eta_0(k + 1)].$$

$$(4.3.57)$$

(ix)   Update $\bar{\mathbf{Z}}_k$:

$$\bar{\mathbf{Z}}_{k+1} = M(k + 1) - \mathbf{B}(k + 1) \cdot \mu(k + 1). \qquad (4.3.58)$$

A proof of this procedure will not be shown here, but it is available from several sources, e.g., [Ljung et al., 1978]. Most of the steps involve either predicting the waveform $x(k)$ or using those predictions to improve the predictors. The last step produces $\bar{\mathbf{Z}}_k$ as a weighted combination of the forward and backward prediction vectors. The theory on which this procedure is based is quite powerful and allows proofs to be given and other algorithms to be developed. It is, however, beyond the scope of this book, and the interested reader is invited to pursue the topic in [Ljung et al., 1978], [Cioffi and Kailath, 1984], [Honig and Messerschmitt, 1984], or [Alexander, 1986].

Given a method for recursively computing $\bar{\mathbf{Z}}_k$, we can now state a "fast" algorithm for recursively computing $\mathbf{W}_{k+1}^o$ from $\mathbf{W}_k^o$.

For each k:

(a) Compute the "a priori" output $y_0(k + 1)$

$$y_0(k + 1) = \mathbf{X}^t(k + 1)\mathbf{W}_k^o. \tag{4.3.59}$$

(b) Form the "a priori" output prediction error $e_0(k + 1)$:

$$e_0(k + 1) = d(k + 1) - y_0(k + 1). \tag{4.3.60}$$

(c) Update $\bar{\mathbf{Z}}_k$ to $\bar{\mathbf{Z}}_{k+1}$.
(d) Compute the updated impulse response vector

$$\mathbf{W}_{k+1}^o = \mathbf{W}_k^o + e_o(k + 1)\bar{\mathbf{Z}}(k + 1). \tag{4.3.61}$$

Here step (c) includes steps (i) through (ix) listed in (4.3.50)–(4.3.58). We note that $\bar{\mathbf{Z}}(k)$ carries most of the needed information about $x(k)$ into the weight vector update. The filter output is computed using $\mathbf{X}(k)$ and compared to the template waveform $d(k + 1)$, producing $e_o(k)$. The vector $\bar{\mathbf{Z}}(k + 1)$ carries all the directional information needed to update $\mathbf{W}_k$.

4.3.3.3 Computational Issues
The principal objective of developing these relatively complicated algorithms was to reduce the amount of computation needed to attain "exact" optimality at each time step. A review of steps (i)–(ix) and (a)–(d) shows that this objective has been met. Steps (i), (ii), (iii), (v), (vii), (viii), and (ix) involve a vector dot or scalar product and require N multiply-adds each. The conversion of $\bar{\mathbf{Z}}_{k+1}$ to $\bar{\mathbf{W}}_{k+1}$ requires two more vector operations, at a cost of $2N$ multiplications and additions. Thus, if we ignore a small number of scalar operations, we see that this particular fast algorithm needs $9N$ multiply-adds, a distinct improvement over the "slow" RLS procedure.

[Cioffi and Kailath, 1984] have analyzed a large variety of fast RLS algorithms and shown that all can be written as recursions on a set of transversal filters by using the generic updating formula

$$\mathbf{C}_{new} = s_1 \cdot \mathbf{C}_{old} + s_2 \cdot \mathbf{D} \tag{4.3.62}$$

and

$$\rho = \mathbf{X}^t(k)\mathbf{E}. \tag{4.3.63}$$

A quick comparison of (i)–(ix) and (a)–(d) with these equations shows this observation to be true for the algorithm stated in Section 4.3.3.2. [Cioffi and Kailath, 1984] further show, however, that the most computationally efficient algorithms attain their efficiency by making s_1 and/or s_2 equal to one as often as possible. Using this approach, algorithms requiring about $7N$ multiply-adds per sample have been developed. Perversely, it appears that this efficiency is attained at the cost of numerical stability. Algorithms with good numerical stability tend to use values of s_1 and s_2 other than unity, and require $11N$ to $20N$ computations per sample. The trade-off between computational efficiency and numerical stability continues to be an active research area.

4.4 AN EXAMPLE USING BOTH LMS AND RLS

To demonstrate the behavior of some of the algorithms described in earlier sections, we now examine a simple filtering problem. Consider the configuration shown in Figure 4.2. The input of the adaptive filter consists of white noise and two real sinusoids of amplitudes a and b and distinct radian frequencies of ω_1 and ω_2. The filter is a real N-tap FIR discrete-time tapped-delay-line structure. The error $e(k)$ is the difference between the filter output and the template waveform $d(k)$, which, for this example, is a real sinusoid of amplitude c and frequency ω_2. Thus the template waveform has exactly the same frequency but possibly a different phase and amplitude than one of the two input sinusoids.

Intuitively we would expect an adaptive filter to reduce the error of its average squared value by adjusting its impulse response to provide a gain of c/b

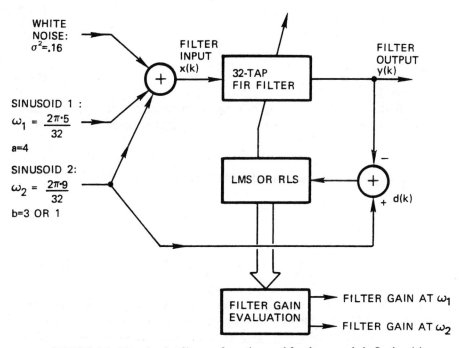

FIGURE 4.2. The adaptive filter configuration used for the example in Section 4.4.

at frequency ω_2, a gain of zero at ω_1, and a generally small gain across the band to suppress the input white noise. If this is done, then the filter removes the sinusoid at ω_1 from its output $y(k)$ and scales the sinusoid at ω_2 to match the amplitude and phase of the template waveform.

In this section we present computer simulation examples of both the LMS algorithm and the RLS algorithm. The algorithms used are exactly those stated in Sections 4.2 and 4.3, and the software implementations of the algorithms are exactly those listed in Appendix A, Sections A.1 and A.5. Before discussing the simulation results, however, we first predict the filter's behavior. To simplify the mathematics we assume that ω_1 and ω_2 are distinct integer multiples of $2\pi/N$. When the input frequencies are selected in this way, the eigenanalysis of Section 3.4 is particularly simple.

Consider the LMS algorithm first. We can predict its convergence rate and determine allowable values of the adaptation constant by determining the eigenvalues of the input data correlation matrix. Following the type of decomposition used in Section 3.4, we find the eigenvalues of R_{ms} to be given by:

$$\lambda_1 = \lambda_2 = \frac{Na^2}{2} + \sigma^2 \tag{4.4.1}$$

and

$$\lambda_3 = \lambda_4 = \frac{Nb^2}{2} + \sigma^2 \tag{4.4.2}$$

and

$$\lambda_n = \sigma^2, \qquad N \geqslant n \geqslant 5, \tag{4.4.3}$$

assuming here that the filter length N equals or exceeds 4, and that the input noise variance is given by σ^2. From (4.2.20), the adaptive constant μ is bounded by

$$0 < \mu < \min\left(\frac{1}{\lambda_i}\right), \qquad i = 1, \ldots, N \tag{4.4.4}$$

or

$$0 < \mu < \frac{2}{N} \min\left(\frac{1}{a^2}, \frac{1}{b^2}\right), \tag{4.4.5}$$

assuming that a^2 and b^2 considerably exceed σ^2. As long as this equation is satisfied, then LMS will theoretically converge. It is common practice, however, to use a value of μ substantially smaller than this theoretical limit, simply to reduce the effects of input noise and gradient estimation error.

The convergence rate of the algorithm is determined by the time constants of the individual adaptive modes. In this case from (4.2.24)

$$\tau_1 = \tau_2 = \frac{1}{\mu\lambda_1} \cong \frac{2}{\mu N a^2} \tag{4.4.6}$$

and

$$\tau_3 = \tau_4 = \frac{1}{\mu\lambda_3} \cong \frac{2}{\mu N b^2}. \tag{4.4.7}$$

If $N > 4$, then the remaining $N - 4$ eigenvalues are all equal σ^2, making the associated time constants equal $1/\mu\sigma^2$.

The RLS algorithm converges very rapidly since it uses the inverse covariance matrix to direct the weight vector update at each step. Because of this, the algorithm can be expected to converge in about N iterations. For the purpose of the example we select $N = 32$ and $\sigma^2 = 0.16$. The first simulation, with results shown in Figure 4.3, was conducted with $a = 4$, $b = 3$, $c = 2$, and $\omega_1 = 2\pi\cdot5/32$ and $\omega_2 = 2\pi\cdot9/32$ rad/sec. From (4.4.5), μ must be less than 0.0078, and has been chosen to be one-tenth that, that is, $\mu = 0.00078$. The adaptive time constants are thus

$$\tau_1 = \tau_2 = 10 \tag{4.4.8}$$

and

$$\tau_3 = \tau_4 = (\tfrac{4}{3})^2 10 = 18, \tag{4.4.9}$$

meaning that the LMS algorithm should be expected to asymptotically converge with a time constant of 18 iterations, the longest of time constants.

Figure 4.3a shows the actual response of the LMS algorithm. The filter gain at frequencies ω_1 and ω_2 are shown versus update iteration, and it is seen that, other than a small amount of jitter due to gradient estimation error, the filter gains converge to the values expected (i.e., zero and c/b) at the rates expected. The RLS results are shown in Figure 4.3b, with the gains converging promptly as expected. Note the large excursions associated with the startup of RLS.

Another aspect of performance can be demonstrated by keeping all parameters the same except the input level of the desired sinusoid. Suppose now that $b = 1$. Thus, instead of the two input sinusoids having roughly equivalent power, they now differ by 12 dB. Retracing our steps, we find that since a still equals 4, μ must still be less than 0.0078. As before, it is chosen to be 0.00078 and $\tau_1 = \tau_2 = 10$. Since b is so much smaller, however, $\tau_3 = \tau_4 = 160$ iterations. This result is borne out in the results shown in Figure 4.4. As before, the filter gains at frequencies ω_1 and ω_2 are plotted versus the iteration number in Figure

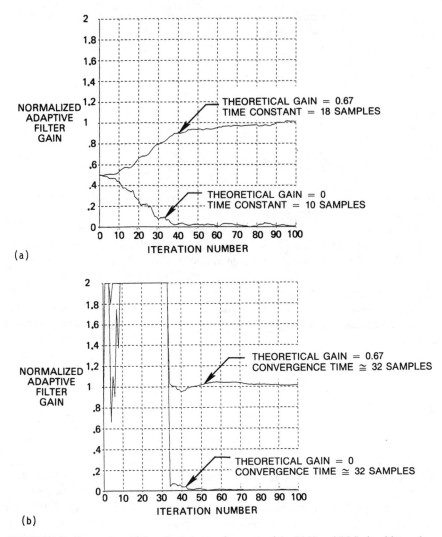

FIGURE 4.3. Comparison of the adaptation performance of the LMS and RLS algorithms where the input sinusoids are nearly matched in power (a) response of the LMS algorithm (b) response of the RLS algorithm.

4.4a. The gain at ω_1 again responds quickly, but the gain at ω_2 takes several hundred samples to converge to c/b. As before, the RLS technique quickly converges for both filter gains, as seen in Figure 4.4b.

With this example we see the algorithms do in fact converge to the expected results and that the rate can be reasonably well predicted. By conducting several trials and averaging the "ensemble," even closer adherence can be attained. For this example, it is also tempting to conclude that the RLS technique is superior

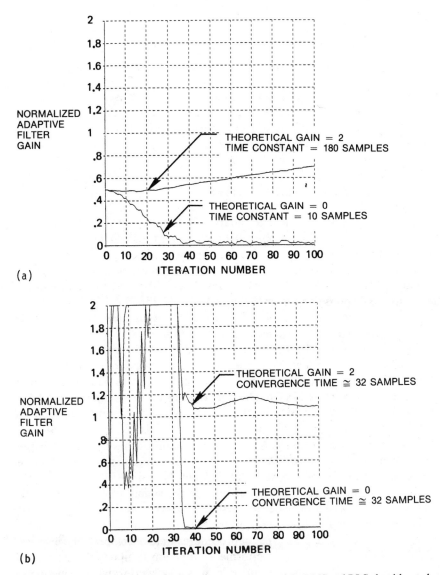

FIGURE 4.4. Comparison of the convergence performance of the LMS and RLS algorithms where the input sinusoids are disparate in power (a) response of the LMS algorithm (b) response of the RLS algorithm.

to gradient search methods because of its prompt convergence. This conclusion must be made carefully, however. While it was not done here in this simple example, careful comparisons of convergence rate are usually done while holding some measure of filter quality constant. The quality measure usually employed is the "filter misadjustment," defined in Section 4.2.7. Also, convergence might be measured in "multiplications" instead of input samples, thus giving the simple gradient algorithms a large advantage. Thus, we see that selection of the "best" algorithm depends on the arithmetic precision available, the rank of the input process, the required output quality, the time-variability of the inputs, and the amount of computation which can be tolerated. In general, RLS and "fast RLS" require fewer data samples to converge, but gradient descent techniques are simpler, more robust, use less computation per iteration, and need less memory. All these issues must be weighed in selecting a filter structure and algorithm for an application.

4.5 SUMMARY AND PERSPECTIVE

This chapter has developed a set of adaptive filtering algorithms based on the idea of optimizing some function of the data waveforms provided to the filter. While the algorithms discussed here are the basis of most adaptive filters in current use, it is useful to remember that Chapter 4 has considered a very specific and quite limited problem. We have assumed that the filter has a discrete-time finite impulse response and is implemented with a tapped-delay-line (transversal) structure. The filter's performance is judged based on some template or reference waveform also provided to the algorithm.

Given these choices for the filter structure and the manner in which the quality of the filter output would be judged, we then turned to the third aspect of adaptive filter design: the development of the algorithms to be used to choose the filter impulse response. In Chapter 2 we did this by developing an intuitively reasonable set of rules for updating an estimate of the desired impulse response in such a way as to improve the match between the filter output and the reference or template waveform. A more rigorous approach was used in Chapter 3. We defined a performance function based on an average of the squared output error, that is, the difference between the filter output and the reference, and then in this chapter developed methodical procedures which allow us to find the impulse response which provides the best performance. We focused principally on the performance function based on the sum of the squared errors over the entire input data record. Two types of procedures were developed. The first used estimates of the gradient of the performance function to update an estimate of the optimal impulse response. In this way the performance function is "searched" to find the best possible operating point. The LMS algorithm and many of its variants stem from this idea. They are relatively low in their computational requirements, work adequately in a variety of signal environments, and are the basis for many existing adaptive processors, such as adaptive equalizers and telephone echo cancellers. The second approach is based on the idea of starting with the optimal solution and then using each input sample to update the impulse response in such a way as to maintain that optimality. This approach yields the recursive least squares (RLS) algorithm and a set of "fast" algorithms that maintain the optimality of the solution with a relative minimum of computation. The normal and "fast" RLS algorithms offer faster convergence than the gradient-search-type algorithms but usually at the cost of more computation per data sample and more numerical difficulties. Adaptive filters employing the RLS approach are coming into practical use but so far only in applications which need fast convergence, such as "fast-startup" modem equalizers and adaptive differential pulse code modulation (ADPCM) encoders.

Given this background and the set of adaptive algorithms that satisfy our limited objectives in this chapter, we expand our horizon in the next two chapters. Chapter 5 treats the case of adapting an infinite-impulse-response filter instead of the transversal filter employed in this chapter. As will be shown, the presence of feedback in the filter itself complicates many aspects of the adaptive algorithm design. Chapter 6 examines yet a different problem, that is, what if there is no reference waveform to guide the algorithm's adaptation? We see there that criteria other than output prediction error can be used to define a performance function, and hence to fashion an adaptive algorithm. Thus, we will have probed into each aspect of the generic adaptive filter cited in the introduction of Chapter 2: filter structure, performance evaluation, and parameter updating.

5
Adaptive Infinite-Impulse-Response (IIR) Filtering

● *PRECIS   Relative to the FIR case, adaptive IIR filter algorithms compensate for the nonunity filtering of the signal-weighted parameter error in the measured prediction error via algorithm modifications and excitation constraints.*

The algorithms developed in Chapter 4 provide a variety of methods for adapting finite-impulse-response (FIR) filters. To date, the overwhelming majority of adaptive filtering problems have been solved with FIR filters, primarily because these algorithms are well understood. Applications for which an infinite-impulse-response (IIR) filter appears to be more appropriate have been dealt with by using an FIR filter which is sufficiently long to adequately model the resonances (poles) and nulls (zeros) of the desired frequency response. While this method is effective and has the advantage of relying on a background of known FIR techniques, it can lead to significant computational costs compared to using the "right" IIR filter. To emphasize the computational impact we reexamine the problem of reducing the data transmission rate needed to send a speech signal over a digital telephone circuit.

In Section 1.2 we examined methods that would allow a speech signal to be transmitted with substantially less than the 12 to 13 bits per sample needed for toll-quality quantization via pulse code modulation (PCM). As shown in Figure 1.5, the basic technique of adaptive differential pulse code modulation (ADPCM) employs an adaptive filter to model the speech generation process of the speaker. This estimate is subtracted from the actual speech, and the difference, presumed to be much smaller in variance than the actual signal, is quantized and transmitted.

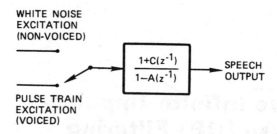

FIGURE 5.1. Simple model of speech generation.

Figure 5.1 shows a simplified model of the human speech generation process. The vocal tract is excited in one of two ways. For "voiced" sounds, the vocal cords vibrate and produce quasi-periodic excitation. For "unvoiced" sounds, air is forced through constrictions in the vocal tract, causing turbulence and a noiselike excitation [Rabiner and Schafer, 1978]. The excitation, whether "voiced" or "unvoiced" in origin, is filtered by the vocal tract, i.e., the throat, mouth, and nasal passages, to modify the spectrum of the speech waveform. The vocal tract has resonances at certain frequencies, modeled by the roots of $1 - A(z^{-1})$, i.e., the vocal tract poles, and nulls at certain frequencies, modeled by the roots of $1 + C(z^{-1})$, i.e., the vocal tract zeros.

Any nonzero choice of the polynomials $A(z^{-1})$ and $C(z^{-1})$ increases the dynamic range, or, equivalently, the variance of the output speech signal over that of the original white excitation signal entering the vocal tract. The filtering also adds correlation to the processed version of the excitation signal, making each speech sample correlated with those before and after it. This correlation and the associated increase in the variance of the speech waveform increase the dynamic range needed to transmit the speech signal over that needed for the excitation alone.

Figure 5.2 shows how ADPCM techniques can be used to reduce the number of bits required for each sample. By developing just the right "decorrelation filter" $H(z)$, the filtering introduced by the vocal tract can be cancelled, thereby removing the sample-to-sample correlation from the waveform and reducing its

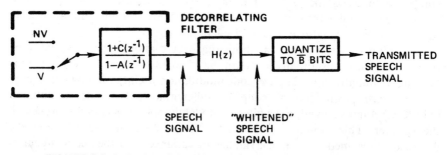

FIGURE 5.2. Reduction of dynamic range by decorrelating the speech signal.

variance. The output of the decorrelation filter can then be quantized to a smaller number of bits, e.g., four, and transmitted. As we will discuss in Section 5.5, this filter can be viewed as a predictor of the speech waveform and may in fact be implemented with a different structure than that in Figure 5.2. In any case, the reduction in the dynamic range of the transmitted signal is obtained by "inverting" the spectral shaping introduced by the vocal tract, and thus decorrelating the speech waveform. An appropriate $H(z)$ will then approximate the inverse of the vocal tract transfer function $[1 - A(z^{-1})]/[1 + C(z^{-1})]$. To adequately capture the unpredictable time-varying character of the human vocal tract during speech, this inverse filter should be time-varying, which suggests its implementation as an adaptive filter.

The required length of the decorrelation/inverse filter impulse response can be determined from a physical analysis of the vocal tract. Such analysis [Rabiner and Schafer, 1978] has shown that the desired inverse filter impulse response sequence $\{h(k)\}$ should be significantly nonzero for approximately 20 msec. Given an 8-kHz sampling frequency, which is standard for the less than 4-kHz frequency bandwidth considered adequate for reconstructed speech, 20 msec represents a vocal tract decorrelator impulse-response duration of 160 samples. Thus, a viable FIR filter should have approximately 160 taps (or delay elements and weights). For operation at 8 kHz, this results in a computation rate of 1.28 megamultiples per second, just for filter output computation without considering the computational requirements of adaptation.

In contrast, the CCITT standard for ADPCM [Jayant and Noll, 1984], which was developed before some of the results in this chapter were, uses an adaptive IIR filter to accomplish the same 4-bit quantization at 8 kHz (for a 32-kbps data transmission rate). This IIR structure uses eight weights for a two-pole, six-zero model. This results in a total of only 8 multiplications per input sample for prediction formation or a computation rate of 0.064 megamultiplies per second. This 20:1 ratio cannot be attained for all applications, but in this ADPCM application it certainly does motivate the search for algorithms capable of adapting IIR digital filters.

We should observe that this considerable reduction in computation comes at some cost. For example, a time-invariant FIR filter with bounded coefficients has all of its poles at the origin of the z-plane and is therefore stable in the sense that any bounded input sequence generates a bounded output sequence. This is not true for IIR filters, where it is comparatively easy to find a choice of bounded filter coefficients which leads to an unbounded output given a bounded input. Any fixed filter with a pole outside the unit circle can produce such an unbounded output. This stability problem is compounded with adaptation. We have already seen that adaptive FIR filter instability can result due to "parameter runaway" if the adaptive step size is chosen inappropriately. With adaptive IIR filters we must be wary of both parameter runaway (which can also happen with FIR adaptive filters) and output explosion without parameter runaway (which happens only with IIR filters).

Essentially, the same practical issues that suggest the use of nonadaptive IIR filters in place of nonadaptive FIR filters also apply to the choice of adaptive IIR filters in place of adaptive FIR filters. Furthermore, the same needs that lead to the use of adaptation in FIR filters, such as lack of adequate a priori information for a fixed filter design and/or the expectation of unpredictable variations over time in the desired filter parameterization, also imply the addition of adaptation to IIR implementations. Despite such practical utility, adaptive IIR filter algorithms are a relatively recent development, with the earliest algorithm publicly suggested as recently as 1975 in [White, 1975]. As a result, far less practical experience has been accumulated with adaptive IIR filters compared to their FIR counterparts. Thus, adaptive IIR filter "theory" lacks the widely tested guidelines for practical applications that are available for adaptive FIR filter use. Given this state of affairs, this chapter will focus on the analytical underpinnings for the derivation and interpretation of adaptive algorithms for IIR implementations. This basis should ultimately be merged with application experience to provide future design guidelines.

In our efforts here to reveal the source and interpretation of adaptive IIR filter algorithms we will follow first the gradient descent approach to cost function minimization so successful and widely adopted in adaptive FIR filter development and characterization, as indicated in Chapter 4. We will quickly discover that adaptive IIR filter algorithms utilize distinctive elaborations on their FIR counterparts. This revelation of algorithmic distinctions will continue as we turn, later in this chapter, to an alternate source for adaptive filter algorithm development and analysis: nonlinear system stability theory. The fundamental form of the adaptive algorithm will remain intact, however, as in (2.3.1), i.e.,

$$
\begin{bmatrix} \text{New} \\ \text{parameter} \\ \text{estimate} \end{bmatrix} = \begin{bmatrix} \text{Old} \\ \text{parameter} \\ \text{estimate} \end{bmatrix} + \begin{bmatrix} \text{Bounded} \\ \text{step} \\ \text{size} \end{bmatrix} \cdot \begin{bmatrix} \text{Function} \\ \text{of} \\ \text{information} \\ \text{vector} \end{bmatrix} \cdot \begin{bmatrix} \text{Function} \\ \text{of} \\ \text{prediction} \\ \text{error} \end{bmatrix} \cdot
$$

This chapter will introduce various adaptive IIR filter algorithms that have an apparent basis from the developments in this chapter but have not yet been individually investigated. Although some of these alternatives, cited in Section 5.3, have not yet been proven analytically to possess desirable behavior in less than ideal circumstances, some of these algorithm modifications have been shown to exhibit attractive features via simulation studies. Although the theoretical approaches of gradient descent and stability theory will be used to justify the algorithm forms considered in Section 5.2, Section 5.4 will tie (at least heuristically) the analysis and behavioral interpretation of all of these candidate adaptive IIR filter algorithms to the expected-moment approach of Chapter 3.

The final section of this chapter will return to the adaptive differential pulse code modulation (ADPCM) example presented in Chapter 1 and discussed in the preceding paragraphs of this chapter. The adaptive IIR filter algorithm candidates of this chapter will be considered for their suitability to this problem. This reveals some of the decisions that must be taken in translating adaptive IIR theory to design.

5.1 GRADIENT DESCENT MINIMIZATION OF SQUARED PREDICTION ERROR

Using an autoregressive, moving-average (ARMA) model, our adaptive IIR filtering task is to update the a_i and b_j parameters in

$$y(k) = \sum_{i=1}^{n} a_i(k)y(k - i) + \sum_{j=0}^{m} b_j(k)x(k - j). \qquad (5.1.1)$$

Note that (5.1.1) possesses an infinite impulse response due to its use of past output estimates in forming its current output estimate $y(k)$. Assuming that we would like to have y emulate a desired signal d in a least squares sense, our objective is to select the a_i and b_j in (5.1.1) such that

$$J(a_1, a_2, \ldots, a_n, b_0, b_1, b_2, \ldots, b_m) = \sum_{k=1}^{L} [d(k) - y(k)]^2 \qquad (5.1.2)$$

is minimized. Obviously, the character of d and x and the choices of n and m will strongly influence the optimal parameterization.

As with the adaptive FIR filter development in Section 4.2.2, we resort to consideration of a gradient descent procedure based on evaluation of the instantaneous squared prediction error gradient as an approximation to the gradient of its "averaged" value in (5.1.2). In other words our chosen algorithm form is

$$a_i(k + 1) = a_i(k) - \mu_i \frac{\partial(\frac{1}{2}[d(k) - y(k)]^2)}{\partial a_i(k)}, \qquad \mu_i > 0 \qquad (5.1.3)$$

$$b_j(k + 1) = b_j(k) - \rho_j \frac{\partial(\frac{1}{2}[d(k) - y(k)]^2)}{\partial b_j(k)}, \qquad \rho_j > 0. \qquad (5.1.4)$$

Since d is not a function of a_i and b_j,

$$\frac{\partial(\frac{1}{2}[d(k) - y(k)]^2)}{\partial a_i(k)} = -[d(k) - y(k)] \frac{\partial y(k)}{\partial a_i(k)}, \qquad 1 \leqslant i \leqslant n \qquad (5.1.5)$$

and

$$\frac{\partial(\frac{1}{2}[d(k) - y(k)]^2)}{\partial b_j(k)} = -[d(k) - y(k)] \frac{\partial y(k)}{\partial b_j(k)}, \qquad 0 \leqslant j \leqslant m. \qquad (5.1.6)$$

Thus, our task is to evaluate the partial derivative of y with respect to the a_i and b_j. Focusing, for the moment, on the derivative with respect to a_i and utilizing (5.1.1) results in

$$\frac{\partial y(k)}{\partial a_i(k)} = y(k - i) + \sum_{s=1}^{n} a_s(k) \frac{\partial y(k - s)}{\partial a_i(k)}. \tag{5.1.7}$$

Note that each a_i is functionally related to all previous y due to the interconnection of (5.1.3), (5.1.5), and (5.1.7) and the causal interdependence of the y in (5.1.1). This functional relationship is complex and forbidding to our ability to evaluate (5.1.7) by determining a closed-form expression for $\partial y(k - s)/\partial a_i(k)$. At the very least, we can assert that this partial derivative of past y with respect to more recent a_i is not to be ignored. It is not necessarily zero.

To further evaluate (5.1.7), we utilize the small step-size convention common in adaptive IIR (and FIR) filter development [White, 1975; Stearns, et al., 1976; Parikh and Ahmed, 1978; Horvath, 1980]. The use of a sufficiently small step size μ_i in (5.1.3) will cause small changes in the $a_i(k)$ over each iteration. We actually assume that each μ_i is sufficiently small such that

$$\frac{\partial y(k - s)}{\partial a_i(k)} \approx \frac{\partial y(k - s)}{\partial a_i(k - s)}, \qquad \text{for } s = 1, 2, \ldots, n. \tag{5.1.8}$$

This assumption converts (5.1.7) to

$$\frac{\partial y(k)}{\partial a_i(k)} \approx y(k - i) + \sum_{s=1}^{n} a_s k) \frac{\partial y(k - s)}{\partial a_i(k - s)}, \qquad 1 \leqslant i \leqslant n. \tag{5.1.9}$$

Note that (5.1.9) provides a recursive formula for approximating $\partial y(k)/\partial a_i(k)$. Similarly, if we assume that the ρ_j in (5.1.4) are sufficiently small such that

$$\frac{\partial y(k - s)}{\partial b_j(k)} \approx \frac{\partial y(k - s)}{\partial b_j(k - s)}, \tag{5.1.10}$$

then using (5.1.10) in (5.1.6) yields

$$\frac{\partial y(k)}{\partial b_j(k)} \approx x(k - j) + \sum_{s=1}^{n} a_s(k) \frac{\partial y(k - s)}{\partial b_j(k - s)}, \qquad 0 \leqslant j \leqslant m. \tag{5.1.11}$$

Concatenating the $n + m + 1$ partial derivative approximations into one vector yields

$$\left[\frac{\partial y(k)}{\partial a_1(k)} \cdots \frac{\partial y(k)}{\partial a_n(k)} \quad \frac{\partial y(k)}{\partial b_0(k)} \cdots \frac{\partial y(k)}{\partial b_m(k)} \right] \approx [y(k - 1) \cdots y(k - n) \quad x(k) \cdots x(k - m)]$$

$$+ \sum_{s=1}^{n} a_s(k) \left[\frac{\partial y(k - s)}{\partial a_1(k - s)} \cdots \frac{\partial y(k - s)}{\partial a_n(k - s)} \quad \frac{\partial y(k - s)}{\partial b_0(k - s)} \cdots \frac{\partial y(k - s)}{\partial b_m(k - s)} \right]. \tag{5.1.12}$$

Define the information vector (also called the regressor)

$$\mathbf{X}(k) = [\, y(k-1) \cdots y(k-n) \quad x(k) \cdots x(k-m)]^t \qquad (5.1.13)$$

and an autoregressively filtered version of $\mathbf{X}$,

$$\psi(k) = \mathbf{X}(k) + \sum_{s=1}^{n} a_s(k)\psi(k-s). \qquad (5.1.14)$$

Given (5.1.12), ψ can be viewed as an approximation of the partial derivative of the current output estimate $y(k)$ with respect to the parameter estimate vector composed of $a_i(k)$ and $b_j(k)$. As a brief aside, note that this approximate gradient descent strategy could be followed for other IIR filter structures than the direct form parameterization of (5.1.1). For example, [Parikh et al., 1980] uses the gradient descent approach to develop a centralized algorithm for IIR lattice parameter adaption.

Using (5.1.14) in the concatenation of (5.1.3)–(5.1.6) yields the adaptive IIR filter algorithm

$$\mathbf{W}(k+1) = \mathbf{W}(k) + \Lambda\psi(k)e(k), \qquad (5.1.15)$$

for updating the parameters in the direct form of (5.1.1), where

$$\mathbf{W}(k) = [a_1(k) \cdots a_n(k) \quad b_0(k) \cdots b_m(k)]^t \qquad (5.1.16)$$

$$\Lambda = \mathrm{diag}[\mu_1 \cdots \mu_n \quad \rho_0 \cdots \rho_m], \qquad \mu_i > 0, \forall i \quad \text{and} \quad \rho_j > 0, \forall j, \quad (5.1.17)$$

$$e(k) = d(k) - \mathbf{X}^t(k)\mathbf{W}(k), \qquad (5.1.18)$$

$\mathbf{X}(k)$ is given by (5.1.13), and $\psi(k)$ by (5.1.14). Note that the algorithm in (5.1.15) provides a causal method for determining y. In (5.1.15), $d(k)$, $x(k)$ and past $x(x(k-j)$ for $j = 1, 2, \ldots, m)$, past y, $\mathbf{W}(k)$, and past ψ are sufficient to compute $\mathbf{W}(k+1)$. Thus, in computing $y(k+1)$ as in the incremented version of (5.1.1), or

$$y(k+1) = \mathbf{X}^t(k+1)\mathbf{W}(k+1), \qquad (5.1.19)$$

$\mathbf{W}(k+1)$ is actually available before the $(k+1)$th sample instant, after measurements from the kth sample and an appropriately brief computation period. Note that (5.1.19) is not strictly causal due to the inclusion of the $b_0(k+1)x(k+1)$ term in the right-side inner product. But $y(k+1)$ from (5.1.19) can essentially be provided as $d(k+1)$ is measured. This same timing issue exists in adaptive FIR filter algorithms, since it hinges on the $b_0(k+1)x(k+1)$ erm also present in the adaptive FIR filter computation of $y(k+1)$, e.g., as $w_0^{(k}+1)x(k+1)$ in the time-incremented version of (4.2.5).

5.1.1 A Less Expensive Gradient Approximation

The form of ψ in (5.1.14) imposes a significant computational and storage burden since n past values of the full $(n + m + 1) \times 1$ ψ vector must be stored and each element of ψ updated independently via an nth-order regression. Recognizing that $X(k)$ in (5.1.13), which drives the propagation of $\psi(k)$ in (5.1.14), is composed of successively delayed versions of y and x suggests [Söderström et al., 1978; Horvath, 1980] a more computationally efficient approximation for $\partial y(k)/\partial W(k)$, the gradient of $y(k)$ with respect to the coefficients of $W(k)$. This alternate approximation uses filtered versions of y and x

$$y^F(k) = y(k) + \sum_{s=1}^{n} a_s(k)y^F(k - s) \tag{5.1.20}$$

$$x^F(k) = x(k) + \sum_{s=1}^{n} a_s(k)x^F(k - s) \tag{5.1.21}$$

and composes ψ^F from past values of y^F and x^F

$$\psi^F(k) = [\, y^F(k - 1) \cdots y^F(k - n) \quad x^F(k) \cdots x^F(k - m)\,]^t. \tag{5.1.22}$$

The updating of ψ^F thus requires only two nth-order autoregression operations rather than the $n + m + 1$ for ψ in (5.1.14). The algorithm of (5.1.15) is unchanged except for the use of ψ^F rather than ψ.

That ψ^F in (5.1.22) is not equivalent to ψ in (5.1.14) can be noted by comparing the second entry ψ_2 in (5.1.14)

$$\psi_2(k) = y(k - 2) + \sum_{s=1}^{n} a_s(k)\psi_2(k - s) \tag{5.1.23}$$

and the second entry ψ_2^F in (5.1.22)

$$\psi_2^F(k) = y^F(k - 2) = y(k - 2) + \sum_{s=1}^{n} a_s(k - 2)\psi_s^F(k - 2). \tag{5.1.24}$$

Even if the past values $\psi_2(k - s)$ and $\psi_s^F(k - 2)$ were equal for $s = 1, 2, \ldots, n$, $\psi_2(k)$ need not equal $\psi_2^F(k)\, (= \psi_1^F(k - 1))$ due to the difference in the time indices on the a_i. However, under the small μ_i assumption used to generate (5.1.8), $a_i(k) \approx a_i(k - i)$ for $i = 1, 2, \ldots, n$, and these two approximations ψ and ψ^F for $\partial y(k)/\partial W(k)$ are essentially interchangeable.

5.1.2 Stability Check and Projection

Compare the adaptive IIR filter algorithm of (5.1.15) and the similar LMS algorithm for FIR filter adaptation in (4.2.7). An obvious difference in the adaptive IIR filter algorithm is the filtering of the information vector before forming the prediction error and filtered information vector product ψe, which is absent in the adaptive FIR filter algorithm update kernel $\mathbf{X}e$. In the adaptive FIR algorithm, the unfiltered information vector is used for the update kernel $\mathbf{X}e$. In the proof of adaptive FIR filter convergence, $\mathbf{X}$ is assumed bounded, which it is since x is bounded. Similarly, for the convergence of the adaptive IIR filter algorithm in (5.1.15), ψ should remain bounded. Establishment of this property is not as immediate since, from (5.1.14) (or (5.1.22)), ψ (or ψ^F) is the output of autoregressions driven by x and y in $\mathbf{X}$ in (5.1.13). Since either parameter explosion or input–output instability of (5.1.1) could result in unbounded y from bounded x, the adaptive IIR filter information vector $\mathbf{X}$ in (5.1.13) need not be bounded simply because x is bounded.

Under the assumption that the μ_i are small and that the a_i are very slowly varying, a stability test for

$$1 - A(q^{-1}, k + 1) = 1 - \sum_{i=1}^{n} a_i(k + 1)q^{-i}, \tag{5.1.25}$$

where q^{-i} is the delay operator, that is, $q^{-i}\mathbf{X}(k) = \mathbf{X}(k - i)$, would be reasonable in checking the stability of ψ generation from $\mathbf{X}$ (and y from x). (The shift operator q^{-i} is used here rather than the more familiar z-transform operator z^{-i} to allow us to distinguish the shift operator from the complex variable in the z-transform.) It turns out that in order to establish the stability and convergence of adaptive IIR filter algorithms, such as (5.1.15), such a test should be frequently performed, and if (5.1.25) is unstable, the $a_i(k + 1)$ should be modified so all the roots of $1 - A(q^{-1}, k + 1)$ are projected to within the stability region of the unit circle in the complex q-plane. As might be suspected, simulations have verified that when the desired roots for this information vector filter are near the unit circle, more frequent projection is required.

One concern with this need for stability check and projection is that a real-time stability test of the time-varying, autoregressive information vector filter can be a significant computational burden in many situations. Another concern is selection of a successful procedure for the associated projection requirement. One simple projection facility, mentioned in [Ljung and Söderström, 1983, pp. 366–368], is to ignore those updatings of the parameter estimate vector that would lead to the instability of the information vector filtering polynomial. With reference to (5.1.14)–(5.1.16) and (5.1.25), the a_i coefficients of $1 - A(q^{-1}, k + 1)$ are the first n entries of $\mathbf{W}(k + 1)$. The procedure would be to form $\mathbf{W}(k + 1)$ via (5.1.15) and check the stability of $1 - A(q^{-1}, k + 1)$ in (5.1.25). If unstable, the $a_i(k + 1)$ portion of $\mathbf{W}(k + 1)$ is simply retained from $\mathbf{W}(k)$. The question is whether or not the approximate instantaneous gradient descent trajectory of $\mathbf{W}(\cdot)$ will require that $1 - A(q^{-1}, \cdot)$ be instantaneously unstable. If so, it seems possible that the a_i portion of $\mathbf{W}$ could lock up with some of the roots of $1 - A(q^{-1}, k)$ near the unit circle and unable to proceed due to the repeated request for an "unstable update." This undesirable stall mechanism has been observed in simulated and real data tests.

An alternate approach to projection, also suggested in [Ljung and Söderström, 1983, pp. 366–368], repeatedly shrinks the adaptive step size until the down-scaled correction term does not result in the instability of $1 - A(q^{-1}, k + 1)$, i.e., the $a_i(k + 1)$ portion of the new parameter estimate $\mathbf{W}(k + 1)$. This technique appears more robust, but not infallible, in simulation.

A suggestion, made in [Ljung, 1977] for any projection scheme, is that it move unstable roots to a decidedly interior point of the stable region. This helps avoid lockup in a nearly unstable parameterization. The same concept could be exploited by performing the "stability" check for a circle of radius slightly less than unity. The problem is to decide how much less so as not to exclude a stable, though very lightly damped, desired parameterization. The same idea is behind the "pull" factor $0 \leqslant \alpha \leqslant 1$ suggested in [Friedlander, 1982a]. The idea is to use $1 - A(\alpha q^{-1}, k)$ rather than $1 - A(q^{-1}, k)$ to autoregressively filter $\mathbf{X}$ to generate ψ. For $\alpha < 1$, the roots of $1 - A(\alpha q^{-1}, k)$ are pulled radially toward the origin relative to those of $1 - A(q^{-1}, k)$. Thus, $1 - A(\alpha q^{-1}, k + 1)$ should require less frequent projection. But, as with all of these various schemes for implementing the necessary projection facility, this pull factor scheme has been evaluated mostly via simulations. Though further examination of projection candidates is needed to determine the best technique, they cannot, in general, be omitted from adaptive IIR filter algorithms using time-varying, autoregressive information vector filtering.

5.1.3 Squared A Posteriori Error Gradient Descent

Examination of (5.1.15) reveals that $\mathbf{W}(k + 1)$ is computable just after measurement of $d(k)$ and $x(k)$. Therefore, between the kth and $(k + 1)$th sample instants an improved estimate of $d(k)$ can be formed over that output by (5.1.1). After

$$y(k) = \mathbf{X}^t(k)\mathbf{W}(k) \qquad (5.1.26)$$

is output by (5.1.1), the improved, a posteriori estimate

$$\bar{y}(k) = \mathbf{X}^t(k)\mathbf{W}(k + 1) \qquad (5.1.27)$$

could be formed. A reasonable suggestion [Ljung and Söderström, 1983] would be to use these a posteriori output estimates in the information vector. This would replace the a priori information vector of (5.1.13) with its a posteriori counterpart

$$\bar{\mathbf{X}}(k) = [\bar{y}(k - 1) \cdots \bar{y}(k - n) \quad x(k) \cdots x(k - m)]^t. \qquad (5.1.28)$$

Using $\bar{\mathbf{X}}$ rather than $\mathbf{X}$ in (5.1.14)–(5.1.18) still results in a causal algorithm. The advantage is that the use of this a posteriori information vector $\bar{\mathbf{X}}$ has been observed to result in faster convergence and greater accuracy [Ljung and Söderström, 1983, sec. 5.11]. As an alternate explanation of this phenomenon we will convert the a priori error gradient of (5.1.3) to a posteriori error gradient. To more clearly describe this distinction, we will return momentarily to the adaptive FIR filter problem.

5.1.3.1 A Posteriori Error Gradient Adaptive FIR Filter

Recall the LMS algorithm from (4.2.7) for adaptive FIR filter parameter updating

$$\mathbf{W}(k + 1) = \mathbf{W}(k) + \Lambda \mathbf{X}(k)e(k), \tag{5.1.29}$$

employing the new notation

$$\mathbf{W}(k) = [b_0(k) \cdots b_m(k)]^t \tag{5.1.30}$$

$$\mathbf{X}(k) = [x(k) \cdots x(k - m)]^t \tag{5.1.31}$$

and

$$e(k) = d(k) - \mathbf{X}^t(k)\mathbf{W}(k). \tag{5.1.32}$$

The a posteriori error gradient formulation is

$$\mathbf{W}(k + 1) = \mathbf{W}(k) - \Lambda[\partial(\tfrac{1}{2}e^2(k))/\partial \mathbf{W}(k)]. \tag{5.1.33}$$

The a posteriori error gradient formulation is

$$\mathbf{W}(k + 1) = \mathbf{W}(k) - \Lambda[\partial(\tfrac{1}{2}\bar{e}^2(k))/\partial \mathbf{W}(k + 1)]. \tag{5.1.34}$$

where the a posteriori prediction error is

$$\bar{e}(k) = d(k) - \sum_{j=0}^{m} b_j(k + 1)x(k - j) = d(k) - \mathbf{X}^t(k)\mathbf{W}(k + 1). \tag{5.1.35}$$

Compare (5.1.32) and (5.1.35). Evaluating the gradient in (5.1.34) yields

$$\partial(\tfrac{1}{2}\bar{e}^2(k))/\partial \mathbf{W}(k + 1) = -\bar{e}(k)\mathbf{X}(k). \tag{5.1.36}$$

Combining (5.1.34)–(5.1.36) yields

$$\mathbf{W}(k + 1) = \mathbf{W}(k) + \Lambda \mathbf{X}(k)[d(k) - \mathbf{X}^t(k)\mathbf{W}(k + 1)], \tag{5.1.37}$$

an implicit formula since $\mathbf{W}(k + 1)$ appears on both sides.

However, the implicit form in (5.1.37) is not a direct implementation. In fact,

(5.1.37) appears noncausal since $W(k + 1)$ is needed to form $\bar{e}(k)$ in order to compute $W(k + 1)$. But note that

$$\bar{e}(k) = d(k) - X^t(k)W(k + 1)$$
$$= d(k) - X^t(k)W(k) + X^t(k)W(k) - X^t(k)W(k + 1). \quad (5.1.38)$$

Using (5.1.32), (5.1.35), and (5.1.37), (5.1.38) can be rewritten as

$$\bar{e}(k) = e(k) - X^t(k)[W(k + 1) - W(k)]$$
$$= e(k) - X^t(k)\Lambda X(k)\bar{e}(k) \quad (5.1.39)$$

or

$$\bar{e}(k) = e(k)/[1 + X^t(k)\Lambda X(k)]. \quad (5.1.40)$$

Note that, with Λ a diagonal matrix with all positive entries as in (5.1.17), the divisor in (5.1.40) cannot be zero. Thus, (5.1.37) can be rewritten in the causally implementable form

$$W(k + 1) = W(k) + \Lambda X(k)[d(k) - X^t(k)W(k)]/[1 + X^t(k)\Lambda X(k)]. \quad (5.1.41)$$

Compare the a posteriori error gradient adaptive FIR filter algorithm of (5.1.41) with the normalized LMS adaptive FIR filter algorithm of (4.2.36). They are equivalent for $\Lambda = 1$ in (5.1.41) and $\alpha = \gamma = 1$ in (4.2.36). An argument for the improved convergence rate of (5.1.41) versus (5.1.29) comes from the approximation by (5.1.41) of a projection procedure (when $X^t\Lambda X \gg 1$) [Goodwin and Sin, 1984, pp. 50–54].

5.1.3.2 Approximate A Posteriori Error Gradient Adaptive IIR Filter
We will now attempt to replicate the preceding a posteriori error gradient reformulation for the adaptive IIR filter problem. The a posteriori error gradient approach alters (5.1.3) to

$$a_i(k + 1) = a_i(k) - \mu_i \partial(\tfrac{1}{2}[d(k) - \bar{y}(k)]^2)/\partial a_i(k + 1), \quad (5.1.42)$$

where

$$\bar{y}(k) = \sum_{i=1}^{n} a_i(k + 1)\bar{y}(k - i) + \sum_{j=0}^{m} b_j(k + 1)x(k - j)$$
$$= \bar{X}^t(k)W(k + 1) \quad (5.1.43)$$

with $\bar{X}$ defined in (5.1.28) and $W(k)$ defined in (5.1.16). As with (5.1.5) and (5.1.7),

$$\partial(\tfrac{1}{2}[d(k) - \bar{y}(k)]^2)/\partial a_i(k + 1) = -[d(k) - \bar{y}(k)][\partial\bar{y}(k)/\partial a_i(k + 1)] \quad (5.1.44)$$

and

$$\frac{\partial \bar{y}(k)}{\partial a_i(k+1)} = \bar{y}(k-i) + \sum_{s=1}^{n} a_s(k+1) \frac{\partial \bar{y}(k-s)}{\partial a_i(k+1)}. \tag{5.1.45}$$

Reusing the small μ_i assumption, (5.1.45) can be approximated as

$$\frac{\partial \bar{y}(k)}{\partial a_i(k+1)} \approx \bar{y}(k-i) + \sum_{s=1}^{n} a_s(k+1) \frac{\partial \bar{y}(k-s)}{\partial a_i(k-s)}. \tag{5.1.46}$$

Thus, (5.1.14) is replaced by

$$\boldsymbol{\Psi}(k) = \bar{\mathbf{X}}(k) + \sum_{s=1}^{n} a_s(k+1)\boldsymbol{\Psi}(k-s) \tag{5.1.47}$$

and (5.1.15) becomes

$$\mathbf{W}(k+1) = \mathbf{W}(k) + \Lambda\boldsymbol{\Psi}(k)\bar{e}(k), \tag{5.1.48}$$

where

$$\bar{e}(k) = d(k) - \bar{\mathbf{X}}^t(k)\mathbf{W}(k+1). \tag{5.1.49}$$

As is (5.1.37), (5.1.48) is implicit. Similar to (5.1.38)–(5.1.40),

$$\begin{aligned}\bar{e}(k) &= d(k) - \bar{\mathbf{X}}^t(k)\mathbf{W}(k) - \bar{\mathbf{X}}^t(k)[\mathbf{W}(k+1) - \mathbf{W}(k)] \\ &= d(k) - \bar{\mathbf{X}}^t\mathbf{W}(k) - \bar{\mathbf{X}}^t(k)\Lambda\boldsymbol{\Psi}(k)\bar{e}(k)\end{aligned} \tag{5.1.50}$$

or

$$\bar{e}(k) = [d(k) - \bar{\mathbf{X}}^t\mathbf{W}(k)]/[1 + \bar{\mathbf{X}}^t(k)\Lambda\boldsymbol{\Psi}(k)]. \tag{5.1.51}$$

Even using (5.1.51) in (5.1.48), the update remains implicit due to the need for $\mathbf{W}(k+1)$ in computing $\boldsymbol{\Psi}(k)$ in (5.1.47). Also the divisor in (5.1.51) could possibly be zero or near zero since the addend $\bar{\mathbf{X}}^t(k)\Lambda\boldsymbol{\Psi}(k)$ is not a positive quadratic form as $\mathbf{X}^t(k)\Lambda\mathbf{X}(k)$ is in (5.1.40). So, to avoid division by zero and the need for $\mathbf{W}(k+1)$ in computing $\boldsymbol{\Psi}(k)$, we will again invoke the smallness of the step size such that $1 + \bar{\mathbf{X}}^t(k)\Lambda\boldsymbol{\Psi}(k) \approx 1$ and $a_s(k) \approx a_s(k-i)$ for $i = 1, 2, \ldots, n$. With (5.1.51) and these approximations (5.1.47)–(5.1.48) become

$$\boldsymbol{\Psi}(k) = \bar{\mathbf{X}}(k) + \sum_{s=1}^{n} a_s(k)\boldsymbol{\Psi}(k-s) \tag{5.1.52}$$

and

$$\mathbf{W}(k+1) = \mathbf{W}(k) + \Lambda\boldsymbol{\Psi}(k)[d(k) - \bar{\mathbf{X}}^t(k)\mathbf{W}(k)]. \tag{5.1.53}$$

Note that (5.1.52) and (5.1.53) are the same as simply using the a posteriori information vector in (5.1.14)–(5.1.18), as suggested earlier for convergence rate improvement, as documented in [Ljung and Söderström, 1983].

5.1.4 "Least Squares" Time-Varying Step-Size Matrix Form

In the development of exact solutions to the least squares problem underlying adaptive FIR filtering in Section 4.3, the step-size scale factor became a time-varying matrix that served as a recursively computed approximation of the inverse of the information vector autocorrelation matrix, that is, R_{ss} in (4.3.20). This interpretation led to claims that, in certain situations, such a time-varying step-size matrix engenders more rapid adaptive filter convergence. The same "improvement" is possible for the results of the preceding gradient descent approach to adaptive IIR filter algorithm development. For example, the algorithm of (5.1.13)–(5.1.18) can be rewritten as [Ljung and Söderström, 1983]

$$\mathbf{W}(k + 1) = \mathbf{W}(k) + P(k)\boldsymbol{\psi}(k)e(k), \tag{5.1.54}$$

where

$$P^{-1}(k) = \lambda(k)P^{-1}(k - 1) + \boldsymbol{\psi}(k)\boldsymbol{\psi}^t(k) \tag{5.1.55}$$

or using the matrix inversion lemma

$$P(k) = \frac{1}{\lambda(k)}\left[P(k - 1) - \frac{P(k - 1)\boldsymbol{\psi}(k)\boldsymbol{\psi}^t(k)P(k - 1)}{\lambda(k) + \boldsymbol{\psi}^t(k)P(k - 1)\boldsymbol{\psi}(k)} \right]. \tag{5.1.56}$$

Note that P^{-1} in (5.1.55) corresponds to R in (4.3.37). In (5.1.55) and (5.1.56), $\lambda(k) \in (0,1]$ is the (possibly time-varying) scalar forgetting factor. This forgetting factor must be less than 1 for this algorithm to retain the ability to track variations in the desired IIR filter parameterization.

Such a time-varying matrix step size can also be readily incorporated in the differently developed adaptive IIR filter algorithms in the following sections. However, as with Sections 5.1.1–5.1.3, we will focus on constant step-size algorithms for pedagogical clarity. Do not fail to recognize that the elaboration to a "least squares" form, as in (5.1.54)–(5.1.56), is possible.

5.2 PARAMETER IDENTIFICATION FORMAT

In the preceding section, no explicit assumptions were made regarding the generation of the desired filter output $d(k)$. Implicitly, we assumed that d was generated from x in such a manner that (5.1.1), if properly parameterized, would offer a reasonable approximation. In this section we will begin by imposing the assumption that the desired signal $d(k)$ is generated by a model with the same structure as (5.1.1),

$$d(k) = \sum_{i=1}^{n} a_i^o d(k - i) + \sum_{j=0}^{m} b_j^o x(k - j).$$ (5.2.1)

To further avoid confusion between the desired a_i^o of (5.2.1) and the adapted $a_i(k)$ of (5.1.1), the adapted parameters will always be cited with the time index included. Later in this section (in Section 5.2.2.1) we will expand (5.2.1) to include more "modes" than are in (5.1.1), which is the practical situation when the desired signal is generated from a physical operation on x that is unlikely to be linear and of known finite dimensions n and m. The analytical approach taken in this undermodeling case will also be shown to apply to analysis of the effect of time variations in the desired signal generator parameters.

For now, the goal of adapting the parameters of (5.1.1) is to precisely identify those in (5.2.1) in order to make y match d for any possible x. This places us in the domain of recursive system identification [Ljung and Söderström, 1983], which is a field that has progressed somewhat separately from that of adaptive filtering. Clearly, with the extra structure that (5.2.1) provides to our problem statement, a more thorough theoretical understanding is possible. One result has been that a wider range of system-theoretic tools has been employed to develop and interpret adaptive algorithms than just the gradient descent approach of the preceding section. In this section we will exploit one of them: stability theory. But before we embark on our utilization of this approach, we will consider some of the algebraic consequences of the precise structural match of the adaptive filter and the desired signal generator.

First, we will reconsider the a posteriori adaptive FIR filter

$$\bar{y}(k) = \sum_{j=0}^{m} b_j^o(k + 1)x(k - j).$$ (5.2.2)

Assuming that the desired signal is similarly composed,

$$d(k) = \sum_{j=0}^{m} b_j^o x(k - j),$$ (5.2.3)

the a posteriori prediction error can be written as

$$\bar{e}(k) = d(k) - \bar{y}(k) = d(k) - \mathbf{X}^t(k)\mathbf{W}(k + 1)$$
$$= \mathbf{X}^t(k)\mathbf{V}(k + 1),$$ (5.2.4)

where

$$\mathbf{V}(k) = \mathbf{W}^o - \mathbf{W}(k)$$ (5.2.5)

$$\mathbf{W}^\circ = [b_0^\circ \quad b_1^\circ \cdots b_m^\circ]^t \tag{5.2.6}$$

$$\mathbf{W}(k) = [b_0(k) \quad b_1(k) \cdots b_m(k)]^t \tag{5.2.7}$$

and

$$\mathbf{X}(k) = [x(k) \quad x(k-1) \cdots x(k-m)]^t. \tag{5.2.8}$$

The pertinent observation is that the prediction error in (5.2.4) is simply the inner product of the information vector and the parameter estimate error vector, that is, $\mathbf{X}^t\mathbf{V}$.

For an adaptive IIR filter the a posteriori output estimate is, as in (5.1.43),

$$\bar{y}(k) = \sum_{i=1}^{n} a_i(k+1)\bar{y}(k-i) + \sum_{j=0}^{m} b_j(k+1)x(k-j) = \bar{\mathbf{X}}^t(k)\mathbf{W}(k+1), \tag{5.2.9}$$

with $\bar{\mathbf{X}}$ defined in (5.1.28) and $\mathbf{W}(k)$ in (5.1.16). Subtracting (5.2.9) from (5.2.1) yields the a posteriori prediction error

$$\bar{e}(k) = d(k) - \bar{y}(k)$$

$$= \sum_{i=1}^{n} a_i^\circ[d(k-i) - \bar{y}(k-i)]^\circ \sum_{i=1}^{n} [a_i^\circ - a_i(k+1)]\bar{y}(k-i)$$

$$+ \sum_{j=0}^{m} [b_j^\circ - b_j(k+1)]x(k-j)$$

$$= \sum_{i=1}^{n} a_i^\circ \bar{e}(k-i) + \bar{\mathbf{X}}^t(k)\mathbf{V}(k+1) \tag{5.2.10}$$

or

$$[1 - A(q^{-1})]\bar{e}(k) = \bar{\mathbf{X}}^t(k)\mathbf{V}(k+1). \tag{5.2.11}$$

In (5.2.11), $\mathbf{V}$ is still the parameter error vector, as in (5.2.5) for the adaptive FIR filter. However, the definition of $\mathbf{W}^\circ$ changes from that of (5.2.6) to a vector of the parameters of (5.2.1) in correspondence with the estimated parameter vector of (5.1.16). The operator notation of $1 - A(q^{-1})$ in (5.2.11) is simply shorthand for

$$[1 - A(q^{-1})]\bar{e}(k) = \left[1 - \sum_{i=1}^{n} a_i^\circ q^{-i}\right]\bar{e}(k) = \bar{e}(k) - \sum_{i=1}^{n} a_i^\circ \bar{e}(k-i). \tag{5.2.12}$$

With inversion of the scalar operator $1 - A(q^{-1})$, (5.2.11) can be written as

$$\bar{e}(k) = [1 - A(q^{-1})]^{-1}\{\bar{\mathbf{X}}^t(k)\mathbf{V}(k+1)\}, \tag{5.2.13}$$

which is simply shorthand for (5.2.10). Recognizing $[1 - A]^{-1}$ as an autoregressive operator, the prediction error for the parameter identification formulation of adaptive IIR filtering is an autoregressively filtered version of the inner product of the information vector and the parameter estimate error vector. Note that this autoregression is the "unknown" autoregression of the desired signal model in (5.2.1).

An obvious question is: Can we use the a posteriori error $\bar{e}$ of (5.2.13), with its inclusion of the filtering by $[1 - A]^{-1}$, to update the parameters of an adaptive IIR filter, much as we used $\bar{e}$ of (5.2.4) in (5.1.37) to adapt the parameters of an FIR filter? The next section develops a stability theory problem reinterpretation of this question that allows us to apply substantial analytical tools for answering it.

5.2.1 Homogeneous Error System Stability Formulation

Return, once again, to the adaptive FIR filter algorithm of (5.1.37) given (5.2.4),

$$\mathbf{W}(k + 1) = \mathbf{W}(k) + \Lambda\mathbf{X}(k)\mathbf{X}^t(k)\mathbf{V}(k + 1). \qquad (5.2.14)$$

Subtract both sides of (5.2.14) from $\mathbf{W}^o$ to yield

$$\mathbf{V}(k + 1) = \mathbf{V}(k) - \Lambda\mathbf{X}(k)\mathbf{X}^t(k)\mathbf{V}(k + 1) \qquad (5.2.15)$$

or

$$\mathbf{V}(k + 1) = [I + \Lambda\mathbf{X}(k)\mathbf{X}^t(k)]^{-1}\mathbf{V}(k). \qquad (5.2.16)$$

The reformulation in (5.2.16) can be viewed as the state equation of a homogeneous (or unforced) system with state $\mathbf{V}$ and time-varying state transition matrix $[I + \Lambda\mathbf{X}\mathbf{X}^t]^{-1}$. Our objective of parameter identification can be stated as a zero-state asymptotic stability problem. In other words, we would like for (5.2.16) to stably return an initial nonzero state to zero as $k \to \infty$. This is due to the fact that if the state $\mathbf{V}$ of (5.2.16) is zero, then from (5.2.5) $\mathbf{W} = \mathbf{W}^o$ (and $\bar{e} = d - \bar{y} = 0$) as desired. It is readily proven that the "energy" of the state of the system in (5.2.16), measured as $\mathbf{V}^t\Lambda^{-1}\mathbf{V}$, is nonincreasing, such that (5.2.16) is globally asymptotically stable when Λ is finite, symmetric, and positive definite [Mendel, 1973]. In fact,

$$\mathbf{V}^t(k + 1)\Lambda^{-1}\mathbf{V}(k + 1) - \mathbf{V}^t(k)\Lambda^{-1}\mathbf{V}(k) = -(2 + \mathbf{X}^t(k)\Lambda\mathbf{X}(k))\bar{e}^2(k). \qquad (5.2.17)$$

With Λ diagonal, the computation of (5.2.17) mimics that of (2.2.24)–(2.2.26) in the nondivergence proof for a number guessing strategy in Chapter 2. It can also be proven that the linear, time-varying system exhibits zero state asymptotic stability if Λ is positive definite; i.e., all eigenvalues of Λ are greater than zero, *and* $\mathbf{X}$ satisfies certain richness requirements [Anderson and Johnson, 1982a].

The condition on $\{X\}$ said to define $\{X\}$ as persistently exciting, and which yields *exponentially fast* convergence of V to zero, is

$$\alpha_1 I > \sum_{k=j}^{j+S} X(k)X^t(k) > \alpha_2 I > 0 \qquad (5.2.18)$$

for some S and all j. As noted in [Anderson and Johnson, 1982a], the sequence $x(k)$, from which $\{X\}$ is composed in (5.2.8), containing a sufficient number $(> m/2)$ of distinct sinusoids satisfies the persistent ecitation condition of (5.2.18). Compare this to the exisence of R^{-1} required in Section 3.2.6. If X does not satisfy these richness requirements, (5.2.16) is still stable but V need not decay to zero. In any event (5.2.16) is stable such that V remains bounded *and* $X^t V(= \bar{e})$ decays to zero if X remains bounded. As noted before, the decay of the prediction error to zero is the sole objective in many adaptive filtering applications.

One might be tempted, as suggested in [Feintuch, 1976] to simply apply an adaptive FIR filter algorithm to an adaptive IIR filter problem. For example, consider updating the parameters in (5.2.9) via

$$W(k + 1) = W(k) + \Lambda \bar{X}(k)\bar{e}(k), \qquad (5.2.19)$$

where $\bar{e} = d - \bar{y}$ as in (5.2.10) and $\bar{X}$ is defined in (5.1.28). A homogeneous error system can also be formed for this candidate adaptive IIR filter algorithm. Subtracting both sides of (5.2.26) from W^o and using (5.2.13) yields the implicit error system

$$V(k + 1) = V(k) - \Lambda \bar{X}(k)\{[1 - A(q^{-1})]^{-1}[\bar{X}^t(k)V(k + 1)]\}. \qquad (5.2.20)$$

The adaptive FIR filter error system of (5.2.15) is diagrammed in Figure 5.3. Note that the right summing junction in Figure 5.3 replicates (5.2.15) premultiplied by $X^t(k)$, while the left summing junction produces (5.2.15). Such a block diagram description of the adaptive filter error system as a feedback system indicates the strong connection of estimate error decay and adaptive filter convergence to the stability analysis of time-varying feedback systems. Feedback system stability analysis is a subject more frequently associated with control systems analysis rather than digital signal processing. The block diagram description of the adaptive IIR filter error system in (5.2.20) is shown in Figure 5.4. Similar to the development of Figure 5.3 from (5.2.15), the left summing junction in Figure 5.4 replicates (5.2.20) premultiplied by $X^t(k)$ while the right summing junction produces (5.2.20). One complication of the adaptive IIR filter error system embodied by (5.2.20) over that associated with the adaptive FIR filter error system of (5.2.15) is that the adaptive IIR filter error system is nonlinear (and time-varying) while that of the adaptive FIR filter is linear (and time-varying). The nonlinearity of (5.2.20) arises since $\bar{X}(k)$ is a

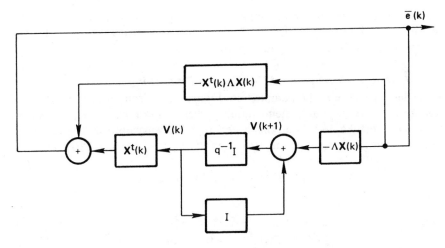

FIGURE 5.3. Stable error system of adaptive FIR filter.

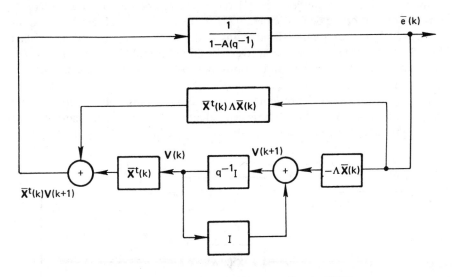

FIGURE 5.4. Error system of adaptive IIR filter in (5.2.20).

function of $\mathbf{V}(k - i + 1)$. Refer to the definition of $\bar{\mathbf{X}}(k)$ in (5.1.28), which shows its inclusion of past $\bar{y}$. For the adaptive FIR filter error system of (5.2.15), $\mathbf{X}$ as defined in (5.2.8) is composed of past values of x, which are clearly not affected by the adaptive mechanism. The more prominent addition in Figure 5.4 relative to Figure 5.3 is a linear, time-invariant, autoregressive transfer function in the forward path of the feedback loop. A reasonable question is: Under what conditions on this transfer function $H(q^{-1}) = [1 - A(q^{-1})]^{-1}$ does (5.2.19) retain the stability properties of (5.2.14) where effectively $H(q^{-1}) = 1$? The remainder of this section will address this question.

5.2.1.1 Strictly Positive Real (SPR) Error Smoothing

Consider the system diagrammed in Figure 5.5. From the application of nonlinear, time-varying system stability theory [Popov, 1973], it can be proven [Tomizuka, 1982] that this system is globally asymptotically stable if the time-invariant transfer function $H(q^{-1})$ is strictly positive real (SPR), i.e.,

$$\text{Re}[H(q^{-1})] > 0 \qquad \text{for all } |q| = 1 \qquad (5.2.21)$$

and

$$\tfrac{1}{2}\bar{\mathbf{X}}'(k)\Lambda\bar{\mathbf{X}}(k) - \lambda \geqslant 0 \qquad \text{for all } k, \qquad (5.2.22)$$

where Λ is a positive definite matrix and λ is a positive scalar. Global asymptotic stability of Figure 5.5 implies that $v(k) \to 0$ as $k \to \infty$ and that the state of $H(q^{-1})$ and the output of the decay element $q^{-1}I$ in the feedback path are bounded for all k, as long as they are initialized at any finite value. Clearly (5.2.21) is satisfied for the error system of the adaptive FIR filter in (5.2.15) since $H(q^{-1}) = 1$. As shown in [Anderson and Johnson, 1982a], a persistent excitation condition on $\{\bar{\mathbf{X}}\}$, the same as that of (5.2.18) with $\mathbf{X}$ replaced by $\bar{\mathbf{X}}$, leads to exponential stability of the system in Figure 5.5. Translating this requirement on $\{\bar{\mathbf{X}}\}$ in (5.1.28) to one on $\{x\}$ alone is dependent on the lack of near-cancellations in the "transfer function" of (5.1.1). Refer to [Taylor and Johnson, 1982] for

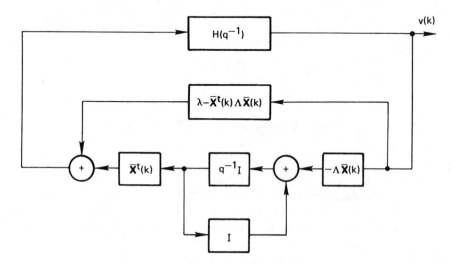

FIGURE 5.5. Stable system.

further comments on this point, which is intimately connected to the practical selection of the orders n and m in (5.1.1).

The SPR requirement of (5.2.21) can be interpreted that the transfer function $H(q^{-1})$, when evaluated as a complex number for all values of q on the unit circle in the q-plane, must have a positive real part. In other words, the angle of the polar coordinate description of the complex-valued $H(q^{-1})$ for all $|q| = 1$ must be less than 90° and greater than $-90°$. Evaluation of the time-invariant $H(q^{-1})$ on the unit circle is equivalent to assessing the magnitude scaling and phase shift of the steady-state sinusoidal response of $H(q^{-1})$. Thus (5.2.21) can also be viewed as a requirement that the Bode phase plot for $H(q^{-1})$ be between 90° and $-90°$ for all frequencies.

As noted, the stability of Figure 5.5 is such that $v(k) \to 0$ as $k \to 0$ and the state of this feedback system remains bounded for all k despite the character of $\bar{\mathbf{X}}$. Compare Figure 5.4 and Figure 5.5. This comparison indicates that the adaptive IIR filter candidate of (5.2.19) is desirably stable, that is, $\bar{e} \to 0$ and $\mathbf{V}$ bounded, if $[1 - A(q^{-1})]^{-1}$ is SPR, since (5.2.22) is trivially satisfied with $\lambda = 0$.

It is reasonable to ask what type of constraint is imposed on the poles of $[1 - A(q^{-1})]^{-1}$ for it to be SPR. We will address this question directly only for a second-order denominator

$$1 - A(q^{-1}) = 1 - a_1^o q^{-1} - a_2^o q^{-2} \qquad (5.2.23)$$

with complex conjugate roots. If we evaluate $H(q) = [1 - A(q^{-1})]^{-1}$ of (5.2.23) for $|q| = 1$, the region of complex conjugate pole pairs that satisfy the SPR condition in (5.2.21) are within the unshaded oval in Figure 5.6. It is interesting to note that this excludes the vicinity near $z = 1$, which is the region where poles from a digital equivalent of a rapidly sampled, continuous-time, second-order plant will cluster. We could alter this situation if we could modify $H(q^{-1})$ to $[1 + D(q^{-1})]/[1 - A(q^{-1})]$. In Figure 5.7 we see how the vicinity near $q = 1$ can be accommodated with only $D(q^{-1}) = d_1 q^{-1}$ and (5.2.23) with complex conjugate roots.

Such modification of the transfer function can be achieved by using a moving-average smoothing of the prediction error

$$v(k) = [1 + D(q^{-1})]\bar{e}(k) = \bar{e}(k) + \sum_{i=1}^{n} d_i \bar{e}(k - i)$$

$$= d(k) - \bar{\mathbf{X}}^t(k)\mathbf{W}(k + 1) + \sum_{i=1}^{n} d_i[d(k - i) - \bar{\mathbf{X}}^t(k - i)\mathbf{W}(k - i + 1)] \quad (5.2.24)$$

in

$$\mathbf{W}(k + 1) = \mathbf{W}(k) + \Lambda\bar{\mathbf{X}}(k)v(k). \qquad (5.2.25)$$

(Do not confuse the coefficients of $D(q^{-1})$ which are denoted as d with a subscript, that is, d_i, and the desired signal sequence values which are denoted by

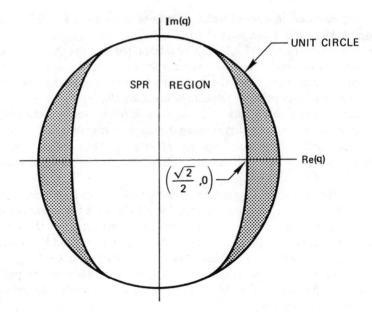

FIGURE 5.6. SPR region (unshaded area within unit circle) for second-order $[1 - A(q^{-1})]^{-1}$ with complex conjugate poles © IEEE.

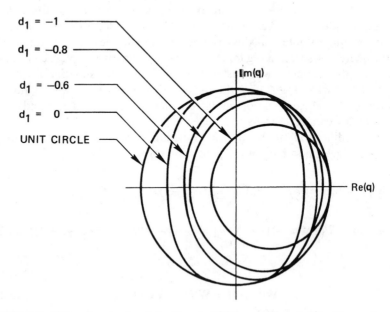

FIGURE 5.7. SPR regions (enclosed) for $[1 + d_1 q^{-1}]/[1 - a_1^o q^{-1} - a_2^o q^{-2}]$ with complex conjugate poles © IEEE.

d with a time argument, that is, $d(k - i)$.) A difficulty with (5.2.25) is that it is implicit, since $\mathbf{W}(k + 1)$ is needed in $v(k)$ on the right of (5.2.25) to compute $\mathbf{W}(k + 1)$ on the left. To generate an explicit formula for $\mathbf{W}(k + 1)$ rewrite (5.2.24) as

$$v(k) = d(k) - \bar{\mathbf{X}}^t(k)\mathbf{W}(k + 1) + \sum_{i=1}^{n} d_i \bar{e}(k - i)$$

$$= d(k) - \bar{\mathbf{X}}^t(k)\mathbf{W}(k) + \bar{\mathbf{X}}^t(k)[\mathbf{W}(k) - \mathbf{W}(k + 1)] + \sum_{i=1}^{n} d_i \bar{e}(k - i). \quad (5.2.26)$$

Using (5.2.25) in (5.2.26) yields

$$v(k) = d(k) - \bar{\mathbf{X}}^t(k)\mathbf{W}(k) - \bar{\mathbf{X}}^t(k)\Lambda\bar{\mathbf{X}}(k)v(k) + \sum_{i=1}^{n} d_i \bar{e}(k - i) \quad (5.2.27)$$

or

$$v(k) = [1 + \bar{\mathbf{X}}^t(k)\Lambda\bar{\mathbf{X}}(k)]^{-1}[d(k) - \bar{\mathbf{X}}^t(k)\mathbf{W}(k) + \sum_{i=1}^{n} d_i \bar{e}(k - i)]. \quad (5.2.28)$$

Using (5.2.28), rather than (5.2.24), in (5.2.25) yields an explicit, causal algorithm. This algorithm is guaranteed to be stable if

$$H(q^{-1}) = \frac{1 + D(q^{-1})}{1 - A(q^{-1})} = \frac{1 + \sum_{i=1}^{n} d_i q^{-i}}{1 - \sum_{i=1}^{n} a_i^o q^{-i}} \quad (5.2.29)$$

is SPR [Landau, 1976; Johnson, 1979]. Its error system is diagrammed in Figure 5.8. Note that it matches Figure 5.5 with $\lambda = 0$. The stability result for Figure 5.5 implies that $v \to 0$ and $\mathbf{V}$ remains bounded in Figure 5.8. Recall that our minimal objective for adaptive filtering use is to prove the convergence of $\bar{e}$ to zero. Note from Figure 5.8 that $\bar{e}$ is the input to $1 + D(q^{-1})$ generating v. Thus, if $[1 + D(q^{-1})]^{-1}$ is stable, which it must be for (5.2.29) to be SPR, then $v \to 0$ implies that $\bar{e} \to 0$.

A minor issue regarding implementation of (5.2.25) is what signal to use as the causal output of the adaptive IIR filter. Obviously, from (5.2.25) and (5.2.28), $\bar{y}(k) = \bar{\mathbf{X}}^t(k)\mathbf{W}(k + 1)$ is not available until a significant computation time after the kth sample. Therefore, $\bar{\mathbf{X}}^t(k)\mathbf{W}(k)$ should be used instead. This output of an a priori estimate and the maintenance of an a posteriori estimate record in $\bar{\mathbf{X}}$ and for the past $\bar{e}$ in v, as well as the multiplications and division for the scale factor in (5.2.28), make the computational load for (5.2.25) and (5.2.28) more significant than that of the comparable a priori error gradient-based scheme in (5.1.53). We can use the small step-size assumption used in the generation of (5.1.13)–(5.1.18)

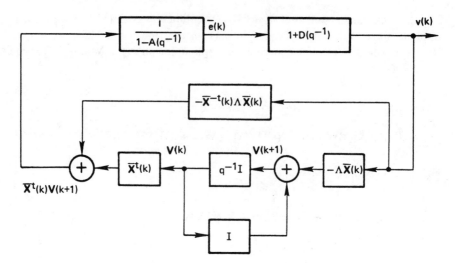

FIGURE 5.8. Error system of adaptive IIR filter in (5.2.24)–(5.2.25).

to reduce (5.2.25) and (5.2.28) to the SHARF algorithm [Larimore et al., 1980; Johnson et al., 1981]

$$\mathbf{W}(k + 1) = \mathbf{W}(k) + \Lambda\mathbf{X}(k)v(k), \tag{5.2.30}$$

where

$$v(k) = e(k) + \sum_{i=1}^{n} d_i e(k - i), \tag{5.2.31}$$

$\mathbf{X}$ is defined in (5.1.13), $\mathbf{W}$ in (5.1.16), Λ in (5.1.17), and e by

$$e(k) = d(k) - y(k) = d(k) - \mathbf{X}^t(k)\mathbf{W}(k), \tag{5.2.32}$$

where d is in (5.2.1) and y in (5.1.1). Now the principal algorithmic difference between (5.1.13)–(5.1.18) and (5.2.30)–(5.2.32) is replacement of the time-varying, autoregressive information vector filtering in the gradient-descent-based algorithm of (5.1.13)–(5.1.18) by a time-invariant, moving-average error smoothing in the stability-based algorithm. Their computational requirements are similar.

The major drawback of this stability-based algorithm is the SPR requirement on $H(q^{-1})$ in (5.2.29). The problem is that the denominator of $H(q^{-1})$ is patently unknown, which is in fact the source of the need for adaptation. An unanswered question is how to select the error smoothing coefficients in $1 + D(q^{-1})$ given partial information on the desired $1 - A(q^{-1})$. In fact, it is not clear what information, short of off-line identification of $1 - A(q^{-1})$, is obtainable in practical situations to guide selection of $1 + D(q^{-1})$. One unfortunate tendency is apparent. As roots $1 - A(q^{-1})$ become more oscillatory, the situation in which an IIR filter could have an advantage over an FIR filter, matching roots in $1 + D(q^{-1})$ must be closer to these oscillatory roots of $(1 - A(q^{-1}))$ for $H(q^{-1})$ to be SPR.

One proposal that has been forwarded for error-smoothing coefficient selection is to adapt them as well. A simple method is to use the estimated $a_i(k)$ to parameterize $D(q^{-1}, k)$ via $d_i(k) = -a_i(k)$, such that, as $a_i(k) \rightarrow a_i^o$, $H(q^{-1})$ in (5.2.29) approaches the SPR value of unity. An elaboration of such an ad hoc procedure has been formalized in [Landau, 1978]. As shown in [Johnson and Taylor, 1980], this adaptation of the error-smoothing coefficients results in the need for a "stability" check on $1 + D(q^{-1}, k + 1)$ and a projection facility for any roots found outside the unit circle. This is to avoid the masking of unbounded $\bar{e}$ with a nonminimum phase $1 + D(q^{-1}, \cdot)$ while $v \rightarrow 0$. A particular (computationally nontrivial) stability check and projection scheme has been shown [Dugard and Goodwin, 1985] to result in the desired global stability with ideal use of this adapted error-smoothing algorithm. It is not yet certain whether a similar projection scheme can be used successfully on the coefficients of related time-varying, autoregressive information vector filters, which also require projection as discussed in Section 5.1.2.

5.2.1.2 Information Vector Sufficient Instantaneous Power
Why not simply ignore the information vector filtering of gradient-descent-based adaptive IIR filters and the error smoothing of stability-theory-based adaptive IIR filters? After all, both modifications have their practical difficulties, reliance either on a successful stability check and projection facility or on a successful operator selection satisfying a seemingly unverifiable condition. In the resulting algorithm, as stated in (5.2.19), the a posteriori prediction error can be modified, as in (5.1.38)–(5.1.40),

$$\bar{e}(k) = d(k) - \bar{\mathbf{X}}^t(k)\mathbf{W}(k + 1)$$
$$= [d(k) - \bar{\mathbf{X}}^t(k)\mathbf{W}(k)]/[1 + \bar{\mathbf{X}}^t(k)\Lambda\bar{\mathbf{X}}(k)], \qquad (5.2.33)$$

to provide a causal implementation. The question is whether or not such an algorithm would converge. We have already seen that, if $[1 - A(q^{-1})]^{-1}$ is SPR, convergence is assured. But satisfaction of this requirement is extremely unlikely, especially as any roots of $1 - A(q^{-1})$ approach the unit circle. Simulated examples of nonconvergent behavior (of a simplified version assuming a small step size), when $[1 - A(q^{-1})]^{-1}$ is not SPR, are provided, e.g., in [Larimore et al., 1980]. Thus, it appears that this algorithm would not always be expected to converge without satisfaction of the SPR condition.

Fortunately, we can use the stability of Figure 5.5 to provide an alternte condition that guarantees the convergence of $\bar{e}$ and the stability of $\mathbf{W}$ in (5.2.15) and (5.2.33). Redraw Figure 5.4 as in Figure 5.9. Note that $\lambda\bar{e}(k)$ is added to the

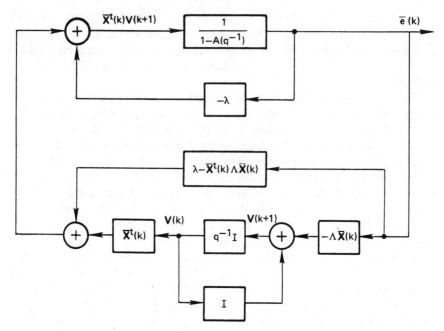

FIGURE 5.9. Reconstruction of Figure 5.4.

feedback path at the left summing junction and is then subtracted before input to $[1 - A(q^{-1})]^{-1}$, thus preserving the equality of Figures 5.4 and 5.9. The transfer function of the forward, time-invariant block in Figure 5.9 is simply

$$H(q^{-1}) = [1 + \lambda - A(q^{-1})]^{-1}. \qquad (5.2.34)$$

If we were to increase the positive λ, a root locus would be sketched with the roots all approaching the origin. Thus, for a sufficiently large λ, $H(q^{-1})$ in (5.2.34) would be SPR. In [Tomizuka, 1982] a sufficiently large λ is derived as

$$\lambda > \sum_{i=1}^{n} |a_i^0| - 1. \qquad (5.2.35)$$

Under the assumption that the poles of the desired filter are all within the unit circle (necessary for convergent adaptive IIR filter stability), a loose upper bound on $\sum_{i=1}^{n} |a_i^0|$ as a function of n is immediate from a binomial expansion [Altay, 1984]. Thus, a sufficient λ can be determined that is dependent only on the order of $1 - A(q^{-1})$ and not on its root locations. Note that with comparison of Figure 5.9 to Figure 5.5, as long as λ is not so large such that (5.2.22) is dissatisfied, the underlying adaptive algorithm would be stable and $\bar{e} \to 0$. For a fixed positive definite Λ (5.2.22) implies that the instaneous "power" of the information vector $\bar{X}$, that is, $\bar{X}^t \bar{X}$, is never too small. Contrast this sufficient power constraint on satisfactory $\{\bar{\mathbf{X}}\}$ sequences with the *different* persistent excitation condition of (5.2.18) on $\{\bar{\mathbf{X}}\}$. With $\bar{\mathbf{X}}$ composed in (5.1.28) from x and $\bar{y}$, the sufficient power condition could be imposed on just a finite window of x values. This is an appealing result. It is apparent from the stability engendered by (5.2.21) and (5.2.22) that mixing the error-smoothing fix of the previous subsection and the sufficient power concept of this subsection can result in adaptive IIR filter stability even in some cases where neither the error-smoothing requirement on the SPRness of (5.2.29) nor the sufficient power condition, reflected in (5.2.35), is satisfied [Lawrence and Johnson, 1986]. Thus we have shown that with modifications, such as information vector and/or prediction error filtering, and/or excitation constraints, such as sufficient power, adaptive FIR filter algorithms can be successfully extended to the adaptive IIR filter problem.

5.2.2 Forced Error System

The parameter identification format of (5.1.1) and (5.2.1) implies that $a_i(k)$ and $b_j(k)$ exist such that $e = d - y$ (and $\bar{e} = d - \bar{y}$) can be zeroed. This is simply impractical in adaptive filtering applications. For example, the prediction error may only be measurable in the presence of noise, such that $e = d - y + n$, where n is a nonzero, albeit potentially small, random signal. With n uncorrelated from x, no fixed choice for the $a_i(k)$ and $b_j(k)$ exists to set $e(k) = d(k) - y(k) + n(k)$ to zero for all k. Additionally, the physical phenomenon modeled by (5.1.1) is not actually linear, time-invariant, and finite-dimensional as implied by (5.2.1). Thus, unavoidable mismodeling in our assumption of the structural equivalence of the adaptive filter output and desired signal generators also leads to circumstances disallowing the presumption that the prediction error can be zeroed for all time by a fixed adaptive filter parameterization. This section will show that this inability to zero the prediction error converts the underlying stability problem from one of global, asymptotic, zero-state stability of a homogeneous, error system to one of local, bounded-input, bounded-state stability of a forced error system. Though still phrasable as a stability theory problem, as we will see, the analytical difficulty in capturing this more practical situation is substantially greater.

5.2.2.1 ARMAX Modeling

A more general difference equation model for a desired signal generator that acknowledges the presence of unmeasurable signals is the autoregressive, moving-average model with exogeneous input (termed an ARMAX model)

$$d(k) = \sum_{i=1}^{n} a_i^o d(k-i) + \sum_{j=0}^{m} b_j^o x(k-j) + \sum_{s=1}^{p} c_s^o w(k-s) + w(k), \qquad (5.2.36)$$

where d is the measurable output, x the measurable input, and w a zero-mean, unmeasurable input. The model of (5.2.36) also accommodates, with parameterization redundancy, the Box–Jenkins model of a desired signal as the sum of an IIR filtering of the measurable x and a different IIR filtering of the unmeasurable w. For a presentation of the suitability of this ARMAX formulation to a number of adaptive filtering applications, refer to [Friedlander, 1982b].

In constructing our adapted model of (5.2.36) we cannot simply append the term $\sum_{s=1}^{p} c_s(k)w(k-s)$ to (5.2.9) since we cannot measure w. But notice that if we could access w *and if* $a_i(k) = a_i^o$, $b_j(k) = b_j^o$, and $c_s(k) = c_s^o$ for all k *and* we used d and not $\bar{y}$ on the right of (5.2.9), then the a posteriori prediction error $\bar{e} = d - \bar{y}$ would equal w. This observation suggests that $\bar{e}$ might be a useful replacement for w. In other words, we will consider estimating (5.2.36) via

$$\bar{y}(k) = \sum_{i=1}^{n} a_i(k+1)d(k-i) + \sum_{j=0}^{m} b_j(k+1)x(k-j) + \sum_{s=1}^{p} c_s(k+1)\bar{e}(k-s)$$

$$= \bar{X}^t(k)W(k+1), \tag{5.2.37}$$

where

$$\bar{e}(k) = d(k) - \bar{y}(k), \tag{5.2.38}$$

$$W(k) = [a_1(k) \cdots a_n(k) \quad b_0(k) \cdots b_m(k) \quad c_1(k) \cdots c_p(k)]^t, \tag{5.2.39}$$

and

$$\bar{X}(k) = [d(k-1) \cdots d(k-n) \quad x(k) \cdots x(k-m) \quad \bar{e}(k-1) \cdots \bar{e}(k-p)]^t \tag{5.2.40}$$

We can rewrite the a posteriori prediction error equation in (5.2.38), given (5.2.36) and (5.2.37), as

$$\bar{e}(k) - w(k) = \sum_{i=1}^{n} [a_i^o - a_i(k+1)]d(k+i) + \sum_{j=0}^{m} [b_j^o - b_j(k+1)]x(k-j)$$

$$+ \sum_{s=1}^{p} [c_s^o - c_s(k+1)]\bar{e}(k-s) - \sum_{s=1}^{p} c_s^o[\bar{e}(k-s) - w(k-s)] \tag{5.2.41}$$

or

$$\bar{e}(k) = [1 + C(q^{-1})]^{-1}[\bar{X}^t(k)V(k+1)] + w(k). \tag{5.2.42}$$

Compare (5.2.42) with (5.2.13). One striking difference is that the autoregression operator on the information vector and parameter estimate error vector inner product is the noise polynomial $1 + C(q^{-1})$ in the desired signal generator of (5.2.36) rather than its autoregression polynomial $1 - A(q^{-1})$. This suggests changing the information vector filtering to $[1 + C(q^{-1}, k)]^{-1}$ in the gradient descent case or relying on the fact that $[1 + C(q^{-1})]^{-1}$ is SPR for a stability-theory-based scheme. These modifications are proven to yield convergent algorithms in [Söderström et al., 1978] and [Solo, 1979], respectively. These convergence theorems rely on vanishing step sizes. For a thorough discussion of the basis of these claims and further elaboration of the associated algorithms see [Ljung and Söderström, 1983].

A second distinction between (5.2.42) and (5.2.13) is that $e \neq 0$ in (5.2.42) when $V = 0$, as it is in (5.2.13). Thus, with a small, but nonvanishing, step size W does not converge to a point. As we will see in the following section, this behavior is described by a forced, rather than unforced, error system.

Another important observation regarding (5.2.37) is to explicitly recognize its IIR form. Given (5.2.38), we can rewrite (5.2.37) as

$$[1 + C(q^{-1}, k + 1)]\bar{y}(k) = [C(q^{-1}, k + 1) + A(q^{-1}, k + 1)]d(k)$$

$$+ [B(q^{-1}, k + 1)]x(k), \qquad (5.2.43)$$

where $A(q^{-1}, k + 1)$, $B(q^{-1}, k + 1)$, and $C(q^{-1}, k + 1)$ are appropriately defined. Thus, we recognize that (5.2.37), as written in (5.2.43), is a two-input (d and x), single-output ($\bar{y}$) IIR filter. Note that if $x(k) \equiv 0$ for all k and $d(k)$ is replaced by the equivalent $\bar{y}(k) + \bar{e}(k)$, (5.2.43) can be written as a single-input ($\bar{e}$), single-output ($\bar{y}$) IIR filter

$$[1 - A(q^{-1}, k + 1)]\bar{y}(k) = [C(q^{-1}, k + 1) + A(q^{-1}, k + 1)]\bar{e}(k). \quad (5.2.44)$$

The IIR form in (5.2.44) is appropriate in a number of signal processing applications, such as adaptive differential pulse code modulation, as discussed in Section 5.5.

5.2.2.2 Mismodeling

As we have noted, a major assumption of the preceding parameter identification formulation in Section 5.2.1 for adaptive filter algorithm establishment is that the model can precisely predict the desired signal with appropriate model parameterization. In comparing (5.1.1), or (5.2.9), and (5.2.1), this assertion appears in the selection of a model order [$\max(n, m)$] that equals that of the desired (minimal) signal generator. This order match allows a single model parameterization to provide the least (actually zero) prediction error for *all* possible input sequences. This independence of the "optimal" adaptive filter parameterization from input signal characteristics is not possible when insufficient model complexity, such as inadequate order, does not permit prediction error zeroing. Refer to the influence of the input signal statistics on the solution for the optimal parameterization of an adaptive FIR filter discussed in Chapters 3 and 4. But, such a structural match is required for the goal of unique parameter identifiability.

The impracticality of this structural completeness assumption is obvious once we acknowledge that the desired signal represents the outcome of a physical operation on the input sequence(s) provided to our adaptive filter. Since linear, time-invariant, finite-dimensional difference equation models are, at best, approximations of nonlinear or time-varying or infinite-dimensional physical phenomena, the model of, e.g., (5.2.9) is doomed to inadequacy. Adding the unavoidable presence of additive noise, as in the ARMAX model of (5.2.36), further exposes the idealized impracticality of (5.2.9).

Having admitted the idealization of the parameter identification format in Section 5.2.1, a reasonable question is how to accommodate the stability theory tools of Section 5.2.1 to situations of desired signal mismodeling as readily as is possible with the gradient descent interpretation of Section 5.1. The major insight is that the underlying stability theory problem is converted from the study of the stability of unforced, nonlinear, time-varying error systems to examination of forced ones as, e.g., in [Desoer and Vidyasagar, 1975]. The remainder of this section will elaborate on this extension.

Consider the more practical desired signal generator model as composed of two terms, one modelable and one not,

$$d(k) = d_M(k) + d_U(k). \tag{5.2.45}$$

The modelable component d_M matches the structure of the adaptive IIR filter, such as (5.2.9), as did (5.2.1),

$$d_M(k) = \sum_{i=1}^{n} a_{Mi} d_M(k - i) + \sum_{j=0}^{m} b_{Mj} x(k - j). \tag{5.2.46}$$

The unmodelable component d_U represents the nearly unavoidable portion of d not predictable by the linear, time-invariant, finite-dimensional model of (5.2.46). Note that (5.2.45) also subsumes the ARMAX modeling case of (5.2.36) and (5.2.37) with $d_U = w$. The a posteriori prediction error obtained by subtracting (5.2.9) from (5.2.45) given (5.2.46) is, paralleling (5.2.10),

$$\bar{e}(k) = \sum_{i=1}^{n} a_{Mi}[d_M(k - i) - \bar{y}(k - i)] + \bar{\mathbf{X}}^t(k)\mathbf{V}(k + 1) + d_U(k)$$

$$= \sum_{i=1}^{n} a_{Mi}[d(k - i) - \bar{y}(k - i)] + \bar{\mathbf{X}}^t(k)\mathbf{V}(k + 1)$$

$$+ \left[d_U(k) - \sum_{i=1}^{n} a_{Mi} d_U(k - i) \right], \tag{5.2.47}$$

where $\bar{\mathbf{X}}$ is as defined in (5.1.28) and $\mathbf{W}$ in (5.1.16). Rewrite (5.2.47) as

$$\bar{e}(k) = [1 - A(q^{-1})]^{-1}[\bar{\mathbf{X}}^t(k)\mathbf{V}(k + 1)] + d_U(k). \tag{5.2.48}$$

Forming v from $\bar{e}$ via (5.2.24) for use in (5.2.25) yields

$$\mathbf{V}(k + 1) = \mathbf{V}(k) - \Lambda\bar{\mathbf{X}}(k)\left\{ \frac{[1 + D(q^{-1})]}{[1 - A(q^{-1})]}[\bar{\mathbf{X}}^t(k)\mathbf{V}(k + 1)] + [1 + D(q^{-1})]d_U(k) \right\}.$$

$$\tag{5.2.49}$$

The error system of (5.2.49) is diagrammed in Figure 5.10. Compare Figure 5.8, which has no input, to Figure 5.10, which does.

A similar forced error system arises if, instead of the reduced-order but still time-invariant desired modeling implied by (5.2.46), the desired signal generator model is time-varying. After all, the supposition that the desired para-meterization is time-varying is a major impetus to the use of an adaptive filter. Subtracting both sides of (5.2.19) from a time-varying desired parameterization at time $k + 1$, $\mathbf{W}^o(k + 1)$, yields

$$\mathbf{W}^o(k + 1) - \mathbf{W}(k + 1) = \mathbf{W}^o(k + 1) - \mathbf{W}(k) - \Lambda\bar{\mathbf{X}}(k)\bar{e}(k)$$
$$= \mathbf{W}^o(k) - \mathbf{W}(k) - \Lambda\bar{\mathbf{X}}(k)\bar{e}(k) + [\mathbf{W}^o(k + 1) - \mathbf{W}^o(k)].$$
$$(5.2.50)$$

With $\mathbf{V}(\cdot)$ appropriately defined as $\mathbf{W}^o(\cdot) - \mathbf{W}(\cdot)$, (5.2.50) adds an input to the error system of the time-invariant case ($\mathbf{W}^o(k) = \mathbf{W}$) in (5.2.20). This input $\mathbf{W}^o(k + 1) - \mathbf{W}^o(k)$ is larger the more rapid the variation in $\mathbf{W}^o(\cdot)$. This error system construction for a time-varying desired parameterization is somewhat more subtle than indicated by (5.2.50), since the $A(q^{-1})$ in (5.2.20) is replaced by the desired signal *time-varying*, autoregressive polynomial. However, the con-version of the homogeneous error system stability problem underlying Section 5.2.1 to a forced error system is undeniable due to mismodeling effects such as order inadequacy *and* desired parameterization time variations.

In terms of the adaptive filtering task, a bounded-input, bounded-state (BIBS) objective relies on the boundedness of d_U or $\mathbf{W}^o(k + 1) - \mathbf{W}^o(k)$. Since the error system states are a vector of past values of prediction errors $\bar{e}$, which we will designate $\bar{\mathbf{E}}$, and the vector of parameter estimate errors $\mathbf{V}$, proving their

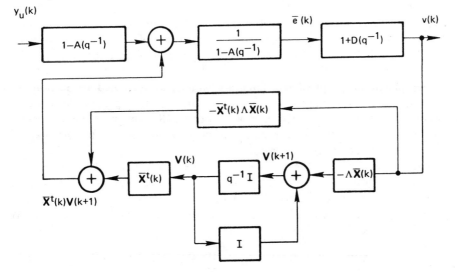

FIGURE 5.10. Forced error system.

subsequent boundedness in terms of the bounds on such forcing inputs similarly bounds the prediction error and parameter estimates of the adaptive IIR filter. Thus, for effective performance we are requiring that some model of known structure but unknown parameterization will provide a close prediction of the desired signal once the model is appropriately parameterized, that is, d_U is reasonably small, and that any variations in the desired parameterization are reasonably small over one sample interval, that is, $\mathbf{W}^o(k+1) - \mathbf{W}^o(k)$ is small.

Recall from linear, time-invariant, finite-dimensional systems theory that exponential, zero-state, homogeneous system stability and BIBS forced system stability both arise with all transfer function poles within the unit circle in the z-plane. This same connection of exponential zero-state homogeneous system stability to BIBS forced system stability exists for nonlinear, time-varying systems, such as that in Figure 5.10, *if* the bounded inputs and initial system states are small enough so that the forced nonlinear, time-varying system state trajectories are not perturbed too far from exponentially stable homogeneous system state trajectories [Desoer and Vidyasagar, 1975, pp. 133–135]. The requirement that both initial states *and* the forcing input remain small is due to their potential interaction in affecting the output of a nonlinear system, unlike the superposition of their separate effects possible in a linear system. For the adaptive IIR filter algorithm of (5.2.24) and (5.2.25), BIBS stability is proven in [Anderson and Johnson, 1982*b*] given exponential stability with ideal use, i.e., of Figure 5.8 and suitably small $\bar{\mathbf{E}}(0)$, $\mathbf{V}(0)$, and d_U. Recall that this exponential stability arises with persistent excitation, as in (5.2.18) with $\bar{\mathbf{X}}$ replacing $\mathbf{X}$. A similar result for suitably small $\mathbf{W}^o(k+1) - \mathbf{W}^o(k)$, $\bar{\mathbf{E}}(0)$, and $\mathbf{V}(0)$ is given in [Anderson and Johnstone, 1983].

5.3 ALGORITHMS

The identification format in Section 5.2 revealed that the significant difference between adaptive FIR and IIR filters is the inclusion of a nonunity operator on the information vector and parameter error vector inner product generating the prediction error. Refer to (5.2.13). This difference substantially impacts the development of adaptive IIR filter algorithms, as indicated in Section 5.2. Essentially, the information vector and prediction error product forming the update kernel in the adaptive law of adaptive FIR filters must be augmented and/or the information vector class restricted to compensate for this adaptive IIR filter prediction error operator.

More specifically, in the case where the difference between the desired signal and the a posteriori adaptive filter output $d - \bar{y}$ is a filtered version of the a posteriori information vector $\bar{\mathbf{X}}$ and the parameter error vector $\mathbf{V}$, that is, $F(q^{-1})\{\bar{\mathbf{X}}^t(k)\mathbf{V}(k+1)\}$, plus an unremovable portion, we have isolated two possible implicit algorithm forms

$$\mathbf{W}(k+1) = \mathbf{W}(k) + \Lambda\{L(q^{-1}, k)[\bar{\mathbf{X}}(k)]\}\{d(k) - \bar{y}(k)\} \qquad (5.3.1)$$

and

$$\mathbf{W}(k + 1) = \mathbf{W}(k) + \Lambda\{\bar{\mathbf{X}}(k)\}\{M(q^{-1}, k)[d(k) - \bar{y}(k)]\}. \qquad (5.3.2)$$

We have also seen that the filter $F(q^{-1})$ in the prediction error is composed of unknown desired signal generator parameters, i.e., a subset of $\mathbf{W}^{\circ}$.

In Section 5.1, using a gradient descent development of a candidate algorithm, we generated a candidate algorithm of the form of (5.3.1), where the appropriate subset of $\mathbf{W}(k)$ was used to parameterize L as our current estimate of $F(q^{-1})$. In other words, convergence ensued as

$$L^{-1}(q^{-1}, k)F(q^{-1}) \to 1, \qquad \text{as } k \to \infty. \qquad (5.3.3)$$

In Section 5.2, using nonlinear system stability theory, we supported convergent behavior when M was fixed in (5.3.2) and

$$M(q^{-1})F(q^{-1}) \approx 1, \qquad (5.3.4)$$

or, more precisely, MF was SPR. This suggests two additional possibilities: selecting a fixed L such that $L^{-1}F$ is SPR, or using $\mathbf{W}(k)$ to appropriately parameterize a time-varying M such that $M(q^{-1}, k)F(q^{-1}) \to 1$.

Using small step sizes, appropriate in most data-rich adaptive signal-processing applications, allows the statement of four simplified candidate algorithms for the updating of the parameters used in forming the adaptive filter output

$$y(k) = \sum_{i=1}^{n} a_i(k)y(k - i) + \sum_{j=0}^{m} b_j(k)x(k - j). \qquad (5.3.5)$$

Recall that, with d formed by a similar model, $F(q^{-1})$ is $[1 - A(q^{-1})]^{-1}$. To write the parameter update laws individually for each $a_i(\cdot)$ and $b_j(\cdot)$, we will select Λ as a diagonal matrix as in (5.1.17), as is commonly done. Thus, the four algorithms are

Algorithm 1:

$$a_i(k + 1) = a_i(k) + \mu_i y^F(k - i)[d(k) - y(k)] \qquad (5.3.6)$$

$$y^F(k) = y(k) + \sum_{i=1}^{n} a_i(k)y^F(k - i) \qquad (5.3.7)$$

$$b_j(k + 1) = b_j(k) + \rho_j x^F(k - j)[d(k) - y(k)] \qquad (5.3.8)$$

$$x^F(k) = x(k) + \sum_{i=1}^{n} a_i(k)x^F(k - i). \qquad (5.3.9)$$

Algorithm 2:

$$a_i(k + 1) = a_i(k) + \mu_i y^F(k - i)[d(k) - y(k)] \qquad (5.3.10)$$

$$y^F(k) = y(k) + \sum_{i=1}^{n} \alpha_i y^F(k - i) \qquad (5.3.11)$$

$$b_j(k + 1) = b_j(k) + \rho_j x^F(k - j)[d(k) - y(k)] \qquad (5.3.12)$$

$$x^F(k) = x(k) + \sum_{i=1}^{n} \alpha_i x^F(k - i). \qquad (5.3.13)$$

Algorithm 3:

$$a_i(k + 1) = a_i(k) + \mu_i y(k - i)v(k) \qquad (5.3.14)$$

$$b_j(k + 1) = b_j(k) + \rho_j x(k - j)v(k) \qquad (5.3.15)$$

$$v(k) = d(k) - y(k) - \sum_{i=1}^{n} \alpha_i[d(k - i) - y(k - i)]. \qquad (5.3.16)$$

Algorithm 4:

$$a_i(k + 1) = a_i(k) + \mu_i y(k - i)v(k) \qquad (5.3.17)$$

$$b_j(k + 1) = b_j(k) + \rho_j x(k - j)v(k) \qquad (5.3.18)$$

$$v(k) = d(k) - y(k) - \sum_{i=1}^{n} a_i(k)[d(k - i) - y(k - i)]. \qquad (5.3.19)$$

Algorithm 1 is the gradient-descent-based algorithm of Section 5.1 using the simplified time-varying autoregressive filtering of the information vector of Section 5.1.1. As noted in Section 5.1.2, to assure local convergence algorithm 1 requires a stability check and projection facility on the polynomial $1 - \sum_{i=1}^{n} a_i(k + 1)q^{-i}$.

Algorithm 2 is a variant of the gradient-descent-based algorithm of Section 5.1, also using a simplified form of the autoregressive filtering of the information vector, but with fixed parameters. Based on the discussion in Section 5.2.1, we speculate that local convergence would result if an SPR condition or a sufficient power condition or a blend of the two is satisfied. These conditions are dependent on the desired autoregressive parameterization a_i^o. For example, the SPR condition is on the transfer function $[1 - \sum_{i=1}^{n} \alpha_i q^{-i}]/[1 - \sum_{i=1}^{n} a_i^o q^{-i}]$, which is satisfied if $\alpha_i \approx a_i^o$.

Algorithm 3 is derived from the nonlinear system stability theory analysis of Section 5.2 and uses a time-invariant moving-average filtering of the prediction error rather than the autoregressive information vector filtering of algorithms 1 and 2. Local convergence should result with the same SPR condition associated with algorithm 2. As discussed in Section 5.2.1.2, a sufficient power condition, blended with the prediction error filtering, should also lead to local convergence. Assuming small step sizes to remove the distinction between a priori and a posteriori prediction error use requires that $X^t \Lambda X \ll 1$. Refer to (5.2.28). This does not permit significant λ in (5.2.22), which reduces our ability to rely on a sufficient power condition with the use of small step sizes.

Algorithm 4 uses a time-varying moving-average filtering of the prediction error that is subject to the same stability check and projection facility as algorithm 1.

If the ARMAX model of (5.2.36) is deemed more appropriate for the desired signal generator, the adaptive filter output could be formed as

$$y(k) = \sum_{i=1}^{n} a_i(k)d(k - i) + \sum_{j=0}^{m} b_j(k)x(k - j) + \sum_{s=1}^{p} c_s(k)e(k - s), \quad (5.3.20)$$

where

$$e(k) = d(k) - y(k). \tag{5.3.21}$$

Note that, with comparison to (5.2.37), (5.3.20) is based on the a priori form of the output and prediction error, which is appropriate with the use of small step sizes. Similar to the formation of algorithms 1–4, for (5.3.20) four additional algorithms are possible given the realization that, with d approximated by a similar model as in (5.2.36), $F(q^{-1})$ is $[1 + C(q^{-1})]^{-1}$.

Algorithm 5:

$$a_i(k + 1) = a_i(k) + \mu_i d^F(k - i)e(k) \tag{5.3.22}$$

$$d^F(k) = d(k) - \sum_{s=1}^{p} c_s(k)d^F(k - s) \tag{5.3.23}$$

$$b_j(k + 1) = b_j(k) + \rho_j x^F(k - j)e(k) \tag{5.3.24}$$

$$x^F(k) = x(k) - \sum_{s=1}^{p} c_s(k)x^F(k - s) \tag{5.3.25}$$

$$c_s(k + 1) = c_s(k) + \delta_s e^F(k - s)e(k) \tag{5.3.26}$$

$$e^F(k) = e(k) - \sum_{s=1}^{p} c_s(k)e^F(k - s). \tag{5.3.27}$$

Algorithm 6:

$$a_i(k + 1) = a_i(k) + \mu_i d^F(k - i)e(k) \tag{5.3.28}$$

$$d^F(k) = d(k) - \sum_{s=1}^{p} \gamma_s d^F(k - s) \tag{5.3.29}$$

$$b_j(k + 1) = b_j(k) + \rho_j x^F(k - j)e(k) \tag{5.3.30}$$

$$x^F(k) = x(k) - \sum_{s=1}^{p} \gamma_s x^F(k - s) \tag{5.3.31}$$

$$c_s(k + 1) = c_s(k) + \delta_s e^F(k - s)e(k) \tag{5.3.32}$$

$$e^F(k) = e(k) - \sum_{s=1}^{p} \gamma_s e^F(k - s). \tag{5.3.33}$$

Algorithm 7:

$$a_i(k + 1) = a_i(k) + \mu_i d(k - i)v(k) \tag{5.3.34}$$

$$b_j(k + 1) = b_j(k) + \rho_j x(k - j)v(k) \tag{5.3.35}$$

$$c_s(k + 1) = c_s(k) + \delta_s e(k - s)v(k) \tag{5.3.36}$$

$$v(k) = e(k) + \sum_{s=1}^{p} \gamma_s e(k - s). \tag{5.3.37}$$

Algorithm 8:

$$a_i(k + 1) = a_i(k) + \mu_i d(k - i)v(k) \tag{5.3.38}$$

$$b_j(k + 1) = b_j(k) + \rho_j x(k - j)v(k) \tag{5.3.39}$$

$$c_s(k + 1) = c_s(k) + \delta_s e(k - s)v(k) \tag{5.3.40}$$

$$v(k) = e(k) + \sum_{s=1}^{p} c_s(k)e(k - s). \tag{5.3.41}$$

Similar comments to those made for algorithms 1–4 apply to algorithms 5–8 regarding stability checks and SPR conditions. The principal difference is that in algorithms 5 and 8 the stability check should be done on $[1 + C(q^{-1}, k + 1)]^{-1}$ rather than $[1 - A(q^{-1}, k + 1)]^{-1}$ as in algorithms 1 and 4. Also, the SPR (or sufficient power) condition for algorithms 6 and 7 should consider

$[1° + °\sum_{s=1}^{p} \gamma_s q^{-s}]/[1 + \sum_{s=1}^{p} c_s^o q^{-s}]$ rather than the transfer function $[1 - \sum_{i=1}^{n} \alpha_i q^{-1}]/[1 - \sum_{i=1}^{n} a_i^o q^{-i}]$ of concern for algorithms 2 and 3.

What should be emphasized here is that these eight algorithms are the logical outgrowth of the pedagogy established in this chapter. This pedagogy blends the gradient descent and average squared error minimization concepts more commonly associated with adaptive filtering, as detailed in Chapters 3 and 4, and the nonlinear system stability theory concepts typically exploited in recursive system identification [Ljung and Söderström, 1983] and adaptive control [Goodwin and Sin, 1984] problems. The next section blends the different analysis tools associated with these respective approaches to reveal an insightful common ground.

5.4 ANALYTICAL INSIGHTS

The principal complication of the adaptive IIR filter problem is its nonlinearity. This is well recognized from attempts at recursive (IIR) filter synthesis [Cadzow, 1976]. The question is if (and if so, how) the mean moment analysis techniques underlying Chapters 3 and 4, and so popular in adaptive FIR filter analysis, can be extended to the analysis of adaptive IIR filters. Only if this extension is possible will the behavioral insights and their associated design guidelines accumulated from abundant adaptive FIR filter application experience be useful in the emerging applications of adaptive IIR filters. Fortunately, under certain conditions of broad practical interest, a qualified extension is possible. This section will sketch this connection.

Recall the two candidate implicit algorithm forms of (5.3.1) and (5.3.2). One uses information vector filtering and the other uses prediction error smoothing to compensate for the filtering $F(q^{-1})$ present in the adaptive IIR filter problem. Blending these two fixes into one algorithm statement and using a small step size to remove the a priori/a posteriori distinctions yield

$$\mathbf{W}(k+1) = \mathbf{W}(k) + \Lambda\{L(q^{-1}, k)[\mathbf{X}(k)]\}\{M(q^{-1}, k)[d(k) - \mathbf{X}^t(k)\mathbf{W}(k)]\}. \quad (5.4.1)$$

For any particular $\mathbf{W}$, denoted here as $\mathbf{W}^*$, we can write

$$d(k) - \mathbf{X}^t(k)\mathbf{W}(k) = F(q^{-1})[\mathbf{X}^t(k)(\mathbf{W}^* - \mathbf{W}(k))] + e^*(k). \quad (5.4.2)$$

Recall (5.2.42) and (5.2.48). If we presume indeed that $\mathbf{W}^*$ is the average convergence point of (5.4.1) then on average $\mathbf{W}(k+1) = \mathbf{W}(k)$ and on average $\{L[\mathbf{X}]\}\{M[F(\mathbf{X}^t(\mathbf{W}^* - \mathbf{W})) + e^*]\}$ must be zero. But, since $\mathbf{W}^*$ is the presumed average convergence point for $\mathbf{W}$, $\text{avrg}[\mathbf{W}^* - \mathbf{W}] = \mathbf{0}$ and from (5.4.2) $\text{avrg}[e^*] = \text{avrg}[d - \mathbf{X}^t\mathbf{W}^*]$, so that (5.4.1) reduces to

$$\text{avrg}[\{L(q^{-1}, k)[\mathbf{X}(k)]\}\{M(q^{-1}, k)[d(k) - \mathbf{X}^t(k)\mathbf{W}^*]\}] = \mathbf{0}, \quad (5.4.3)$$

where "avrg" indicates an averaging operation. Conversely, (5.4.3) can be thought of as a logical condition for average (point) convergence of (5.4.1). In addition, solving (5.4.3) for $\mathbf{W}^*$ produces candidate average stationary points.

In the LMS adaptive FIR filter algorithm central to Chapter 4, (5.4.3) also applies but with L and M equal to unity operators. Thus, for LMS, (5.4.3) reduces to $\text{avrg}\{[\mathbf{X}d] - [\mathbf{X}\mathbf{X}^t]\mathbf{W}^*\} = \mathbf{0}$. Replacing the averaging operator by an "equivalent" expectation operator converts this to $\mathbf{W}^* = \{E[\mathbf{X}\mathbf{X}^t]\}^{-1}\{E[\mathbf{X}d]\}$. This is exactly the representation of the optimal parameterization being sought by LMS, as discussed in Chapter 4. Thus, (5.4.3) represents the elaboration of the central equation of Chapter 4 to the adaptive IIR filter problem.

One difficulty in using (5.4.3) to derive a closed-form expression for the average convergent parameterization of particular adaptive IIR filters using various L and M is that (5.4.3) is strongly nonlinear in $\mathbf{W}^*$. In some adaptive IIR filter algorithms, e.g., algorithms 1, 4, 5, and 8 of the preceding section, L and M are functions of $\mathbf{W}$, and thus $\mathbf{W}^*$ under the presumption that $\text{avrg}[\mathbf{W}^*] = \text{avrg}[\mathbf{W}]$. Similarly, a portion of $\mathbf{X}$ is composed from the adaptive IIR filter output, which is dependent on the particular value of $\mathbf{W}$, or $\mathbf{W}^*$. Refer to (5.1.13) and (5.2.40). Thus, the analytical solution of (5.4.3) for $\mathbf{W}^*$ is an unresolved problem. But (5.4.3) can be solved numerically for a particular problem by testing various $\mathbf{W}^*$ until finding one that satisfies (5.4.3). This was effectively done in the adaptive IIR filter examples in [Johnson and Larimore, 1977]. These examples also display the possibility of multiple solutions satisfying (5.4.3) expected due to its nonlinearity. This multimodality of adaptive IIR filters is discussed further in [Stearns, 1981]. Furthermore, if the averaged expression on the left of (5.4.3) is nonzero, it describes the average incremental motion that $\mathbf{W}$ in (5.4.1) will follow. This approach was effectively taken in [Johnson et al., 1981] for evaluating algorithm 3 of (5.3.14)–(5.3.16). This method of trajectory tracing is formalized via the stochastic ODE (ordinary differential equation) analysis in [Ljung and Söderström, 1983].

Given the potential multimodality of adaptive IIR filters due to the nonlinearity of (5.4.1), it is only logical to investigate its *local* stability properties. This was recognized in Section 5.2.2. Building on the concept of local stability induced by homogeneous system exponential stability, discussed in Section 5.2.2.2, and the averaging in (5.4.3), [Anderson et al., 1986] formalize the following heuristic insight. For local stability of the small-step-size version of (5.4.1), the partial derivative with respect to $\mathbf{W}$ of the average update term about a candidate stationary point $\mathbf{W}^*$ should be negative definite. Use a scalar example to convince yourself of the reasonableness of this statement. This suggests for (5.4.1) that the matrix

$$\frac{\partial\{\text{avrg}(\mathbf{W}(k+1) - \mathbf{W}(k))\}}{\partial\mathbf{W}(k)}\bigg|_{\mathbf{W}(k)=\mathbf{W}^*}$$

$$= \frac{\partial}{\partial\mathbf{W}}\{\text{avrg}(L[\mathbf{X}]M[F(\mathbf{X}^t(\mathbf{W}^* - \mathbf{W}) + e^*)])\}|_{\mathbf{W}=\mathbf{W}^*} \qquad (5.4.4)$$

is negative definite.

Admittedly, (5.4.4) is nontrivial to interpret. Consider the special case where L and M are fixed transfer function operators and not functions of $\mathbf{W}$, that is, as in algorithms 2, 3, 6, and 7. We will also assume that e^* is adequately small, as it should hopefully be at a local equilibrium point. Note that when the algorithm step size is small enough, $\mathbf{W}$ is so slowly time-varying that for a lowpass operator $F, F(\mathbf{X}^t\mathbf{W}) \approx F(\mathbf{X}^t)\mathbf{W}$. This reduces (5.4.4) to the requirement that

$$\mathrm{avrg}\{L[\mathbf{X}^*]MF[\mathbf{X}^{*t}]\} \qquad (5.4.5)$$

is positive definite, where $\mathbf{X}^*$ is the information vector that would result if the adaptive IIR filter where parameterized with the fixed $\mathbf{W}^*$. Note that if L, M, and F are unity operators, as in the equation error case, (5.4.5) reduces to the persistent excitation condition of (5.2.18). When $\mathbf{X}^*$ is persistently exciting and F is nonunity but strictly positive real (SPR) it is easy to visualize (and prove) that (5.4.5) will be satisfied. This corresponds to the conditions in Section 5.2.1.1. If L, M, and F are all fixed nonunity transfer functions and MF/L is SPR then (5.4.5) reduces to a persistent excitation condition on $L[\mathbf{X}^*]$. Note how this differentiates the desirability of various excitation classes for algorithms with different nonunity L. If MF/L is not strictly positive real, (5.4.5) will still be satisfied if the "Re$[MF/L]$-weighted" energy in $\{\mathbf{X}^*\}$ at frequencies for which Re$[MF/L] > 0$ is sufficiently greater than that over frequencies for which Re$[MF/L] < 0$. If L or MF are functions of $\mathbf{W}$, for this local stability analysis they can be considered functions of $\mathbf{W}^*$. Refer to [Anderson et al., 1986] for further details.

Thus, we can see that analysis of the behavior of adaptive IIR filter algorithms can proceed from somewhat the same philosophy as in Chapter 3, as has proven so powerful in adaptive FIR filtering, as shown in Chapter 4. The increased complexity of the analytical machinery that must be applied to the more complex adaptive IIR filter problem is undeniable. The characterization of the average behavior of adaptive IIR filters also reflects this increase in complexity relative to that of adaptive FIR filters.

5.5 AN EXAMPLE OF ADAPTIVE IIR FILTER APPLICATION TO ADPCM

To help illustrate the connection of the theory of this chapter to adaptive IIR filter design, this section abstracts and examines the prediction problem central to adaptive differential pulse code modulation (ADPCM) as an adaptive IIR filtering problem. For a more complete communications systems context for this ADPCM problem, refer to Section 1.2 or [Jayant and Noll, 1984]. In this section, we will focus on a formulation of this problem that reveals its appropriateness for adaptive IIR filter application.

Reconsider the signal-processing task introduced in Section 1.2. A signal is presumed to contain some redundancy that allows a significant portion of its current value to be predicted from its past values. The source signal can be accurately approximated by

$$d(k) = \sum_{i=1}^{n} a_i^o d(k - i) + \sum_{s=1}^{p} c_s^o w(k - s) + w(k), \qquad (5.5.1)$$

where w is a "fictitious," white, zero-mean signal driving the source model. Thus, if we knew the past d's *and* w's we could remove further redundancy from d. The reason for wishing to model and predict the predictable component of d is that a lower-dynamic-range, whitened, unpredictable component, effectively w in (5.5.1), can be quantized with fewer bits than the source itself. Transmission of only this unpredictable component and reconstruction of the entire source, via recombination of the unpredictable component with the component predictable from this transmitted residual sequence, more efficiently utilizes communication channel capacity. The problem at this point is how to model the predictable component from only the received unpredictable component sequence. This seems to be impossible with direct use of (5.5.1), which combines past values of the original source sequence d and the fictitious, patently unmeasurable w. Both of these signals are unavailable at the receiver.

Consider forming

$$y(k) = \sum_{i=1}^{n} a_i(k)[y(k - i) + e(k - i)] + \sum_{s=1}^{p} c_s(k)e(k - s), \qquad (5.5.2)$$

where e is the transmitted signal. In our problem, the prediction error $d - y$ is not the transmitted signal. Actually, the quantized prediction error sequence is transmitted. Thus, the signal transmitted is

$$e(k) = d(k) - y(k) + e_q(k), \qquad (5.5.3)$$

where e_q is the error in quantizing $d - y$. Given (5.5.1) and (5.5.2), (5.5.3) becomes

$$e(k) = \sum_{i=1}^{n} a_i^o d(k - i) + \sum_{s=1}^{p} c_s^o w(k - s) + w(k)$$

$$- \sum_{i=1}^{n} a_i(k)[y(k - i) + e(k - i)] - \sum_{s=1}^{p} c_s(k)e(k - s) + e_q(k). \qquad (5.5.4)$$

Since $y + e = d + e_q$, (5.5.4) can be written as

$$e(k) - w(k) = \sum_{i=1}^{n} [a_i^o - a_i(k)][y(k - i) + e(k - i)] + \sum_{s=1}^{p} [c_s^o - c_s(k)]e(k - s)$$

$$+ e_q(k) - \sum_{i=1}^{n} a_i^o e_q(k - i) - \sum_{s=1}^{p} c_s^o [e(k - s) - w(k - s)] \qquad (5.5.5)$$

or

$$e(k) = \left[1 + \sum_{s=1}^{p} c_s^o q^{-s} \right]^{-1} \left\{ \sum_{i=1}^{n} \left[a_i^o - a_i(k) \right] \left[y(k-i) + e(k-i) \right] \right.$$

$$\left. + \sum_{s=1}^{p} \left[c_s^o - c_s(k) \right] e(k-s) \right\} + w(k)$$

$$+ \left\{ \left[1 + \sum_{s=1}^{p} c_s^o q^{-s} \right]^{-1} \left[1 - \sum_{i=1}^{n} a_i^o q^{-i} \right] \right\} e_q(k). \qquad (5.5.6)$$

If the quantization error were negligible and the $a_i(k)$ and $c_s(k)$ matched the a_i^o and c_s^o, respectively, (5.5.6) would reduce to $e \approx w$ and $y + e \approx d$. This explains the use of (5.5.2) to model the predictable component of (5.5.1). With $e_q \approx 0$, (5.5.6) compares favorably to an a priori version of (5.2.42), which establishes the connection of the problem of (5.5.1) and (5.5.2) to the ARMAX modeling formulation of adaptive IIR filtering of Section 5.3.

This suggests mimicking algorithms 5–8 in (5.3.22)–(5.3.41) for establishment of adaptive algorithms updating the parameters in (5.5.2). Recall that algorithms 5–8, and thus the following four algorithms, are based on the use of small, positive step sizes, that is, μ_i and δ_s.

Algorithm 9:

$$a_i(k+1) = a_i(k) + \mu_i [y^F(k-i) + e^F(k-i)] e(k) \qquad (5.5.7)$$

$$y^F(k) = y(k) - \sum_{s=1}^{p} c_s(k) y^F(k-i) \qquad (5.5.8)$$

$$c_s(k+1) = c_s(k) + \delta_s e^F(k-s) e(k) \qquad (5.5.9)$$

$$e^F(k) = e(k) - \sum_{s=1}^{p} c_s(k) e^F(k-s). \qquad (5.5.10)$$

Algorithm 10:

$$a_i(k+1) = a_i(k) + \mu_i [y^F(k-i) + e^F(k-i)] e(k) \qquad (5.5.11)$$

$$y^F(k-i) = y(k) - \sum_{s=1}^{p} \gamma_s y^F(k-s) \qquad (5.5.12)$$

$$c_s(k+1) = c_s(k) + \delta_s e^F(k-s) e(k) \qquad (5.5.13)$$

$$e^F(k) = e(k) - \sum_{s=1}^{p} \gamma_s e^F(k-s). \qquad (5.5.14)$$

Algorithm 11:

$$a_i(k + 1) = a_i(k) + \mu_i[y(k - i) + e(k - i)]v(k) \tag{5.5.15}$$

$$c_s(k + 1) = c_s(k) + \delta_s e(k - s)v(k) \tag{5.5.16}$$

$$v(k) = e(k) + \sum_{s=1}^{p} \delta_s e(k - s). \tag{5.5.17}$$

Algorithm 12:

$$a_i(k + 1) = a_i(k) + \mu_i[y(k - i) + e(k - i)]v(k) \tag{5.5.18}$$

$$c_s(k + 1) = c_s(k) + \delta_s e(k - s)v(k) \tag{5.5.19}$$

$$v(k) = e(k) + \sum_{s=1}^{p} c_s(k)e(k - s). \tag{5.5.20}$$

Note that each of these algorithms needs only an initialization of $a_i(k)$, $c_s(k)$, and past y and e (or past y^F and e^F) and the ensuing e-sequence to perform their updating. Thus, only the transmitted quantized prediction error sequence in (5.5.3) is needed to update the predictor parameters in (5.5.2) *and* to form, via (5.5.2), the same estimate of the predictable component of the source simultaneously at the transmitter *and* receiver. The resulting ADPCM scheme is illustrated in Figure 5.11. Compare Figure 5.11 to Figure 5.2. Note that (5.2.44) describes the encoder prediction filter in Figure 1.5. Thus, if $a_i(k)$ and $c_s(k)$ are properly convergent, block diagram reduction reveals that the y to e transfer function (ignoring e_q) would be $(1 - A)/(1 + C)$, which is the inverse of the source model. This is the objective cited at the beginning of this chapter.

At this point we will raise an important pragmatic concern regarding direct transcription of generic, theoretically established adaptive filter algorithms into physical problems with seemingly appropriate abstractions. Practicalities outside the scope of the problem abstraction will typically impact the selection of an algorithm for actual use from those applicable to the abstraction. One such practicality in a telephony ADPCM system is that for the speech signal to be transmitted the source model transfer function $[1 + C]/[1 - A]$ is not time-invariant. In fact, this time-varying source model justifies the consideration of an adaptive predictor in DPCM rather than a fixed predictor. The assertion is that an adaptive predictor, if properly operating by closely tracking such source model parameter variations, could outperform any fixed predictor. Recall that the convergence/stability arguments for algorithms 6 and 7, on which 10 and 11 are based, rely on designer parameter γ_s selection in a fixed polynomial delay operator as the numerator of a transfer function such that it is strictly positive real (SPR) or nearly so. Review the remarks below (5.3.41). The unknown source model transfer function numerator is the denominator of this

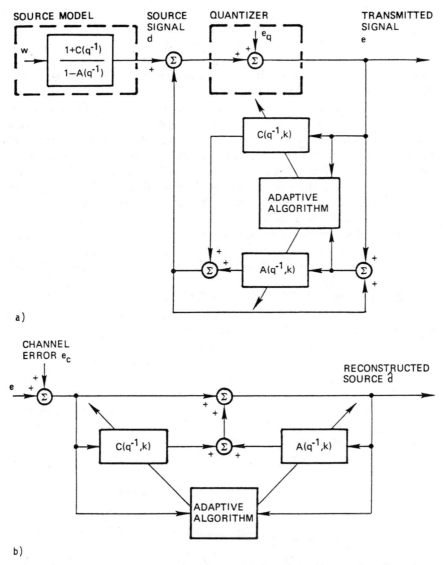

FIGURE 5.11. Adaptive differential pulse code modulation scheme (a) transmitter (b) receiver.

hopefully SPR transfer function. Since the source model is expected to vary extensively, though slowly, over time, no single set of γ_s values would likely satisfy this SPR condition. Thus, the use of algorithms 10 and 11 would seem questionable. Recall that a sufficient power condition may be invoked on the source in conjunction with a potentially non-SPR transfer function and stability retained. But this still does not assure good performance for all possible source sequences. This suggests consideration of Algorithms 9 and 12, as will be done in

Section 8.3. Recall that Algorithms 5 and 8, on which Algorithms 9 and 12 are based, rely on maintenance of the stability of $[1 + C]^{-1}$.

Another practicality raising performance questions is the undeniable possibility of channel errors. Note that the receiver illustrated in Figure 5.11b incorporates possible channel errors e_c. Given such an error, the propagation of the receiver adaptive algorithms will now differ from the transmitter algorithms because e in the potential transmitter algorithms of (5.5.7)–(5.5.20) is replaced by $e + e_c$. Thus, a channel error will separate what may have been previously coincident transmitter and receiver adapted parameter trajectories. Since this will cause the receiver source prediction to differ from the transmitter source prediction, the receiver source reconstruction will be degraded. Of concern is whether or not the transmitter and receiver predictors will quickly resynchronize to reduce this reconstruction degradation. A leakage factor ρ slightly less than one should be added, e.g., in (5.5.7) as $a_i(k + 1) = \rho a_i(k) + \cdots$, to the parameter update equations of algorithms 9–12 to encourage resynchronization after channel errors [Cohn and Melsa, 1975]. Recall Section 4.2.6.2. Chapter 7 will discuss this leakage modification and other practical modifications in a more general setting.

With this example we have shown how a particular adaptive filtering/prediction problem can be cast in a form to which the adaptive IIR filter algorithms developed in this chapter are immediately applicable. This ADPCM example was chosen for discussion here (and in Chapters 1 and 8) because it is apparently the first application of adaptive IIR filtering on a large scale. Our discussion of this example has also indicated how practicalities of an application can suggest potential algorithm performance deficiencies, or, viewed differently, gaps in the theory verifying and quantifying the retention of satisfactory behavior in the face of certain "nonideal" circumstances. Counteraction of such deficiencies in a candidate algorithm may be achieved by a more appropriate algorithm selection (from derivatives of algorithms 1–8) or particular algorithm modifications, such as the introduction of leakage. As noted before, Chapter 7 will present a number of practical modifications to the generic adaptive FIR and IIR filters of this and the preceding chapter.

6

Adaptive Algorithms for Restoring Signal Properties

●*PRECIS Adaptive filtering algorithms can be developed that estimate some property of a signal and use it to guide the "training" of the filter instead of using a reference or template waveform. Such "property-restoring" algorithms can be very useful in applications in which a reference waveform is not available.*

6.1 INTRODUCTION

In Chapters 4 and 5, we developed a variety of algorithms for adapting the coefficients of a digital filter. These algorithms considered different filter structures (e.g., transversal FIR and direct-form IIR), used different performance objectives (e.g., summed squared error reduction and error system stability), and employed different update schemes (e.g., with and without regressor or prediction error filtering). The only aspect constant in the development of these algorithms was that of using the difference between the filter output and the desired or template signal as the prediction error, as illustrated in Figure 6.1. The adaptive filter in Figure 6.1 contains three components. The first filters the input, the second forms the error $e(k)$, and the last adapts the filter coefficients. In the adaptive filters examined thus far, only the first and third components varied. The filter structure and the coefficient-updating algorithms were allowed to change, but the simple form for the system error $e(k)$ as $d(k) - y(k)$ was maintained.

While the model assumed in Figure 6.1 allows an intuitively reasonable formulation for those algorithms, it suffers from the practical problem that a suitable reference signal $d(k)$ is often unavailable. This can be illustrated by

157

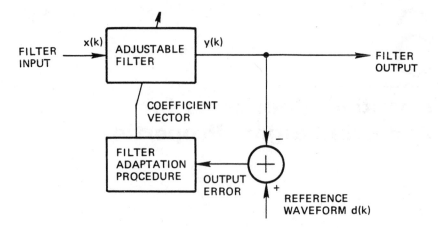

FIGURE 6.1. An adaptive filter employing reference matching as its assessment of performance.

reconsidering the troposcatter communications problem examined in Section 1.3. A signal is sent via the "tropospheric channel" and is degraded in the process. Suppose we consider using an adaptive filter at the receiver to correct this degradation and hence improve the signal quality. The received signal, perhaps after tuning and filtering, is the input to the filter, and the filter output is applied to the demodulator. Refer to Figure 1.9. What, then, is used for the reference waveform $d(k)$? The signal we wish to have as the reference is the transmitted signal, a signal generated hundreds of miles away and in general unknown at the receiver. How, then, do we proceed? In telephone data modems, a roughly equivalent theoretical problem, two procedures are used. In some cases, a specific signal is sent by the transmitter at prearranged times. The receiver keeps this prearranged sequence in local storage and uses it at the appropriate times to initialize or "train" the filter. This special sequence is referred to as the "training sequence," and clearly no other information can be passed while the filter is being trained [Gersho, 1969]. A second technique [Salz, 1973] is "decision direction," which assumes that the system is running well enough that most of the demodulator output decisions are being made properly. The demodulator output is then fed back to form the desired signal $d(k)$. Since most of the decisions are being properly made, the demodulator output closely approximates the transmitted signal. This technique has the advantage that data is always being transmitted, but it can fail catastrophically if the error rate goes too high.

Neither of these techniques is actually practical for the FM troposcatter application. Since the propagation channel changes considerably over even a second, the correction filter must be retrained at least that often, with the associated loss of signal while that training occurs. The decision feedback technique, while useful for tropo data transmissions, cannot be successfully applied to analog FM transmission without the danger of filter divergence

[Treichler, 1980]. Thus, for the particular communications problem examined here the required desired signal is not available nor can it be reasonably inferred from signals available at the receiver.

It turns out that the unavailability of the reference waveform is more the rule than the exception in the adaptive filtering field and that many techniques have been developed to circumvent the problem. In this chapter, we will develop a class of algorithms that avoid a dependence on the "output error" $e(k)$ by judging the algorithm's performance in a more general way. This approach is called "property restoral" and leads to algorithms that adapt the filter in such a way as to restore some property of the signal lost before reaching the adaptive filter. This property, rather than the desired waveform itself, serves to guide the adaptation of the filter.

6.2 THE PROPERTY-RESTORAL CONCEPT

The property-restoral approach exploits the fact that many signals, particularly those used in communications systems, have certain invariant properties that can be sensed and then used as the basis for adapting a filter. For example, many communications signals in common use employ transmitted waveforms with constant envelopes. The use of constant-envelope signals, such as in frequency modulation (FM) and phase modulation (PM), was originally motivated by its noise immunity and its resistance to intermodulation distortion. However, this property can be used as the basis for developing an adaptive equalizer or interference suppressor that does not require explicit knowledge of $d(k)$, which in many cases is the original transmitted waveform.

Conceptually, the idea is as follows. A signal may be designed with some invariant property, e.g., the constant envelope of an angle-modulated signal. If propagation effects or interference that degrade the receiver output also disturb this otherwise invariant property, then an algorithm can often be developed that senses the disturbance and adjusts the coefficients in such a way as to restore the should-be-invariant property. If the algorithm accomplishes this by notching the interference and/or equalizing the channel, then the signal too is corrected, not just the property, and the quality of the receiver output will be improved. It can frequently be shown that restoring the property of the signal is tantamount to correcting the signal itself. When this is true, an adaptive filtering algorithm based on property restoral can be developed and specific knowledge of the desired waveform $d(k)$ is not required.

The adaptive processor resulting from this approach has the structure shown in Figure 6.2. Two of the blocks, the filter and the coefficient adaptation, were shown in Figure 6.1. The third portion is now called "property measurement" and is used to make any measurements needed to sense the degree to which the selected signal property is attained. This property measurement block uses the filter output as its input and produces a sequence $p(k)$, which reflects the

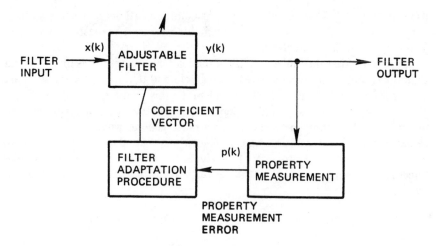

FIGURE 6.2. A general structure that uses property restoral to direct filter adaptation.

difference between the intended state of the property and its actual instantaneous value. This sequence $p(k)$ has the characteristics of an error signal and can be used to drive an adaptive algorithm that adjusts the coefficients of a filter.

Before proceeding with a specific example, we can first make two observations. The first is that this concept is quite general and subsumes a number of existing algorithm design approaches, including the prediction error techniques of Chapters 4 and 5. The second is that in practice the resulting algorithms are quite signal-specific; i.e., an adaptive filter based on restoring the envelope properties of an FM signal may work very poorly on an AM signal. Examples of both of these points will be provided later in the chapter.

6.3 THE CONSTANT-MODULUS ALGORITHM

The discussion of the previous section can be made more concrete by considering a specific example, the "constant-modulus" algorithm (CMA) [Treichler and Agee, 1983]. This algorithm was developed to perform the equalization and interference reduction functions for constant-envelope signals, such as in the troposcatter problem discussed in Chapter 1. If we view the transmitted constant-envelope signal as a complex phasor, the information is contained purely in the phasor angle while the modulus, or instantaneous amplitude, is fixed at some value A. This constant modulus is the invariant property for such angle-modulated signals and is in fact corrupted by the presence of multipath propagation and/or interference.

Following the structure shown in Figure 6.2, we need to define the following in order to specify a property-restoral adaptive algorithm for such signals:

(a) The structure and parameters of the filter;

(b) The error-like signal $p(k)$, which indicates the degree of nonattainment of the constant envelope property; and

(c) An adaptive algorithm for adjusting the filter's coefficients based on some function of the error $p(k)$.

While many choices are available for each of these items, we will examine those made in [Treichler and Agee, 1983].

6.3.1 The Filter Structure

An N-tap transversal FIR filter was assumed as the basis for this adaptive filter. The value of N is determined by practical considerations. Refer to Chapter 8 for an example. An FIR filter was chosen because of its stability with any bounded parameterization. The use of a transversal structure also allows for relatively straightforward construction of the filter.

6.3.2 The Error Signal

If the input and output of the filter are assumed to be complex-valued, then the natural choice for the property measurement is the modulus, or instantaneous amplitude, itself. If $y(k)$ is the complex-valued filter output, then $|y(k)|$ denotes the amplitude. Since we expect the amplitude to have a value of A in the absence of signal degradations, it is equally natural to define the error signal $p(k)$ as

$$p(k) = |y(k)|^2 - A^2. \tag{6.3.1}$$

The error $p(k)$ has the desired features since it is zero when the envelope has the proper value, is nonzero otherwise, and carries sign information to indicate which direction the envelope is in error. Note, however, that other functions with the same qualifications are available, and some may in fact ultimately turn out to work better. Two of these include

$$p_2(k) = |y(k)| - A \tag{6.3.2}$$

and

$$p_3(k) = \ln(|y(k)|/A). \tag{6.3.3}$$

6.3.3 The Adaptive Algorithm

Following steps laid out in the previous chapters, we define the adaptive algorithm by specifying a performance function based on the error $p(k)$ and then developing a procedure that adjusts the filter impulse response so as to minimize that performance function. Suppose we define J, the performance function, by the expression

$$J = E\{p^2(k)\}, \tag{6.3.4}$$

where E is the statistical expectation operator. Thus J is the mean square of the error sensed in measuring adherence to the constant envelope property.

Given this definition for J, many procedures can be developed for adapting the filter. The constant-modulus algorithm reported in [Treichler and Agee, 1983] employed the approximate gradient descent method presented in Section 4.2. The true gradient at time index k is first approximated by its instantaneous value, which can be shown to equal

$$\hat{\nabla} J = \nabla_W\{p^2(k)\} = 2 \cdot p(k) \cdot \nabla_W\{p(k)\} = 4 \cdot p(k) \cdot y(k) \cdot \mathbf{X}^*(k), \tag{6.3.5}$$

where, as before, $\mathbf{X}(k)$ is the vector of current and $N-1$ past input samples. The approximate gradient is used by updating the impulse response vector according to the rule

$$\mathbf{W}(k+1) = \mathbf{W}(k) - \mu \cdot \hat{\nabla}_W J. \tag{6.3.6}$$

Substituting (6.3.5) into (6.3.6) and suitably redefining the adaptation constant μ, we obtain the constant-modulus algorithm

$$\mathbf{W}(k+1) = \mathbf{W}(k) - \mu\{|y(k)|^2 - A^2\} \cdot y(k) \cdot \mathbf{X}^*(k). \tag{6.3.7}$$

We shall show an example of this algorithm's performance in the next section, but we should note here that it has been tested with real and simulated data and shown to operate very well.

The similarity between this gradient descent algorithm and others, such as LMS, can be underscored by defining the term $\tilde{e}(k)$ with the expression

$$\tilde{e}(k) = y(k)\{|(y(k)|^2 - A^2\}. \tag{6.3.8}$$

With $\tilde{e}(k)$ written this way, the CMA recursion expression becomes

$$\mathbf{W}(k+1) = \mathbf{W}(k) - \mu\tilde{e}(k)\mathbf{X}^*(k). \tag{6.3.9}$$

Comparison of this expression and (4.2.7) shows that both have the same form. Only the error term itself indicates the type of performance function being minimized.

It should be emphasized that other "constant-modulus" algorithms can be developed by making different choices for the filter structure, the error signal, the performance function, and the actual updating procedure. The particular choice of the appropriate gradient search technique used here was motivated originally by the need for a simple algorithm that could operate at high clock rates (e.g., 10 megasamples/sec) and for which "slow" convergence (e.g., a 10,000 sample time constant of 1 msec) was not a problem. When these conditions are not the case, e.g., when rapid convergence is required, then more complicated procedures may be applied to advantage.

5.3.4 An Example Using the Gradient Descent Version of CMA

To illustrate the behavior and effectiveness of a property-restoring algorithm, we will examine how CMA responds to an FM signal degraded by multipath propagation. The FM signal itself is assumed to be of the form

$$s(k) = A \ e^{j\{\omega_0 kT + \theta(k)\}}, \tag{6.3.10}$$

where $s(k)$ is complex-valued, A is the real amplitude, ω_o is the center frequency, T is the sampling interval, and $\theta(k)$ is the integrated message waveform. Clearly $s(k)$ has a constant envelope since A is a constant and the exponential has a modulus of unity.

We assume that the multipath propagation path can be modeled by the impulse response $g(k)$, where

$$g(k) = \delta(k) + a \cdot e^{j\phi} \cdot \delta(k - \tau). \tag{6.3.11}$$

This model describes "specular" multipath in which a single delayed, scaled, and phase-shifted version of the transmitted signal is added to the uncorrupted signal that would otherwise be observed at the receiving antenna. The terms a, ϕ, and τ are the multipath attenuation, phase shift, and delay, respectively. The frequency response of this propagation channel can be found by evaluating the Fourier transform of $g(k)$; the resulting power transfer function is given by

$$|G(\omega)|^2 = 1 + a^2 + 2a \cos(\phi - \omega\tau). \tag{6.3.12}$$

This transfer function is periodic in frequency with a period of $1/\tau$ Hz, thus attaining a maximum gain of $(1 + a)^2$ and a minimum gain of $(1 - a)^2$.

The transmitted FM arrives at the receiving antenna after passing through the propagation channel. Convolving $s(k)$ with $g(k)$ produces the received signal $x(k)$:

$$x(k) = A \ e^{j\{\omega_0 kT + \theta(k)\}} + Aa \ e^{j\{\omega_0(k-\tau)T + \theta(k-\tau)\}}. \tag{6.3.13}$$

It is easy to confirm that except for very special cases (such as $\theta(k) = bkT$), the modulus of $x(k)$ varies with time, and therefore the propagation process destroys the constant-envelope property of the transmitted signal.

Figure 6.3 shows the power spectrum of an FM signal generated with a computer simulation. The time waveform closely approximates a 120-channel frequency-division-multiplex/frequency-modulated (FDM/FM) signal and was

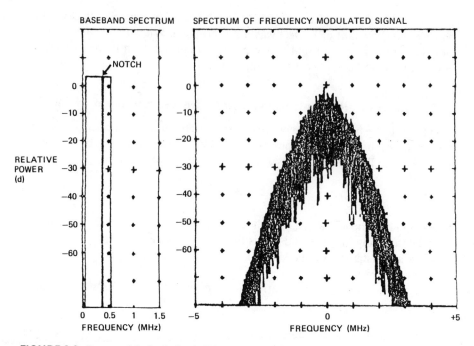

FIGURE 6.3. Spectra of the "noise loaded" baseband test signal and the resulting output of an FM modulator © IEEE.

generated with a 10-MHz sampling rate. The channel transfer function and the resulting received-signal spectra are shown in Figure 6.4a. The multipath attenuation and delay are 0.6 and 2 μsec, respectively, and the phase shift is chosen to set a null on the carrier frequency (that is, $a = 0.6$, $\tau = 20$, and $\phi = \pi$). Note the highly frequency-selective nature of this multipath; the 12-dB nulls are spaced only 500 Hz apart, well within the signal passband. The effect of the channel is clearly evident in the scalloping of the received signal spectrum. A somewhat more realistic situation is shown in Figure 6.4b. Here the signal spectrum is shown with both the effects of multipath and the effect of additive Gaussian noise. The carrier-to-noise ratio (CNR) here is 30 dB, not atypical of practical situations. Note that the scalloping affects the FM signal and not the noise, since the noise is added at the receiver terminals and does not pass through the propagation path.

An adaptive filter using the gradient descent version of the constant-modulus algorithm, as in (6.3.7), was employed to process this simulated signal. A filter length of 256 was chosen for this experiment, and the adaptation constant μ was set to equal 2^{-10}. The filter was allowed to run for 800,000 time samples, the equivalent of 80 msec of real time.

Figure 6.5 shows the "learning curve" for the algorithm, i.e., a plot of the estimate of the performance function $J(k) = p^2(k)$ versus the iteration number. If

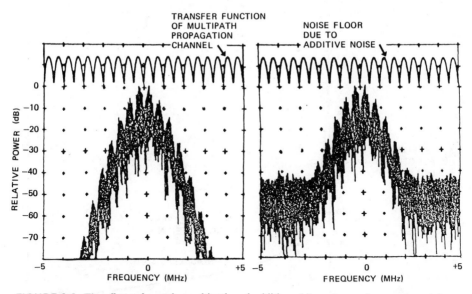

FIGURE 6.4. The effects of specular multipath and additive white noise on the spectrum of the transmitted FM signal © IEEE (a) noiseless FM signal in the presence of specular multipath (b) FM signal in multipath and noise (CNR = 30 dB).

the algorithm is in fact altering the filter impulse response so as to reduce and ultimately minimize the performance function, then we would expect to see the learning curve descend with time and then level out at this minimum. The time constant or, more generally, the convergence time of the algorithm is indicated by the rate at which the performance function is reduced. Reference to Figure 6.5 shows that indeed CMA does reduce $J(k)$ and that most of the convergence occurs within 2 msec of the time when adaptation begins. In the noiseless case, the algorithm tends to continue improving the filter's performance even beyond the 80-msec point. This is also true in the noisy case, but as a practical matter, the performance flattens out to a level constrained by the presence of the additive noise.

Figure 6.6 shows the spectra of the noiseless and noisy signals after passing through the CMA-directed adaptive filter. Also shown with each spectrum is the channel transfer function of the multipath in tandem with the adaptive filter. In considering Figure 6.6a first, we notice that the adaptive filter tends to flatten the propagation-induced spectral shaping *within* the signal passband, but little effort was made in the portions of the band unoccupied by the signal. In the noisy case, however, shown in Figure 6.6b, not only did the filter correct for the spectral shaping of the channel, but it used its remaining degrees of freedom to reduce the noise by developing a bandpass filter response. Note that in both Figures 6.6a and 6.6b, the scalloping effects of the multipath on the signal spectra have been essentially removed.

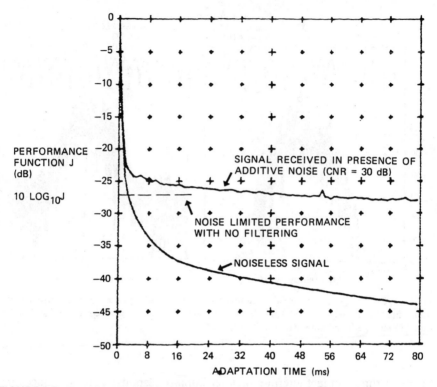

FIGURE 6.5. "Learning curve" of the CM algorithm for a specific signal and multipath condition with 10^4 iterations per millisecond © IEEE.

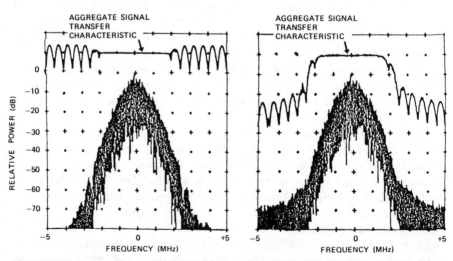

FIGURE 6.6. Spectra of "corrected" FM signals at the output of the CMA adaptive filter after 80 ms of adaptation © IEEE (a) spectrum of the corrected noiseless signal (b) spectrum of the corrected noisy FM signal.

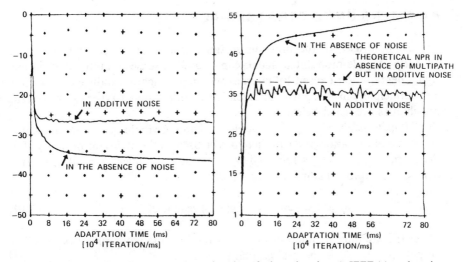

FIGURE 6.7. Signal quality measured as a function of adaptation time © IEEE (a) predetection mean-squared error in dB (b) postdetection noise power ratio (NPR) in dB (notch at 400 kHz).

The effect that the adaptive filter has on signal quality are shown in Figure 6.7. The average-squared difference between the adaptive filter output and the transmitted signal is plotted in Figure 6.7a. As before, the performance improves dramatically over the first 2 msec and then flattens out. The total performance improvement is about 27 dB in the noisy environment and about 37 dB in the noiseless case. The quality of the FDM baseband voice channels is quantified by a measure known as the noise power ratio (NPR), which is roughly analogous to the channel signal-to-noise ratio [Tant, 1974]. This differs from the mean-squared error measurement shown in Figure 6.7a since the NPR is measured after the nonlinear frequency-demodulation process. A plot of the baseband's NPR is shown in Figure 6.7b. Rapid convergence is still evident, and substantial improvement can be observed. In the noiseless case, the performance continues to improve with time, but it levels out in the noisy case. Given the carrier-to-noise ratio and the FM deviation, the theoretical NPR of the signal *in the absence of multipath* can be evaluated and shown to be 37.1 dB. From Figure 6.7b, we can conclude that the CMA-directed adaptive filter removed the effects of the multipath well enough, even in the presence of noise, to return the output quality (NPR) to within a decibel or two of the theoretical level.

From this example, it is clear that the adaptive property-restoral filter performed the desired function. Sensing the envelope variations of the received signal, it directed the adaptation of the filter toward an impulse response that desirably reduced the performance function, equalized the channel distortion, and, most importantly, lifted the quality of the output signal to virtually its noise-limited level.

6.4 EXTENSION TO OTHER SIGNAL TYPES

The example used in the previous section shows that the constant envelope property of angle-modulated signals can be used to sense and ultimately correct signal degradation caused by interference or propagation-induced dispersion. This principle can of course be extended to other signal types. A few of the possibilities are listed below.

1. *Dougle-sideband AM.* In double-sideband AM, the phase of the signal carries no information and can therefore be used as the invariant property after the carrier term $\{\omega_0 t\}$ is removed by the receiver's tuner. A candidate error function to exploit this invariant property is

$$p_{AM}(k) = [\tan^{-1}\{\text{imag}(y(k))/\text{real}(y(k))\} - \phi_0], \tag{6.4.1}$$

where the arctangent function measures the instantaneous phase of the filter output $y(k)$ and ϕ_0 is the desired phase angle.

2. *Bauded signal with pulse shaping.* Concern over spectral occupancy and efficiency has led to the use of carefully controlled pulse shaping in bauded data signals (e.g., QPSK), which would otherwise have a constant envelope. This type of variation is easily accommodated by using a time-varying template of the baud shape, call it $m(k)$, to compare with the instantaneous amplitude. An example of the error signal that might be used in this case is given for the "known modulus algorithm (KMA)" by

$$p_{KMA} = |y(k)|^2 - A^2 m^2(k), \tag{6.4.2}$$

where $p(k)$ deviates from zero only if $y(k)$ deviates from the baud shaping impressed at the transmitter. Algorithms based on this "known modulus" approach are described in [Treichler and Larimore, 1985a].

3. *Quadrature amplitude modulation (QAM).* When sampled at exactly the "slicing instant," complex data signals such as QAM have only a few possible amplitudes. This information constitutes the invariant for such signals and can be exploited. An algorithm for equalizing telephone channels for data modem signals can be interpreted in this way and uses as a performance function of the form

$$J_{QAM}(j) = \prod_{i=1} \{|y(j)| - a_i\}^2, \tag{6.4.3}$$

where j indicates the sample associated with the sampling instant and a_i is the amplitude associated with the ith possible state. If $y(j)$ is reasonably close to one of the possible amplitudes at each slicing instant, then the product in (6.4.3) would be near zero, thus providing an indication of the adherence to the invariant property. Godard [Godard, 1980] also suggested the use of (6.3.7) for equalizing QAM signals, even though the signal modulus cannot be expected to be constant.

4. *Mean Squared Error (MSE) and Sum-Squared-Error (SSE) Techniques.* We note that the more common prediction error techniques discussed in Chapters 4 and 5 can be viewed as a subclass of the property restoral methods. For example, in the MSE case of Sections 4.2 and 5.2, we consider adherence to some template $d(k)$ to be the invariant of the system. The error of interest then becomes

$$p_{\mathrm{MSE}}(k) = d(k) - y(k). \qquad (6.4.4)$$

This output or "prediction" error of course was a critical element in the algorithms of Chapters 4 and 5.

6.5 ANALYTICAL CONSIDERATIONS IN THE DESIGN OF PROPERTY-RESTORAL ALGORITHMS

Several basic conditions must be satisfied to make a specific property-restoral algorithm technically viable and practically useful. They include the following:

1. The signal of interest must have an invariant property that can be measured.
2. The invariant property must be sensitive to (i.e., disrupted by) the type of degradation expected to affect the signal's quality.
3. A filter with the structure selected, adaptive or otherwise, must be able to adequately solve the problem. For example, generally speaking, a filter cannot remove an uncorrelated interferer that spectrally overlaps a signal of interest.
4. The resulting adaptive algorithm must reduce the performance function with an adequate convergence rate.
5. It must be shown that restoring the invariant property of a signal is tantamount to restoring the signal itself to its original uncorrupted state.

The analysis required to examine these points has been performed in some depth for CMA [Treichler and Agee, 1983; Treichler and Larimore, 1985a; Treichler and Larimore, 1985b.], for decision direction [Monson, 1977], and of course, for the prediction error techniques of Chapters 4 and 5. It remains to be carried out for the AM and known modulus algorithms, and in general must be reexamined for each processing problem.

6.6 SUMMARY

This chapter has considered the development of adaptive filtering algorithms that use a property of the signal of interest rather than a template of the desired signal itself to direct the adaptation. Instead of needing the desired signal itself to direct the algorithm, we use some a priori knowledge about the signal. While a specific algorithm developed with this concept may not be as generically applicable as the algorithms of Chapters 4 and 5, they serve in applications where a reference waveform is not available, making it impossible to use the conventional approach.

The property-restoral concept hinges on the identification of an invariant property of the particular signal of interest, a property which carries no intentional information or modulation but is in fact degraded by the same propagation or interference problem that degrades the quality of the demodulated signal. If this property (e.g., the amplitude of an angle-modulated signal) can be measured and compared to its proper value, then an error signal can be formed that can be used to direct the adaptive filter. Once the error has been formed, the remaining aspects of the adaptive filter design are the same as those considered in Chapters 4, 5, and 7: a filter structure must be decided upon, a performance function or objective defined, and a procedure developed for adjusting the impulse response in such a way as to minimize the performance function or achieve the performance objective.

7

Implementation Issues

●*PRECIS Adaptive filters can take several forms, ranging from pure software implementation on a general-purpose computer to a variety of hardware configurations. Concerns regarding complexity, cost, bandwidth, and performance dictate the choices available to the designer.*

To this point, we have dealt principally with the mathematical development of adaptive filtering and not really addressed many practical details. In this chapter we touch on some of the considerations required for design of useful adaptive filters, ranging from common-sense algorithmic efficiencies to the strengths and weaknesses of candidate architectures and technologies. In addition, means of assessing the performance of an implementation are discussed whereby a designer can intelligently select appropriate precision for data and weight storage.

Consider the possible forms that an adaptive filter may take. First, a general-purpose computer may be programmed for off-line processing, where digitized and recorded data of any conceivable bandwidth may be filtered in non-real-time. Sample FORTRAN implementations are found in Appendix A. A pure software implementation, especially using a high-level language, has distinct advantages in that debugging, modifications, and input/output operations are inherently simple. The principal drawback is that the implementation is constrained to off-line applications; the flexibility of a general-purpose computer is invariably accompanied by a sacrifice in execution speed of operations critical to adaptive filtering. Even dedicated central processing units (CPUs) efficiently programmed in assembly code are normally limited for real-time use to applications having relatively low bandwidth, e.g., under 100 Hz. The addition of a general-purpose array processor can boost the throughput of a computer-based implementation, but may call for some degree of machine-dependent "microcoding" for real-time applications with higher bandwidths. The cost of such an adaptive filter implementation may range from routine CPU charges for off-line use to a relatively high investment for a dedicated computer/array processor system.

For real-time applications where cost and size are critical factors, adaptive filters can be implemented using dedicated hardware specially designed for signal-processing functions. In recent years, the "single-chip signal processor" has been developed and refined to the point that it can well provide the computing power for most medium-bandwidth real-time applications; e.g., up to 50 kHz. This class of device is actually a self-contained microprocessor whose instruction set and support circuitry are carefully designed for common signal-processing functions; therefore, it is preferable to other general-purpose micro-processors for signal-processing applications. With suitable off-line development tools, this device becomes as simple to program and configure as a general-purpose CPU, but at a fraction of the cost and size, and operates at significantly higher throughput rates. The class of single-chip signal processors is typified by the Texas Instruments TMS320 family, which includes a number of innovative features for adaptive filtering applications. Typically, these include 16-bit arithmetic, extended precision accumulators, cache memory, delay line shifters, and so on.

Refinements in digital technology improve the speed, cost, and size of digital circuitry almost yearly. Yet for now the single-chip signal processor is limited in terms of its real-time bandwidth. For situations requiring faster digital processing, the designer must resort to less flexible implementations. That is, discrete LSI and VLSI building blocks, e.g., multiplier/accumulators and register files, hardwired in a given configuration can service bandwidths beyond those of the single-chip processor. Use of emitter-coupled logic devices accommodates bandwidths nearing 50 MHz. Naturally, cost, flexibility, and power are sacrificed for such high-bandwidth designs.

Lastly, for bandwidths above the reaches of digital technology, say, above 50 MHz, current technology calls for the use of analog components. Analog systems are as inherently inflexible as digital circuitry based on discrete devices, and in addition, possess nonideal and sometimes unpredictable qualities that make them ill-suited to critical filtering applications. Fortunately, an adaptive structure with its performance feedback serves to correct most linear distortion and aging effects. While the purely analog design has found use at such high bandwidths, filter performance can be greatly enhanced by means of digital circuitry applied at critical points, i.e., forming a hybrid system.

The thread of this discussion is one of trade-offs between flexibility and bandwidth: As bandwidth demands increase, a design typically becomes more complex, less flexible, and more costly. This forms the basis for much of this chapter; algorithmic shortcuts are selected and discussed as means of extending bandwidth without resorting to the costs and complexities of the next level of implementation. It is important to note that the bandwidth boundaries ranking the useful limits of technology alternatives are blurred by cost, size, and power concerns. Naturally, the boundaries are dynamic and expanding, but will most likely always exist in some form; special-purpose digital hardware will always serve the applications beyond the reach of the microprocessor system, just as the analog system will be the only means of processing signals with extremely high bandwidths. However, the low- to medium-bandwidth realm best addressed by the single-chip signal processor will grow as the underlying semiconductor technology evolves.

The first part of this chapter deals with the more mundane, but important, efficiencies that a designer must be familiar with. Included are discussions of complexity and hardware impact, as well as common-sense design steps useful in both hardware and firmware. In conjunction with complexity concerns, we present some means of increasing bandwidth (clockrate) by architectures using distributed processing. Precision of time samples and filter weights has special importance when dealing with adaptive filters, and some guidelines are in order.

Next, we consider specific shortcuts as they can be applied to various algorithms of Chapters 4 and 5. Also, we present arguments on reducing the rate of adaptation in the interest of efficiency. This leads to the concept of dual-mode processing, i.e., a "foreground" signal calculation and "background" adaptation processing. These concepts then serve as a basis for discussions of more block-oriented processing and associated architectures. We conclude by mentioning further alternative implementations and some of their advantages and limitations.

7.1 SIZING CONSIDERATIONS

When presented with a signal-processing requirement, the designer must first assess the value of an adaptive filtering solution. Those not familiar with the inherent limitations often consider an adaptive filter a cure-all, able to design itself and remove most of the burden from the designer. We have seen that there are definite costs, both in complexity and performance, when using an adaptive processor. Whenever possible, a fixed or programmable filter is far preferable, i.e., when enough is known a priori about the environment and when time variation is a secondary concern. Consequently, the designer must first objectively answer the "necessity and sufficiency" question: Will an adaptive filter satisfy the need, and is adaptation necessary? Chapter 1 has provided insight into answering this question for typical applications.

Given that an adaptive filter is indeed called for, the next question to be addressed concerns feasibility. Can a unit be built within size and cost constraints? Again, the answer to this question rests largely with the bandwidth requirement. As an extreme example, equalization of voice channel modem signals (4-kHz bandwidth) has been possible by all-digital adaptive processing for the last two decades. But as yet, similar equalization of radio-frequency PCM signals (with bandwidths about 25 MHz) is only now being attempted, and at present requires often impractical power levels. (This particular problem has been handled more cost-effectively using alternative analog means, to be addressed later in this chapter.)

Designers active in this field rely on experience to answer feasibility concerns. Yet the novice can approach the problem using common sense and knowledge of the capabilities of current technology. One measure of complexity of a signal-processing circuit is the multiplication throughput rate. The multiply operation, a critical step in all digital signal-processing functions, represents an elementary unit of complexity, and per operation it is responsible for more time, area, and heat than other simpler functions (e.g., add, compare, I/O, etc.). For the purposes of this chapter, complexity will be equated with a filter's multiplication requirements.

Let us illustrate by using a concrete but simplified example. Suppose we were to design a 128-tap adaptive echo canceler for use on a 4-kHz telephone voice channel. We assume an FIR architecture adapted by means of (4.2.7). The sampling rate for our signal is 8 kHz. The filtering operation requires $128 \times 8\,\text{kHz} = 1.024$ megamultiplies/sec (M*/s); the update equation similarly requires 1.024 M*/s. As a first-order approximation, we may conclude that our hardware requires enough computational horsepower to perform 2M*/s; a single chip signal processor would certainly suffice, requiring a small space ($3'' \times 4''$) and low power (3 W) for the complete circuit. As an alternative, an LSI multiplier chip would provide an overkill solution; with associated "glue-logic" (memory, registers, I/O, buffers) this might occupy one board, say $6'' \times 10''$, and draw perhaps 10 W.

Had our example involved a 100-kHz sampling rate, our multiply count would have been on the order of 26 M*/s. A naive, yet often adequate, extrapolation would call for about 13 times the silicon, power, and size. If the scaled physical characteristics were unsuitable, a designer could draw on more exotic technologics for faster multipliers to reduce size, or to lower power (usually not both). Needless to say, the concrete numbers in this paragraph reflect current technology at time of publication. Higher multiply rates and lower power are evolving yearly.

For a first cut at sizing, this is usually adequate. However, if a more accurate answer were needed, there are additional subtleties that must be considered. First, implicit in the latter case is a rudimentary type of distributed processing; the net multiply throughput must be spread among several processors or multipliers. As might be suspected, the complexities of distributing data and orchestrating their remultiplexing can account for a fair amount of circuitry. Rather than linear scaling of complexity and circuitry with bandwidth, it can be argued that a quadratic factor appears at bandwidths where distributed processing enters; that is, at a 1-MHz sampling rate, a brute force scaling of the architecture might very well approach a rack of hardware. The bandwidth at which multiplexing effects become significant is naturally dependent on architecture.

Interestingly, an analogous effect can be seen for bandwidth reductions as well. Given a lower clock rate, circuit complexity is reduced according to our earlier definition. However, hardware savings can be realized only when a single processor or multiplier can take on the duties of other circuit elements. Returning to the example of a 128-tap echo canceller, were multiple channels involved, each sampled at 8 kHz, the LSI single-chip multiplier chip with its 10– 20 M*/s rate could service on the order of eight voice channels and becomes a solution competitive with the multiple, independent single-chip signal processor approach. However, this requires multiplexing and control hardware to "funnel" data into the multiplier from several sources and to commutate its output port among subsequent parallel stages, implying an effective increase in complexity and development cost. When multiplexing a processor, this complexity similarly translates into more involved microcode. Such complexity increase resulting from multiplexing is tolerable when cost or size savings through elimination of large or expensive components offsets the impact on design.

The qualitative effect of bandwidth on complexity is shown in Figure 7.1 for a typical digital signal-processing architecture. Reiterating, a region exists for which frequency scaling is essentially linear. Adapting this same architecture to extremes of bandwidth by means of distributed or consolidated processing adversely affects complexity. These end effects serve as sufficient motivation for a designer to consider alternative architectures that better match the particular bandwidth requirement.

A second subtlety of our first-order complexity assessment involves the type of multiplier used in the function. This is of particular importance for adaptive processing, where there are two distinct contexts for multiply operations. Recall that an adaptive FIR filter involves (1) an output calculation and (2) a weight update. In the first, N multiplies are done, and the results accumulated (i.e., forming a convolutional sum). This operation can be made particularly

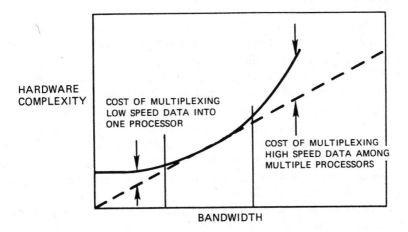

FIGURE 7.1. Effect of bandwidth on circuit complexity for a given architecture.

concise in hardware and has led to the signal processor building block, the multiplier/accumulator (MAC), which includes a pipelined multiplier and on-chip integral accumulator. For convolution operations, such a multiplier can be run at nearly a 100 percent duty cycle. In the latter case, for each multiply operation (e.g., error sample times a delayed input sample), there is again an add operation, but also a memory read (old weight) and a memory write (new updated weight). Consequently, the input/output (I/O) of the multiplier can slow the throughput by a significant fraction. For our original example, while a single-chip signal processor could easily sustain a 2-M*/s rate for convolutional operations, the update operation with its memory I/O when performed at the sample rate may represent a sizable and inefficient burden on the resource. This serves as sufficient motivation to find means of reducing adaptive filter complexity, specifically of the update computation.

Clearly, this discussion has provided only general statements regarding complexity estimates and the assessment of feasibility. We only wish to indicate here that a designer should be aware of certain facets of adaptive filter "lore" and consider seeking an implementation to fit the application's performance and speed requirements.

7.2 REAL-TIME OPERATION

Before investigating specific "shortcuts," it is instructive to review the necessary steps involved in a typical digital adaptive filter implementation. The list of steps involved has special significance in real-time operation, where the "loop" must be completed in one sampling interval. While our procedure breakdown is based on the algorithm of (4.2.7), it is easily generalized. For the analog alternative, the enumerated items correspond to continuous-time calculations and are no longer discrete events.

(1) *Get input sample.* Read the analog-to-digital converter (ADC) port to bring in the next input sample. An interrupt of a time source or ADC has brought us to this point. If an external "desired" waveform or training signal is required by the algorithm, it will also be measured at this point.

(2) *Inject sample into delay line.* This newest sample must be saved for future use in the convolution and update calculations. Storage may be a hardware shift register or a queued random access memory (RAM).

(3) *Compute output.* The convolutional sum of N taps and N input samples is performed using a series of multiply/accumulate operations. Depending on the specific implementation, the accumulator may be implemented using up to full double-word precision.

(4) *Output result.* The output as accumulated is sent out to the appropriate port. Under normal circumstances, rounding or truncation of the accumulator assures that only the most significant part of the convolutional sum is transferred (e.g., 8 to 16 bits).

(5) *Compute error.* A subtraction involving desired and computed output is performed. Normally, a means of monitoring error performance is desirable; i.e., a smoothed error "power" provides a quantitative history of adaptive performance.

(6) *Compute update.* For each weight, an update term is computed from the product of error (from step (5)) and appropriately delayed input (step (2)). According to (4.2.7), this operation must be repeated for each weight, and as mentioned, this operation becomes the dominant burden in the processing loop. It is here that algorithmic sacrifices can make the difference between loop completion and processor overload. Among other things, we will show that the precision of this product can be as coarse as a single sign bit, requiring no multiplication at all.

(7) *Wait state.* Upon completion, the processor can enter an idle state, awaiting the next sampling instant. It's here also that lower-priority "background" processing can be carried out. For example, performance measures (filter frequency response, error power estimate, etc.) can be computed and formatted for display as time permits.

For batch or off-line applications, the timing of this basic loop simply translates into the net "execution" time. Yet in real-time applications, the designer must carefully tabulate the execution time for each step. Neglecting a certain margin for safety, the maximum sampling rate becomes the inverse of the basic loop interval. We next discuss specific means that can contribute to a reduction of the net computation time.

7.3 IMPLEMENTATION EFFICIENCIES

7.3.1 Data Storage

Data samples must be retained for use in the weight update as well as the output convolution sum. For digital implementation this can be done by means of a hardware shift register or delay line; analog filters may use broadband, passive delay elements, i.e., *L-C* sections, surface acoustic wave (SAW) devices, or charge-coupled device (CCD) delay lines. For the remainder of this discussion we concentrate on digital architectures. As a sample enters the processor, the delay line is clocked, physically advancing past samples. It should not be surprising, that while this form of storage arises naturally from the filtering block diagrams, it has severe limitations for long filters. Propagation delays can skew data arrivals at nodes of the delay line. More importantly, component count can become unwieldy.

An alternative approach that can be attractive from both efficiency and timing viewpoints is a queued random access memory (RAM). Samples arriving at the filter are written into a RAM, where an address register keeps track of the next available memory location. Data need not shift down the delay line, but rather simply overwrite data samples that are no longer needed. Of course, advancing the pointer register will require "wraparound" logic when the memory size is exceeded, so that the register points back to the beginning of the memory. The single-chip processor is often configured in such a way that this delay line bookkeeping is handled by special instructions. Each memory location corresponds to a node of the delay line, with the pointer register indicating the most recent arrival. As memory can be had in extremely efficient packaging, discrete delay line hardware can be minimal. For example, two 20-pin DIP circuits can satisfy needs for filters up to 4096 taps long. When using this storage scheme, the maximum filter length is limited by the memory size, a power of two, e.g. 2^M. The M-bit address register is simply allowed to increment with no regard to overflow conditions. A separate read-address register is maintained for output of data in correct sequence for convolution and update operations. Care must be exercised for the synchronization of write and read operations, and access time must be consistent with the loop timing requirements. That is to say, an N-tap filter must complete $2N$ delay line read operations for every sampling interval, N for the convolution and N for the update. A 100-tap filter using a single memory chip with 100 nsec cycle time would be restricted to under 50-kHz operation based solely on memory I/O limitations. Higher rates could be achieved by algorithmic shortcuts, as well as improved or distributed hardware.

Analogous storage efficiencies can be achieved in software implementations. Most novice programmers choose to emulate the shift register operation. Incoming data samples are stored into an array, and as new samples enter, the entire array is shifted one element at a time. Needless to say, this can cause a considerable drain on the execution time, no matter how powerful the CPU. Therefore, for longer filters, say, more than 20 taps, it is usually more efficient to use a queue structure of the array and a pointer variable indicating the top of the queue. The examples in Appendix A make use of this form of data storage.

7.3.2 Passband or Complex Filtering

There are many applications which involve signals concentrated in a narrow band, i.e., with bandwidth much less than its "center frequency." As an example, consider common voice channel data traffic carried on a two-tone frequency-shift keyed (FSK) transmission. The "space" and "mark" (0 and 1) frequencies for the Bell 103 modem are 1070 Hz and 1270 Hz, respectively. During typical modulation intervals the energy of the signal is concentrated above 1000 Hz, in a relatively narrow band. In principle, all discussions in preceding chapters regarding adaptive digital filtering of such signals apply. However, by taking advantage of the narrow-band character of the signal, we can achieve significant efficiencies for discrete time processing. This in turn translates into reduced hardware or execution time. To realize this reduction of signal bandwidth

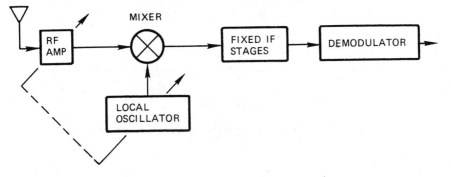

FIGURE 7.2. Conventional communications receiver.

requires an understanding of the concept of complex waveforms and complex filtering, motivated by the following digression. The subtle implications on adaptation then become clear.

Consider a conventional communications receiver, shown in Figure 7.2. The "front-end" stages are used for initial isolation of a narrow signal of interest using variable tuned filtering. Then, a down-conversion or heterodyning operation brings this analog signal to a fixed "intermediate frequency" (IF). At this point, carefully designed fixed filters can isolate the signal from its spectral neighbors. It is at the IF output where digital signal processing can best be applied, providing a convenient means for interference reduction and signal restoration. Clearly, sampling this waveform according to the Nyquist criterion of twice the highest frequency would involve an inordinate data rate and a highly complex filter response with many taps.

There are two basic means of reducing the sampling rate to a more manageable speed. A theoretically sound technique is "bandpass sampling," i.e., undersampling at a rate on the order of the information bandwidth, but chosen such that aliased images do not overlap and that the image nearest DC is properly positioned in the baseband. The details of this technique are beyond the scope of this discussion. However, it is worth noting that high center frequencies place severe practical limitations on sampling accuracy, and as a result bandpass sampling must be carefully implemented. See Figure 7.3 for a graphic explanation of the principle. For a detailed explanation of the bandpass sampling theorem, the reader is referred to [Tretter, 1976].

A second approach is fairly unimaginative but quite effective; this requires a second heterodyning operation to a lower IF frequency, preferably abutting the lower band edge with DC. Then sampling can be done at a rate more in line with twice the signal's IF bandwidth. The spectral layout is seen in Figure 7.4.

As we know, this represents a fundamental lower limit to the sampling rate under the conditions of Nyquist's theorem. However, by exploiting certain mathematical signal properties, the energy of a bandpass signal can be

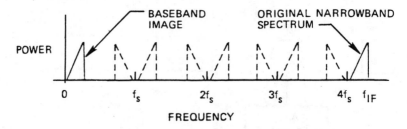

FIGURE 7.3. Bandpass sampling of narrowband signal with $f_s \ll f_{\text{max}}$.

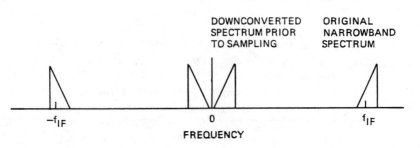

FIGURE 7.4. Downconversion of narrowband signal prior to sampling.

"condensed" even further, leading to additional processing advantages. The complex analytic representation of a real signal is widely used in coherent communication applications, such as modem processing, and properties of linear filtering can likewise be extended to the complex time domain. The following discussion leads to an adaptation algorithm having complex coefficients and certain architectural advantages.

Recall that all real waveforms have an interesting spectral property: Their Fourier transforms are Hermitian, with symmetric magnitude and antisymmetric phase. Now, from an information standpoint, a real signal takes up too much bandwidth; the negative frequency side of its spectrum is redundant. Getting rid of it would allow us to sample at half the rate required by the Nyquist criterion. However, in doing so our signal becomes complex, since its symmetry has been destroyed.

Before discussing means for suppressing half the signal band, it is useful to review the meaning of a complex-time waveform. The concept can be somewhat unnerving to all but mathematicians and communication engineers. In its most general form, a bandpass signal can be represented as the real part of a complex expression. Most communication texts treat single-sideband modulation from this viewpoint; see, for example, [Carlson, 1975]. Thus, the narrowband signal centered at f_c might be written as the real part of a complex analytic time function,

$$x(t) = \text{Re}[(x_I(t) + jx_Q(t))(\cos 2\pi f_c t + j \sin 2\pi f_c t)]$$

$$= \text{Re}[x_B(t)e^{j2\pi f_c t}]. \tag{7.3.1}$$

In this form we see that the bandpass signal is a complex baseband function $x_B(t)$, modulated up to the center frequency. Taking the real part of the complex product generates the spectral symmetry, with the reflected image at $-f_c$.

Mathematically suppressing the negative spectral image can be done in two ways:

Complex down-conversion. The entire spectrum is shifted to the left, centering the positive spectral band at DC. This is done by multiplying by $e^{-j2\pi f_c t} = \cos(2\pi f_c t) - j\sin(2\pi f_c t)$, generating a complex time function. A lowpass filter applied to the both real and imaginary components then eliminates the negative image. See Figure 7.5. This type of operation can be done using either analog or digital processing. For analog demodulation, two matched mixers are used prior to sampling to down-convert with an "in-phase" oscillator (cosine) and a "quadrature" oscillator (sine). The outputs are labeled "real" and "imaginary" time functions, respectively. After appropriate matched antialiasing filters, both can be sampled simultaneously and combined for storage as a complex time sample. At times, practical limitations of analog mixers and filters make such an implementation cumbersome. Instead, the bandpass signal can be sampled while still in its real form, followed by a digital complex conversion. The operations parallel those of the analog conversion, requiring multiplication of each real sample by a "complex oscillator," involving the trigonometric sequence

$$e^{-j2\pi f_c k/f_s} = \cos(2\pi f_c k/f_s) - j \sin(2\pi f_c k/f_s). \tag{7.3.2}$$

This is followed by a digital lowpass filter. Note that after the lowpass filtering operation, the signal bandwidth has been halved; consequently, a decimation-by-two operation can be incorporated.

Hilbert transformation. The second approach to complex conversion may seem somewhat more abstract, but effectively amounts to the same operation. The Hilbert transform is a linear filtering operation that has unity magnitude response but a 90° phase shift at all frequencies. Applying the real-time waveform to such a device generates the "imaginary" half of its complex analytic expression. The Hilbert transformer is a nonrealizable filter, but can be approximated arbitrarily well. While analog realizations are used routinely in single-sideband modulators, their bandwidths are fairly restrictive, limiting their accuracy for baseband spectra. However, a digital filter can provide a very close

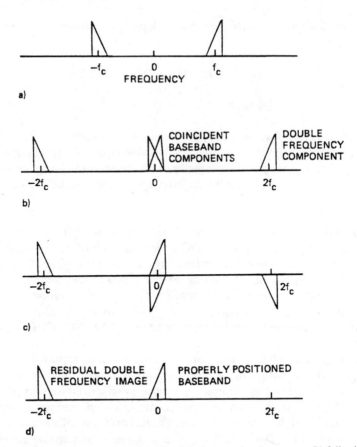

FIGURE 7.5. Complex downconversion (a) original narrowband spectrum (b) following cosine downconversion (c) following $-j$ sine downconversion (d) effective composite spectrum.

approximation to the ideal Hilbert transformer if allowed a noncausal impulse response. Therefore, once again we can sample the incoming real waveform at $2B$ samples/second and apply the sequence to a Hilbert transform filter. Its output is labeled the "imaginary" component, and the original input samples become the "real component" when delayed by the corresponding group delay of the Hilbert transformer. Each pair of values is stored as the complex time sample. The complex Fourier transform of this sequence shows a suppressed negative frequency band. Once again, the rate can be reduced by a factor of 2 as part of the transformation.

At first glance, one might question the wisdom of working with a complex time series. After all, all subsequent processing must be done on both real and imaginary parts, essentially doubling the computation. Remember, though, the sampling rate is likewise halved, so the processing requires slower logic. In a sense, a sampling rate that might have required commutating the processing between two chips has been split into two parallel and "independent" streams, each requiring one chip. The control logic required for overseeing the distribution and remultiplexing can therefore be eliminated.

To complete our digression on complex sampled data, we examine the impact on filter design. One implication of processing complex data can be seen in the spectral domain. Because the signal no longer possesses spectral symmetry, we can see from Figure 7.6 that a filter, fixed as well as adaptive, may require an asymmetric frequency response. For example, a tone interfering with the upper band edge of the signal would require a filter with a one-sided notch response. Therefore, the filter itself must have a complex time response or, in the case of an FIR implementation, complex taps. That is, the delay line has a real and imaginary part, weighted by a real and imaginary component at each node.

It may appear that use of complex data has a hidden cost in computation. Each tap involves a complex multiply, i.e., the equivalent of four real multiplies. Thus, even with the halved sampling rate, net computation is doubled!

Actually, this is true for a given number of filter taps. However, by halving the sampling rate, a given frequency response requires only half the number of taps. Stated in another way, for a given resolution Δf in the frequency domain, determined by (sampling frequency/number of taps), use of complex processing at half-speed requires only half the number of taps. The net multiply rate is then

$$B \text{ samp/sec} \times N/2 \text{ taps} \times 4*/\text{tap} = 2BN*/\text{sec}. \qquad (7.3.3)$$

For a real filter operating at twice the sampling frequency,

$$2B \text{ samp/sec} \times N \text{ taps} \times 1*/\text{tap} = 2BN*/\text{sec}. \qquad (7.3.4)$$

Therefore, after all is accounted for, complex processing requires exactly the same multiplier throughput as equivalent real processing. (Note, however, that the actual operations for complex conversion may account for a fair burden.)

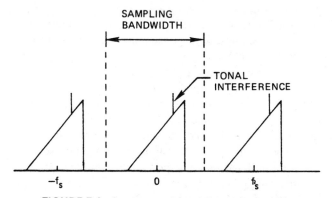

FIGURE 7.6. Asymmetry of sampled complex baseband.

Extrapolation of our previous discussions regarding real adaptive filtering is fairly straightforward. The input and weight vectors as defined earlier now become complex, and the scalar output and its error sequence are likewise complex-valued. A complex weight recursion was developed in Section 4.2.5.1, and was essentially the same as its real counterpart. However, there is a subtle difference in the actual adaptation expression, (4.2.32), that can be easily overlooked. Rather than a squared error cost function, we base optimization on a magnitude-squared error, that is, $e(k)e^*(k) = e_R^2(k) + e_I^2(k)$. This necessitates developing a separate gradient expression with respect to the real component of the weight vector, and a second with respect to the imaginary component:

$$\nabla_{W_R}|e(k)|^2 = 2e_R(k)\mathbf{X}_R - 2e_I(k)\mathbf{X}_I(k)$$

$$= -(e(k)\mathbf{X}^*(k) + e^*(k)\mathbf{X}(k)) \qquad (7.3.5)$$

$$\nabla_{W_I}|e(k)|^2 = 2e_R(k)\mathbf{X}_I(k) - 2e_I(k)\mathbf{X}_R(k)$$

$$= j(e(k)\mathbf{X}^*(k) + e^*(k)\mathbf{X}(k)). \qquad (7.3.6)$$

Details are available in [Widrow et al. 1975a]. Recombining the update vectors into complex form leads to the complex LMS algorithm, as in (4.2.32); i.e.,

$$\mathbf{W}(k + 1) = \mathbf{W}(k) + \mu e(k)\mathbf{X}^*(k). \qquad (7.3.7)$$

Note the presence of the conjugation on the complex input vector. Omission of the conjugate is equivalent to a minus sign on the step size of the imaginary part of the weight vector and leads to a somewhat peculiar divergent mode.

In this section, the subtle efficiencies of complex signal representation and complex filtering have been presented. The associated adaptive algorithm results from a derivation parallelling the conventional real arithmetic version.

7.3.3 Symmetric Filtering

Situations arise where it is very desirable to maintain symmetry of an FIR filter. By this we mean that the impulse response reflects about the center tap, so that for an N-tap filter, $h(0) = h(N - 1)$, $h(1) = h(N - 2)$, etc. The possibility of exact symmetry in the impulse response is one of the valuable strengths of the FIR filter structure. A filter with time domain symmetry has a "linear phase response" in the frequency domain; i.e., a sinusoidal component experiences a phase shift proportional to its frequency. In more practical terms, everything passing through the filter is delayed the same amount in time.

Where might such a property be of interest? Signals sensitive to phase distortion, that is, a nonlinear phase response, often must be processed with due attention to maintaining phase relationships. As a common example, quality of visual image is degraded by smearing caused by phase distortion; in contrast, the insensitivity of the human ear to phase distortion means that audio signals often can undergo serious but imperceptible phase disturbances. In the area of digital communications, phase distortion can be responsible for serious pulse deformation.

At any rate, applications arise where it may be advantageous to maintain the linearity of a filter's phase response, so as to reject or shape energy bands of the spectrum, while maintaining the basic pulse integrity with a constant filter group delay. For this reason, design of nonadaptive symmetric, linear phase filters has received a good deal of attention and can be found in most filter design software packages.

This symmetry can fortunately be exploited to reduce computational complexity, defined earlier in terms of multiplication rate. That is to say, such a filter requires only half the multiply operations of the general filter. To take advantage of this structure, we simply "fold" the delay line as shown in Figure 7.7 and add values appearing at corresponding nodes prior to the convolutional accumulation. Values $x(k)$ and $x(k - N + 1)$ both multiply $h(0) = h(N - 1)$ in the calculation of the output sample; adding them first and then multiplying by $h(0)$ eliminates one multiply

$$y(k) = \sum_{l=0}^{N/2-1} h(l)[x(k - l) + x(k - N + l + 1)]. \qquad (7.3.8)$$

Note that the storage requirement for filter coefficients is likewise halved. However, also note that the introduction of the add prior to the multiply/accumulation may actually not reduce complexity in a processor intended for general impulse responses.

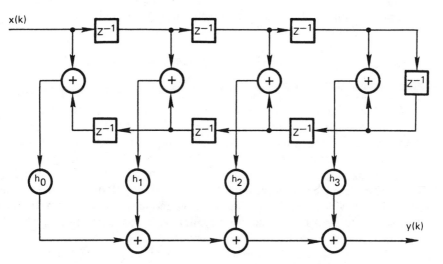

FIGURE 7.7. Implementation of 8-tap symmetric filter with folded delay line.

Adaptation of filter coefficients can be done in such a way as to maintain a linear phase response. For example, a pulsed signal in need of isolation from independent, additive interference might require an adaptive filter to track any time variations in the environment. To prevent arbitrary phase shaping by the filter, its impulse response (or its taps) must be constrained to be symmetric. Constrained adaptation performed on only half the weights reduces the complexity by almost half.

At first glance, it might seem that we must simply perform the update of (4.2.7) on the first half of the weight vector, i.e., using samples $x(k)$ through $x(k - N/2)$. However, this ignores the contribution of the "folded" samples on the output. Rather, noting the form of the output calculation for the symmetric filter above, the derivative of $e(k)$ with respect to w_j is

$$\frac{\partial e(k)}{\partial w_j} = -[x(k-j) + x(k - N + 1 + j)]. \tag{7.3.9}$$

Consequently, (4.2.7) is modified to become

$$\mathbf{W}(k + 1) = \mathbf{W}(k) + \mu e(k)[x(k - j) + x(k - N + j + 1)], \tag{7.3.10}$$

requiring $N/2$ multiplies and N add operations.

7.3.4 Algorithmic Shortcuts

When operating at high sampling rates, it is often necessary to reduce the complexity of calculations to allow completion of the steps enumerated at the beginning of this section. This could mean skipping operations outright when loop timing requires, or perhaps reducing precision to simplify data manipulations. In this section we discuss the general effects of these alternatives and their impact on implementation.

Elimination of operations is a natural consequence of insufficient time for completion of all steps enumerated in Section 7.2. Omission of operations may not be as serious as we might imagine. For adaptive filtering applications, we can classify operations into two distinct categories: those associated with output calculation (the signal flowpath), and those associated with weight update (the performance feedback path). Calculation of the output clearly is of highest priority; that is, we must produce an accurate response value by the end of each sampling instant. The rule is: If the output cannot be calculated at the sampling rate, abandon the current implementation. For a digital design, this means one must either provide more computing power or reduce the filter length. In cases where neither is a satisfactory solution, an analog implementation may provide an alternative with the associated cost to performance, as detailed in Section 7.6.

Fortunately, such a grave verdict is not necessary for the second class of operation, the updating of the filter weights. Clearly, adjustment of a weight is of lesser priority. If an update is lost every now and then, the principal cost is a slower convergence rate. That is, if theory predicts L samples for convergence of the weights of an N-tap filter, to a first order there need be NL individual weight update calculations completed. If only one weight can be updated per sample, we might expect a convergence time of NL samples, an increase by a factor of N. We can view the update operation as a "background" computation, sandwiched in between the "foreground" output calculations. Thus, each weight is updated in turn, as many per sampling instant as possible.

Of course, asynchronous computation can be awkward for a small processor, so it is usually advantageous to structure the firmware/software to update given weights at specific times. For example, we might update one weight per sampling instant, once the output convolution has been completed; we leave enough idle time to assure completion of this task. The next sampling instant, the next weight would be updated, and so on. In effect, the limited computing power of the single processor is being distributed among the N taps of the filter. This could be extended, time permitting, to update groups of two or more during each sampling instant, with a suitable increase in complexity. In higher-speed cases where the entire update cannot be completed during the "background" fraction of the sampling loop, the update operations can be split; i.e., the error-data multiply can be completed during one sampling period, and the add during the next. This again slows response time by roughly another factor of 2.

In such cases, the original adaptation algorithm has effectively overloaded the processing hardware, unfavorably impacting filter evolution rate. Under certain circumstances where tracking ability of the filter is important, it may be necessary to seek an alternative to update skipping. Clearly, this requires introduction of additional hardware. What are the best ways to structure the adaptive filter to use a second processor? Recall that the output convolution and the weight updating involve roughly the same complexity, i.e., the same number of multiplies, but more data manipulations in the additions. One division of labor might be to dedicate one processor to the output calculation, and a second to the update calculation. The former computes output and error terms and passes the error to the update processor. The second processor reads out the delay line data and performs the product and updating for each weight. In this way the updating is completed in parallel with the calculation of the next output. The two processors share the same data and weight memory, as shown in Figure 7.8, so access must be synchronized. Global memory access associated with single-chip signal processors makes such I/O operations simple. This structure allows adaptation in accordance with the original algorithm, with one subtle difference. The update of weights lags one sample; i.e., output at time k is computed using weights not yet incorporating the output performance at time $k - 1$. As we have come to expect when small adaptation step size is involved, such a small time lag is insignificant when compared to the overall time constants of the filter and makes little practical difference to filter evolution.

Sharing memory resources provides certain practical limitations on this dual processor structure; after all, we have effectively doubled the number of memory

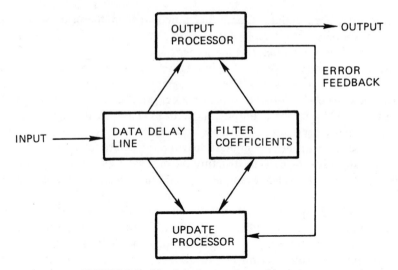

FIGURE 7.8. Distributed processing architecture.

accesses. By further divorcing output and update processors, this concept can be extended to even faster sampling rates. Suppose that each processor were provided its own data and coefficient memory, as shown in Figure 7.9. Note that the output processor draws its coefficients from one of two RAM sections; the second serves as the destination of updated weights. As before, the error sample would be passed from processor 1 to processor 2. Then, each processor operates in parallel without memory access complications. As an update is completed, not only is the new weight written back into the RAM from where it came, but also into the inactive RAM of the output processor. Once the update cycle is completed, the output processor "swaps" the role of its two coefficient RAM sections and uses the new weights for its next output calculation. Again, global memory available to both processors can make interchip communication trivial.

This concept has several ramifications. First, this allows a convenient way to physically segment the hardware, since communications between processors is limited. The FIR filter itself can reside on one board (such units can be purchased off-the-shelf), and the weight calculation can be done on a second. New weights are transferred en bloc.

Secondly, as mentioned before, there is no reason that updated weights need be transferred every sampling period. The output processor can very well use fixed weights for an interval and later accept a new set of improved weights. This leads us to the concept of *block adaptation*. Recall that when speed of update processing was a concern, adaptations were simply skipped, lengthening the convergence rate. With block adaptation, the problem is more one of interboard communication. To minimize the transfer of updated weights between processors, we can keep updating the weights locally, but not pass them on to the filter processor. In effect, subsequent updates are accumulated before an

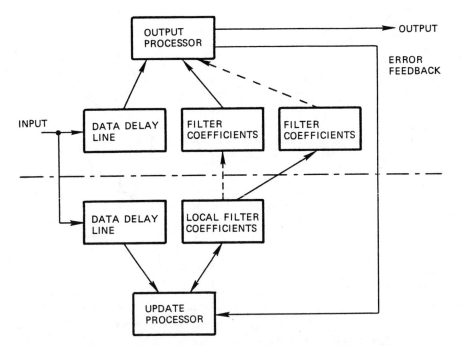

FIGURE 7.9. Divorced output and update processor architecture.

adjustment is seen by the output processor. In terms of the gradient approximation discussion of Section 4.2.2, accumulation of update terms effectively improves the accuracy of the estimate of the error gradient. Consequently, the net update has a lower variance and represents a more accurate adjustment. The rate of convergence is not affected to first order.

Use of block adaptation is especially well-suited to the reverse form of the FIR digital filter, shown in Figure 7.10. Using this architecture allows a convenient means of applying multiple processors to the output convolution, and is thereby useful at very high sampling rates. This form has weights advancing through a delay line, while only a single input sample is present; the N weights multiply the input sample, N multiplies as before, but the accumulation of results is entirely different. Rather than accumulating all N partial products as a single sum, an accumulator is maintained at each node, summing a single product at each sampling period. At any given instant, one of the accumulators holds a complete convolution sum; its result is read as the output sample and is reset to zero. Subsequent output samples commutate from the set of N accumulators. Clearly, this is a convenient means of decoupling the operations of N multiplier/accumulators and of distributing computation. The throughput rate of the filter is then essentially that of the individual processor, e.g., 5–10 MHz for a single-chip signal processor, 10–50 MHz for the LSI multiplier/accumulator.

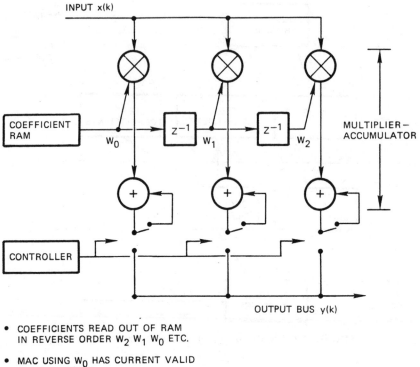

INPUT x(k)

COEFFICIENT RAM

w_0 z^{-1} w_1 z^{-1} w_2

MULTIPLIER − ACCUMULATOR

CONTROLLER

OUTPUT BUS y(k)

- COEFFICIENTS READ OUT OF RAM IN REVERSE ORDER w_2 w_1 w_0 ETC.

- MAC USING w_0 HAS CURRENT VALID OUTPUT SAMPLE

FIGURE 7.10. Reverse form FIR filter.

The same basic architecture also serves the update operation, as shown in Figure 7.11. Thus, a board with modified I/O and control firmware acts as both output processor and update processor, suitable for very high sampling rates. In the update mode, the error signal fed back from the output processor enters to multiply data samples advancing along a delay line. The resulting error-data products are accumulated at each node; each accumulator then contains the respective weight update prior to step size scaling. Step size is incorporated either by scaling of multiplier inputs or scaling outputs from the accumulators. During each sampling period, one weight is read out and passed to the output processor, where it enters the weight delay line. Note that this architecture relies on block adaptation on a weight-by-weight basis; once an updated weight is sent to the output processor, it does not change for N samples. During this interval N updates are accumulated.

These architectural concepts represent a sampling of ways that an adaptive filter may be configured for speed or size efficiency. Note that as general rule, efficiency costs not only filter bandwidth, but convergence rate as well. What else may a designer try to reduce computation time and/or hardware? Often gains can be made by resorting to lower-precision data and coefficients, with the

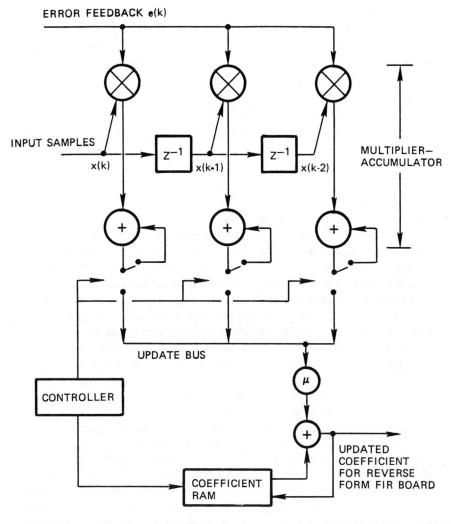

ERROR FEEDBACK e(k)

INPUT SAMPLES

x(k)   x(k-1)   x(k-2)

MULTIPLIER–ACCUMULATOR

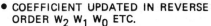

UPDATE BUS

CONTROLLER

μ

+

UPDATED
COEFFICIENT
FOR REVERSE
FORM FIR BOARD

COEFFICIENT
RAM

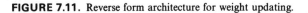

● COEFFICIENT UPDATED IN REVERSE
ORDER W_2 W_1 W_0 ETC.

FIGURE 7.11. Reverse form architecture for weight updating.

obvious cost to output distortion and quantization noise level. Not only does coefficient and data storage require fewer memory chips, but a reduction in multiplier width can result in faster throughput per unit of board area. Depending on the I/O bus of a processor, wider data samples may require multiple accesses; for example, they may allow only one byte at a time.

A discussion of performance and precision issues is found in Section 7.5. Here we are concerned specifically with the effect of precision on update computation.

While modest gains may be possible by use of a lower-precision product of error and data, significant simplification results from use of single-bit update operations, i.e., reduction of precision to the sign bit only. Chapter 4 enumerated the algorithms using "signum-based" adaptation, where the update might involve the terms

$$\text{sgn}(x(k-j))e(k), \tag{7.3.11}$$

$$\text{sgn}(e(k))x(k-j), \tag{7.3.12}$$

or

$$\text{sgn}(e(k)x(k-j)). \tag{7.3.13}$$

In all cases, there is no multiply as such; in the first two forms, the update reduces to an add/subtract. The last form is particularly convenient; the update is simply an increment or decrement operation. Architectures for these algorithms are shown in Figure 7.12. Note that multipliers and multiplier/accumulators have been replaced by signed adders, or in the last case, combinatorial logic driving an up/down counter. Such algorithms may be of particular value for high sampling rates, where use of a full multiply might involve considerable expense. An example of a custom device based on (7.3.13) can be seen in Figure 7.13; it contains a 256 tap adaptive filter using the "sign–sign" update rule, intended for use in automatic echo cancellation.

Not surprisingly, convergence behavior of these forms is degraded by the lower precision. Note that only for the first does the "size" of the error enter into the update, thereby providing finer adjustment as convergence is approached; this retains the flavor of a quadratic cost function with its exponential weight evolution. In contrast, the update associated with the last two algorithms is of constant "size"; near convergence, it is the sign of the error sample that keeps the weight jittering around its steady-state value. This behavior can be likened to that seen with a "linear" cost function, e.g., an absolute value function, where slope magnitude (convergence rate) is constant. Using this intuitive character-ization, we might expect that acceptable behavior at convergence (low-weight jitter) would require reduced step size and an overall loss in convergence rate. Therefore, while "signum-based" adaptation is convenient and efficient in its hardware requirements, it may be of limited use in highly dynamic environments.

7.3.5 Cyclostationary Filtering

Most of the discussion to this point has dealt with signals having a stationary character. Mathematically, this means that the statistics of the signal are constant over time. Of particular interest for the adaptive filtering problem are signal power and time correlation, i.e., the second-order statistics. One of the major attractions of adaptive filters is their ability to track variations in the

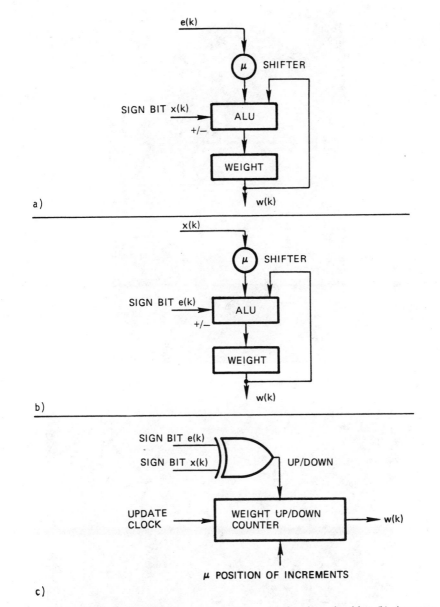

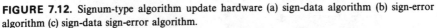

FIGURE 7.12. Signum-type algorithm update hardware (a) sign-data algorithm (b) sign-error algorithm (c) sign-data sign-error algorithm.

signal. We usually assume that such variations occur slowly when compared to convergence modes of the filter, defining the so-called quasi-stationary signal. Such a characterization is suitable for numerous applications involving random continuous waveforms, as found in communication problems.

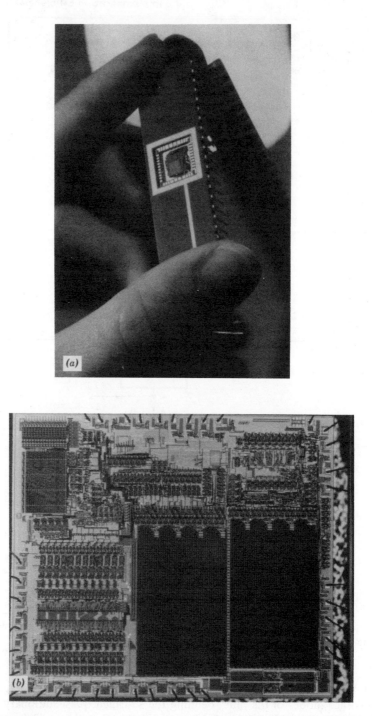

FIGURE 7.13. Adaptive echo cancellation chip (courtesy of Tellabs, Inc.) (a) packaged device (b) microphotograph.

There is a large class of problems involving signals that are not strictly stationary, but with statistics that vary in a periodic or cyclic manner. During a "cycle," the statistics may vary widely, but the variations are the same from cycle to cycle. Sparing any further mathematical qualifications, such signals are referred to as *cyclostationary*. The standard example concerns a periodic pulsed waveform in stationary noise, as depicted in Figure 7.14; measured biological phenomena such as noisy voiced speech are often sources of such data. The wave has a basic period, and at any given instant its variance is constant; the mean traces out the underlying pulse shape.

When the rate of statistical variation is slow compared to the adaptive filter convergence rate, the signal can be treated as a quasi-stationary process. In such cases the filter can track the optimum setting that traces out a periodic trajectory. However, when the signal changes at a rate comparable to the filter time constant, we can no longer expect the normal adaptive algorithm to track accurately the optimal filter settings. In such cases, the filter resorts to some "average" response which typically degrades the quality of the signal.

Fortunately, such problems can be countered by means of *cyclostationary adaptive filtering* [Ferrara, 1981]; we offer an intuitive treatment in this section. For the moment assume that the statistics cycle over a well-defined period. Consider the pulsed signal shown in Figure 7.14. If the filter's task is to isolate the signal and reduce the noise, then it is clear that the "optimum" filter involves some spectral shaping during the "on" time and a zero response during the "off" time. In other words, at a given instant the optimum filter setting is determined by the input statistics at that instant. This implies that in steady state, the filter taps should cycle among a set of weight vectors, each one optimum for a relative position within the period of the statistics. Therefore for a cycle of M samples, there are M vectors of N filter taps each necessary to provide optimal signal recovery. For a given relative position within the period, the best output is computed by using one set of weights, and subsequent outputs by using other sets of weights.

Each such weight vector can be adapted independently, using a cyclic update. That is, each vector is adapted once per cycle following computation of its output. The converged result is dependent only on the statistics of the input at that instant, an invariant. Thus, each of the M filter responses converges to a

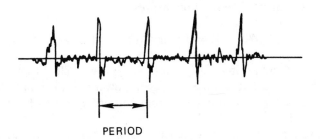

PERIOD

FIGURE 7.14. Cyclostationary process.

unique vector. Mathematically, we define a doubly indexed weight vector, the time index as before, and a subscript to denote the filter position within the cycle:

$$\mathbf{W}_j(kM + j), \qquad 0 \leqslant j \leqslant M - 1. \tag{7.3.14}$$

Note that the time index contains the number of cycles plus an offset within the cycle. As each vector is adapted only once per cycle, we might correctly assume that convergence time is roughly proportional to the length of the cycle time M.

The implementation of a cyclostationary adaptive filter requires only slight modification of the conventional structure. As before, the output is computed every sample, but must involve a different set of weights. These weights are normally stored in small addressable segments of a single RAM, where firmware enables access of the correct segment at the appropriate time. Following calculation of the output error, the update operation reads the correct weight, makes the weight adjustment, and writes back into the same segment of RAM. For the next sampling instant, firmware increments access to the next weight RAM segment.

Use of cyclostationary filters can greatly enhance performance for applications involving periodic signal or noise waveforms. The basic implementational cost is minimal, but does require measurement and tracking of cycle synchronization information.

7.3.6 Algorithmic Leaking

There are certain conditions under which adaptive strategies may prove unsatisfactory, despite the reasonable nature of the environment. Divergence or stalling of weights can often be avoided by means of a "leaking" mechanism used during the update calculation. The benefits of adding a noise to the input of an adaptive filter has been alluded to in Chapter 4; the equivalent effect can be achieved indirectly by an algorithmic adjustment, avoiding unnecessary corruption of the input environment. Specifically, rather than simply adding the update term $\mu e(k)\mathbf{X}(k)$ to the previous weight vector, we first reduce the old weights by a small amount. This action effectively simulates a small "leak" in the integration operation, borrowing from terminology associated with analog operational amplifiers. The resulting expression is

$$\mathbf{W}(k + 1) = (1 - \mu\gamma)\mathbf{W}(k) + \mu e(k)\mathbf{X}(k), \tag{7.3.15}$$

where γ is small and positive such that $(1 - \mu\gamma)$ is very close to unity. This has the same effect as adding a white noise with power γ to the input. Biases away from the optimal filter solution result from nonzero γ, yet can be minimized by specifying $\gamma \ll 1$.

Use of this technique can be beneficial in a number of situations. Conditions where conventional iteration may be inadequate are the following:

Insufficient spectral excitation. When the input spectrum exhibits a wide range of energy densities, the convergence time constants will be correspondingly disparate. In particular, when there are regions for which energy is nearly negligible, adaptive algorithms may respond in odd ways. Absence of energy at a given frequency in the input, and consequently in the output, implies that the filter has no feedback mechanism by which to sense its response at that frequency. Its growth at that frequency may continue unchecked and result in weight divergence. Such divergence is associated with the inversion of the ill-conditioned input correlation matrix.

Finite precision effects. A related effect concerns the quantization of the algorithm computations. The resulting quantization effects may serve as a low-level noise that appears at the output, with no corresponding stimulus at the input. Depending on the specific algorithm, the weights may react in an undesirable fashion.

Stalling or biasing. Weight recursions of Chapter 4 through 6 serve to "integrate" or sum all update contributions, which are derived by means of a performance feedback path. Disturbance of this path (noise, finite precision calculation and storage, transmission errors) may introduce a bias or "glitch" to the weight vector components. Recovery is dependent on the modes of the system, often involving the eigenstructure of the input, and in some cases may come at an insufficient rate. In the case of the ADPCM decoder, which has no true feedback path, recovery will not occur at all; see Sections 5.5 and 8.2. A simple leak factor assures that such errors will decay at an adequate rate.

7.4 FREQUENCY DOMAIN IMPLEMENTATIONS

To this point, all discussion has been limited to the implementation of adaptive filters using time domain techniques. There is a certain consistency and simplicity of the operations that is comforting: a sample enters a delay line, all delay line contents are multiplied in turn by appropriate impulse response or tap values, and the contributing products are accumulated to form the output sample. As we have pointed out along the way, filter performance improves with the number of taps, at a cost in complexity that increases approximately linearly with filter length. For a given bandwidth, long filters may well lead to impractical hardware sizing. It turns out that significant processing efficiencies can be realized by using block-oriented frequency domain operations. The computational efficiency of the fast Fourier transform algorithm, the FFT, reduces the overall complexity of the convolutional sum as well as the adaptation cycle. These ideas were first formalized in [Dentino et al, 1978] and [Ferrara, 1980]. In this section we present an intuitive approach to the development of frequency domain adaptation.

7.4.1 Fast Output Convolution

Recall that we found the bulk of computation is associated with the calculation of filter output. That is, adaptation of weights need be done only as time permits, and stretching out update operations need only affect net convergence rate. For an N-tap filter, the convolution itself involves N multiplies and $N - 1$ adds. (This assumes real data and weights; complex operations implies a fourfold increase in multiplies and doubling of adds.) Therefore, processing N contiguous input samples and generating the N corresponding output samples involves net computation of N^2 multiplies.

In applications involving fixed filtering, significant reduction in hardware can be achieved by using a block-oriented architecture instead of more simple and concise single-sample processing. The costs of block processing, i.e., the addition of memory buffers and more involved control functions, are more than outweighed by the reduction of multiply and accumulate operations. The basic "fast convolution" implementation of a digital filter is shown in Figure 7.15. Details of fast convolution in Fourier analysis are treated in many texts on digital signal processing; see, for example, [Burrus and Parks, 1985]. Here the convolution sum is replaced by a product in the frequency domain, exploiting the property

$$\mathscr{F}[y(k)] = Y(f) = X(f)W(f). \tag{7.4.1}$$

Note that the samples still enter and leave in time sequence form. Incoming samples are written into half of a "double buffer" memory; when a block of N samples is filled, an FFT processor operates on the block to generate complex frequency samples. During this computation, the incoming samples are directed into the second half of the buffer. The filter response is likewise specified in the frequency domain and is applied as a point-for-point complex product with the input spectrum. The block of complex results are then inverse-transformed by another FFT processor and written into an output buffer. At appropriate sample times, each real value is read from the buffer as an output sample. Many

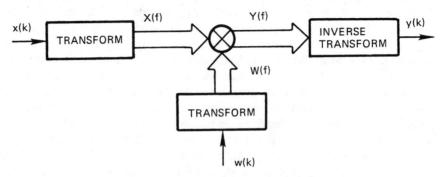

FIGURE 7.15. Concept of fast convolution.

of the architectural features of single-chip signal processors are motivated by the operations associated with the fast Fourier transform. Consequently, very concise transform implementations can be realized with dedicated chips.

It is important to realize that despite its internal differences, this processor is identical to the time domain implementation from an input/output point of view, with one exception. The block-oriented nature of the processor means that there is an inherent propagation delay of at least N samples from input to output port. Consequently, such a filter would normally be inappropriate for use in a feedback loop, as might appear in certain control system functions. On the positive side, the filter response itself is provided in the frequency domain, where its specifications usually originate. In other words, the "filter synthesis" step in the design of a fast convolutional digital filter is trivial. Were the filter's time response designated, we would simply perform a one-time, off-line transform to generate its complex frequency response.

For purposes of this discussion we cite certain key properties of the fast convolution concept. We then elaborate on the importance of these properties in the context of adaptive filtering. First, as a basis for complexity comparison, we note that the number of real multiplication operations involved in an N-point radix 2 complex FFT is $2N \log_2(N)$. (This includes "degenerate multiplies" by ± 1 in one stage.) Secondly, it is important to note that the FFT of an N-point block of real samples can be achieved by a complex $N/2$-point FFT, plus $2N$ additional multiplies. Lastly, we must be aware of the peculiarities of "circular convolution" associated with the periodic nature of the discrete Fourier transform. Avoiding the distortion effects can be assured by proper blocking of input data.

Suppose we wish to apply an N-point digital filter to an input sequence. We might be tempted to partition the data into blocks that are N points in length; that is, the input blocks would be transformed into N complex spectral values, multiplied by N complex frequency response values, and inverse-transformed into N real output values. Unfortunately, the result of this operation is a so-called *circular convolution*; rather than providing the response of the filter taps to an isolated block of N points as it passes down the delay line, this process yields the response to a repeating or periodic sequence of the N points. That is, as point $x(0)$ emerges from the end of the delay line, it reappears at the input. This periodicity is a direct result of the discrete Fourier transform representation; the equations for the inverse transform actually define a repeating time sequence with period N samples.

In order to compute the actual convolutional values, this circular effect must be circumvented by means of double-length transformations. The output of an N-tap filter must be computed using blocks of input $2N$ samples. This follows intuitively since the generation of N sequential output samples requires that the filter have access to $2N$ input values, due to its internal memory. Therefore, the data samples are segmented into blocks of length $2N$ and transformed; the filter impulse response is zero-filled from N to $2N$ values and likewise transformed. The point-by-point product of input transform and filter frequency response is formed, and the resulting $2N$ complex values are inverse-transformed. The emerging time sequence of $2N$ values contains only N valid output samples, that is, the last N points. The first N points, on the other hand, are artifacts of the circular convolution and must be discarded; they represent the convolutional sums had the $2N$ input samples been circularly shifted. To compute the next valid N output points, the next block of $2N$ input samples is defined by shifting over N points, that is, by using blocks of $2N$ samples that overlap by N points. Notationally, this might be expressed as

$$W(f) = \mathscr{F}_{2N}[w_0 \; w_1 \; \cdots \; w_{N-1} 0 \cdots 0]$$

$$X(f) = \mathscr{F}_{2N}[x((k-1)N) \quad x((k-1)N+1) \cdots x((k-1)N+2N-1)] \qquad (7.4.2)$$

$$Y(f) = \mathscr{F}_{2N}[N \text{ discarded values}: \; y(kN) \quad y(kN+1) \cdots y((k-1)N+2N-1)].$$

This is the so-called "overlap and save" fast convolution technique, whereby $2N$ input samples are used to compute N output values via the transform domain, and is diagrammed in Figure 7.16.

A second alternative involves the "overlap and add" technique, where N input values are padded by N zeros, and instead the output values are formed by overlapping and combining sequential blocks of $2N$-point inverse-transformed results. While this technique can be used, it does not fit precisely the adaptive filtering context, where the filter coefficients change from block to block. Again, for further reading on fast convolutional details, the reader is referred to such texts as [Burrus and Parks (1985)] and [Rabiner and Gold (1975)].

OUTPUT BLOCK AT TIME k

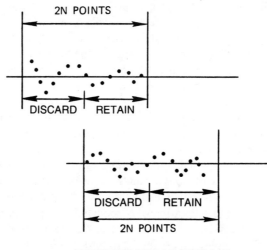

OUTPUT BLOCK AT TIME k+N

FIGURE 7.16. Overlap and save fast convolution.

Consider the net complexity of the fast convolutional filter. For an N-point impulse response, the real input data is blocked into $2N$ samples and transformed. The product with the filter frequency response is performed, and the output sequence is recovered by a $2N$-point inverse FFT; the proper N output time samples are retained. As a basis for comparison, this requires a net $[4N \log_2(N) + 12N]$ multiplies, assuming spectral symmetry is exploited to the full extent. The time domain computation of the same N output samples corresponding to a block of N input samples is N^2; based solely on the number of multiplies, a complexity ratio for radix 2 block sizes can be tabulated:

N	Complexity Ratio (FFT/Time Domain)
32	1
64	0.56
128	0.31
256	0.17
512	0.09
1024	0.05

Thus, for fixed FIR filtering, we see a dramatic reduction of complexity for large filters. Naturally, this disregards the one-time transform associated with the filter frequency response.

Therefore, for sampling rates where the output computation is a significant "bottleneck," replacement of the FIR time domain filter by a frequency domain implementation may be warranted, as shown in Figure 7.17. Note that this exploits the division of labor mentioned in the previous section and allows the slower, less critical update computation to remain in the time domain. This requires that the update processor have access to the input data with sufficient memory to account for the bulk delay in the output calculation. Updated weights are again accumulated by the update processor and passed to the output processor. Each time weights are adjusted, a new frequency response is required, involving another $2N$ point transform.

7.4.2 Fast Update Calculation

Further exploitation of frequency domain techniques allows virtually all operations to be done at a reduced total computational burden. In other words, in situations where the rate of adaptation is critical and loss of update steps is unacceptable, frequency domain implementation of the update processor results in further savings for large filters. Of course, update calculations made in the frequency domain are a form of block adaptation, where N filter taps are adjusted once every N samples.

For each filter update, there are N real multiplies required, that is, the error times the N input samples in the filter delay line. Thus, for a block of N samples,

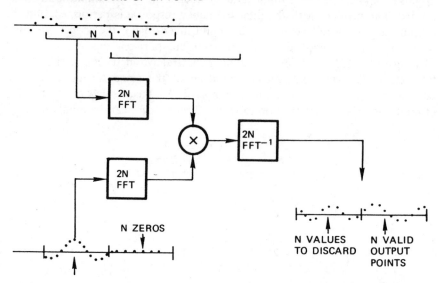

INPUT SEQUENCE
OVERLAPPING BLOCKS OF 2N POINTS

N

N

2N
FFT

×

2N
FFT^{-1}

2N
FFT

N ZEROS

N VALUES
TO DISCARD

N VALID
OUTPUT
POINTS

N-POINT FILTER IMPULSE RESPONSE

FIGURE 7.17. Adaptive filter using fast convolution.

the update process involves N^2 multiplies, assuming that no updates are skipped; this is true for both sample-by-sample updating and block adaptation. Properly viewed, the block adaptation calculation is analogous to the output convolution calculation,

Output at time $kN + i$: $y(kN + i) = \sum_{l=0}^{N-1} w_l x(kN + i - l)$

Block update of lth weight following time $kN + N - 1$:

$kN + N - 1$: $\nabla_l(kN + N - 1) = \sum_{i=0}^{N-1} e(kN + i)x^*(kN + i - l)$;

appropriate substitution of transforms into the previous discussion on fast convolution defines the FFT-based block adaptation of the filter frequency response. In terms of (7.4.2), we have

$E(f) = \mathscr{F}_{2N}[e(kN)\ e(kN + 1)\cdots e(kN + N - 1)\ 0\cdots 0]$

$\tilde{X}(f) = \mathscr{F}_{2N}[x^*(kN + N - 1)\ \ x^*(kN + N - 2)\cdots x^*((k-1)N)]$ (7.4.3)

$\mathbf{V}(f) = \mathscr{F}_{2N}[N \text{ discarded values} : \nabla_0\ \nabla_1\ \cdots\ \nabla_{N-1}].$

Furthermore, use of the Fourier transform conjugation/time-reversal property allows us to note that

$$\tilde{X}(f) = X^*(f)$$

and that

$$V(f) = E(f)X^*(f).$$

This allows use of the input data transform already available, and requires only transformation of the error sequence. The vector to adapt the filter weight values is simply the last N points of the inverse transform:

$$V(kN + N - 1) = \mathscr{F}_{2N}^{-1}[E(f)X^*(f)]. \tag{7.4.4}$$

This N-point vector can be scaled by the constant μ and accumulated into the weight vector as before.

However, rather than updating the weights in the time domain, it is advantageous to apply adjustment in the frequency domain directly. (Clearly in an implementation using the frequency domain output calculation, the weights themselves may need never revert to the time domain.) The gradient can be transformed using a zero-padded $2N$ point FFT and then accumulated according to

$$W(f, k + 1) = W(f, k) + \mu\mathscr{F}_{2N}[\nabla_0 \ \nabla_1 \ \cdots \ \nabla_{N-1} \ 0 \cdots 0]. \tag{7.4.5}$$

This expression is nothing more than a spectral domain version of (4.2.7).

Using this frequency domain implementation, concrete savings in complexity and hardware can be realized. Referring to Figure 7.18, we see that five $2N$-point real FFTs are required, as well as two N-point complex vector multiplications. Net computational complexity for the frequency domain version of the adaptive filter is

$$5 \times [2N \log_2 N + 4N] + 2 \times 4N = 10N \log_2 N + 28N \tag{7.4.6}$$

real multiplies per N-point block. The equivalent complexity of the time domain filter involves simply $2N^2$ real multiplies, leading to a complexity ratio of $(5 \log_2 N + 14)/N$. Tabulated for radix-two block sizes, this implies

N	Complexity Ratio (FFT/Time Domain)
32	1.2
64	0.69
128	0.38
256	0.21
512	0.12
1024	0.063

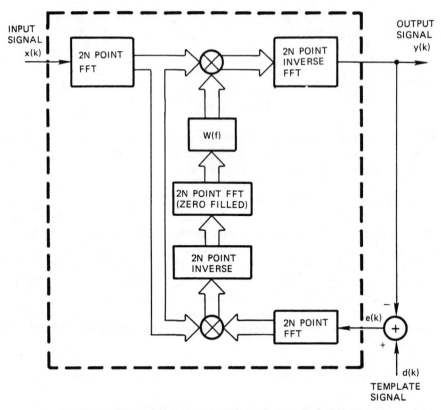

FIGURE 7.18. Frequency domain implementation of adaptive filter.

Note that there is a slight loss over the previous table, due to the fifth transform linking the update and output calculations. While there are subtle costs associated with frequency domain implementations not quantified by this measure (e.g., lack of flexibility, time delays, memory requirements), it is clear that filters requiring in excess of 100 taps may often require simpler hardware using the architecture of Figure 7.18. As a variety of hardware is available that can implement the FFT over a wide range of real-time bandwidths, such an architecture may provide benefits that far outweigh its limitations.

It is important to note that implementation of Figure 7.18 is quite general; it represents the equivalent of (4.2.7), which is itself a general recursive expression. In Chapter 6 it is stated that suitable redefinition of the performance cost function could result in other adaptive algorithms having this general form. Calculation of an alternative performance "error" quantity $e(k)$ can still be easily incorporated in this more efficient implementation. That is, the summing junction in Figure 7.18 where the error $e(k)$ is computed can be replaced by a far more general operation.

7.4.3 Channelized Adaptation

Earlier, it was mentioned that the frequency response of a gradient-based adaptive filter evolves fastest in bands of highest energy concentration. Examination of the frequency domain implementation reveals this phenomenon directly. That is, bands where input energy correlates with error receive the largest perturbations. For example, were a strong undesired narrowband component present in the input spectrum, the filter's response in the appropriate frequency band would be rapidly reduced until the error energy were nulled at that frequency, i.e., $E(f_0)X(f_0) = 0$. Conversely, a low level of broadband noise at the input would be reduced by decay of the filter frequency response in the noise-only band, but at a relatively slow rate. The disparity of time constants (i.e., correlation eigenvalues) describing the convergence of adaptive filter parameters often limits the level of performance that can be achieved. In other words, initial convergence rate and misadjustment effects are dominated by fast time constants, while overall rate of convergence is dictated by the slow time constants. For the most part, the effect of disparate time constants can only be reduced by techniques accelerating convergence by varying the adaptive step size over time.

However, once we resort to the frequency domain implementation, there is an obvious means of improving convergence behavior. The input spectrum has now been "channelized" (i.e., divided into spectral bands), each with its own convergence rate. Now each band can possess its own adaptive step size, dependent on its relative input energy level. By incorporating any a priori information about the energy distribution of the input and desired signal, the convergence modes of the adaptive filter can be compressed to a more reasonable range, thereby improving convergence properties.

Mathematically, the transformation to the frequency domain can tend to decouple the adaptation modes. Recall that in the time domain, convergence dynamics were governed by the input correlation matrix. A set of orthogonal eigenvectors exists that spans the space of all input data vectors. (See Sec. 3.4.4.) A suitable linear similarity transformation of the input, as shown in Figure 7.19, will result in a matrix with the eigenvalues along the diagonal. Once in this new space, each tap corresponds to an eigenvector of the input matrix and converges according to its associated eigenvalue. The filter operates on the transformed data as before, and the output is retransformed back to the original space, if necessary.

Note the similarity of Figure 7.19 to Figure 7.15; i.e., in both cases the data

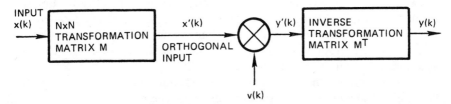

FIGURE 7.19. Orthogonalization of input.

sequence is blocked into N samples, and a transformation is performed. Following filtering and adaptation in this new space, the output is blocked and returned to the original domain. The FFT is merely one means of "orthogonalizing" the input data. It is far preferable to linear algebraic alternatives, which involve not only eigenanalysis of the input correlation matrix, but accurate measurement of the correlation samples making up the matrix as well; it is interesting to note that as the filter length gets large, the Fourier transformation asymptotically approaches a true eigenanalysis of the input correlation matrix [Gray, 1972]. Nevertheless, when input statistics are known a priori, there are adaptive algorithms which approximately orthogonalize the data during the adaptation update; Section 4.3 dealt with a recursive least squares technique using an orthogonalized update calculation.

However, it must be recognized that while the discrete Fourier transform produces a convenient and intuitive dissection of the input signal space (i.e., into frequency bins of the FFT), transformation of any finite record or block of data results in energy leakage between bins. As a simple example, a sinusoid input will result in energy appearing in all bins of the FFT rather than the single bin containing its frequency. (Leakage will not occur only when the sinusoid frequency falls exactly at the center of a bin.) Such leakage represents a small coupling of the dynamics of the adaptive filter in the frequency domain. In certain cases, this small coupling can have a significant impact on convergence modes. For example, consider the same sinusoid component serving as an undesired element of a broadband input. Were we using the frequency domain implementation as described earlier, we might be tempted to assign a small step size to the bin containing the tone and larger step sizes to the rest of the frequency bins. However, leakage from the interfering sinusoid component into adjacent bins may dominate the broadband signal component; this extra energy may induce significant variance of the weight governing this neighboring bin, i.e., performance degradation. In other words, while the frequency domain implementation does provide an efficient architecture, weight behavior is not perfectly decoupled.

For many applications this residual coupling is of minor concern. However, in cases where leakage effects result in a performance degradation, e.g., an array of densely spaced strong sinusoids corrupting a broadband signal, it is often necessary to provide further bin isolation. Essentially, we wish to isolate bands of the input spectrum, passing all energy within the band with no frequency dependence. One simple means of performing this ideal orthogonalization is to use a "comb" of N bandpass filters spanning the input bandwidth. For each band, the magnitude and phase (relative to the filter's center frequency) is determined, weighted by a complex "tap," and used to "remodulate" a corresponding output band of energy. The sum of all such output bands generates the composite time domain output.

It can be seen that the FFT is simply an efficient means of doing the bandpass filtering and reconstruction. However, the equivalent "frequency response" of the FFT is not flat across the bin and does not provide a great deal of band

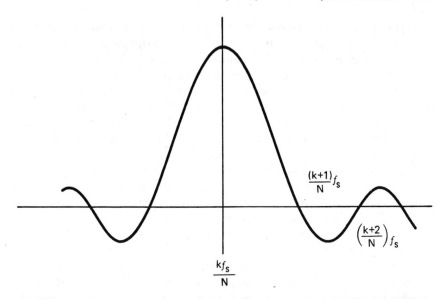

FIGURE 7.20. Approximate spectral response for a bin of an unwindowed DFT: (sin x/x) function.

isolation. The response for the unwindowed FFT is seen in Figure 7.20, i.e. closely approximated by the familiar (sin x/x) function; windowing the data provides some improvement, but may still be inadequate. Fortunately, a means does exist to improve the effective filter response while still exploiting the efficiencies of the FFT. This technique has been popularized by use in the generation and demodulation of frequency division multiplexed basebands, i.e., independent channels stacked in frequency. In the telephone industry, conversion between time and frequency-division-multiplexed formats has become known as "transmultiplexing" and has become an important application of digital signal processing. The details of transmultiplexer theory can be found in numerous publications [e.g., Freeny, 1980; Narasimha and Peterson, 1979; Yam and Redman, 1983].

Suppose that our input were sampled at f_s, and as before we wish to apply an adaptive filter of N weights. To perform the fast convolution, we must block the samples into $2N$ points and transform; the N filter weights are zero-padded to $2N$ values and likewise transformed. Subsequent transformations involve input data blocks overlapping the previous block by N samples. The net bin resolution of this filtering operation is $f_s/2N$. But suppose that the channelizing or bin shaping of the input provided by the $2N$ point FFT were inadequate, and that we insist on lower levels of interbin leakage, while preserving the basic resolution afforded by $2N$ points. To improve the isolation we might work backwards, specifying the bin shape in the frequency domain, and then inverse transforming to determine the weighting sequence necessary to apply to the input prior to the FFT, that is, a frequency-sampling design of the window.

Specifying the shape of the bin response with fine-grain resolution results in a weighting sequence that exceeds $2N$ values, instead spanning $L = 2NQ$ samples, where Q is the integer number of specified frequency samples per $f_s/2N$ interval. Thus, our desire for better channelization has translated directly into a higher resolution spectral analysis, that is, an L-point transform of the input data applied to an L-point frequency response. Again, subsequent transforms must overlap the previous data block by N values. When inverse transformed, N points of the output are valid, as before in the "overlap and save" method of fast convolution.

Therefore, with this initial explanation, the processing burden has gone up with the need for a longer FFT. However, recall that the intent was to channelize the input into only $2N$ bins spaced at $f_s/2N$, not L higher resolution bins spaced at $f_s/2NQ$. The longer transform provides an excess of information, an unnecessary $Q - 1$ bin measurements or channels between each desired channel; this information can simply be discarded. With this in mind, the operations involved in the L-point FFT can be examined and "pruned" to eliminate the unnecessary calculations. It turns out that the longer FFT reduces to the $2N$ point FFT, preceded by the appropriate weighting of L input points, overlapped or "folded" into a single $2N$ point block. The resulting transform provides a channelized version of the input with bin measurements properly decoupled, occurring at a rate of $f_S/2N$.

Once properly channelized, the adaptive filtering operation proceeds as shown in Figure 7.21. An analogous inversion operation recovers the output time domain data from the adaptively weighted, channelized representation. Q sequential time samples emerging from the $2N$-point inverse FFT are interpolated for commutation into an output data stream at the original sampling rate. The interpolation filters applied to each leg of the FFT output must have a "polyphase" relationship. That is, they have identical all-pass characteristics and a flat delay response at exact multiples of $1/f_s$. (For example, the filter for bin 0 data has no delay, and for bin i data has delay i/f_s, etc.) The filter is often derived by decimating the weighting impulse response used to condition data for the input FFT, as shown in Figure 7.22.

Incorporation of data conditioning with each FFT results in an additional QN real multiplies per block of N points. The value of Q is determined by the quality of bin isolation required, i.e. preconditioning filter length, and may range from 3 for an isolation adequate for adaptive filtering applications to 22 for critical telephone transmultiplexer applications. Clearly, this data conditioning can be costly not only in terms of operation complexity, but in terms of memory requirements as well. Nevertheless, it is a powerful means of enhancing resolution of an adaptive filter without increasing its number of adaptive parameters.

A fitting closing comment to this section would note that orthogonalization of the input may have great benefit to convergence properties. We have seen the spectral decoupling using block processing with the FFT, perhaps enhanced by transmultiplexer weighting, allows the disparity in convergence modes to be resolved. That is, each time constant can be controlled by assigning an

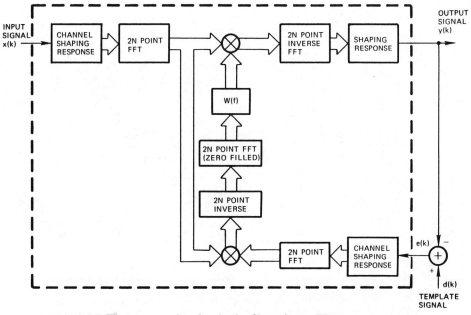

FIGURE 7.21. Frequency domain adaptive filter using an FFT-based filter bank.

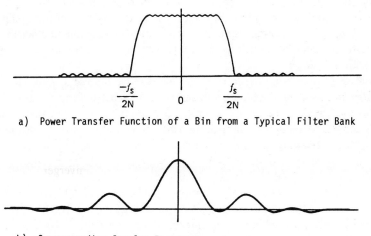

a)   Power Transfer Function of a Bin from a Typical Filter Bank

b)   Corresponding Impulse Response

FIGURE 7.22. Typical shaping function for the FFT-based filter bank and its corresponding impulse response.

individual step size. Mathematical alternatives exist, of course, but are not always practical for the general case.

One very important means of decoupling adaptive modes does lend itself to practice. This concerns an alternative implementation, the so-called lattice or

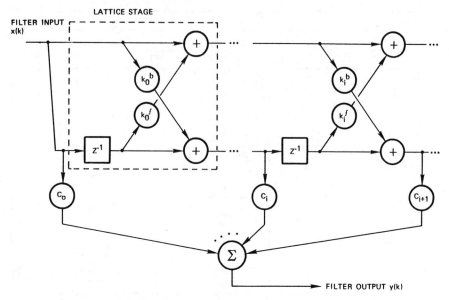

FIGURE 7.23. Structure of an FIR lattice filter.

ladder form for the filter, shown in Figure 7.23. This configuration was shown in [Gray and Markel, 1973] to possess certain favorable numerical properties in terms of coefficient sensitivity. Furthermore, in the context of modeling voice generation, or in autoregressive processes in general, the coefficients may equate to parameters describing energy reflection, and therefore have physical significance.

Adaptive forms using the lattice filter evolved from a recursive least squares solution. In this form, the filter produces the exact least squares solution at each time instant, as described in Chapter 4. To simplify the update calculation, the step size can be made constant as in LMS-type algorithms. By virtue of its architecture involving a cascade of identical sections, the adjustment of each coefficient depends on the output of its section alone; there is no feedback mechanism relating performance of later sections to earlier sections. In a sense, each section acts as a one-tap filter removing the dominant correlation term and orthogonalizing the input one dimension at a time; its convergence rate is not dictated by disparity of input correlation modes. Of course, a section cannot converge until those preceding it have converged. Yet such an architecture has demonstrated favorable convergence behavior compared to equivalent FIR-LMS adaptive filters. [Satorius and Alexander, 1979] discuss a comparison in the adaptive equalization context.

It is worth noting also that this configuration has great appeal for sliced architectures, where a filter can be lengthened by simply adding another section. This is in contrast to FIR implementations, where additional weights impact both delay line and adder tree.

7.5 PRECISION EFFECTS

Arithmetic precision has a profound impact on complexity. For a given processor and memory configuration, increasing precision of data or filter coefficients can often only be accomplished by multiple memory accesses and storage of intermediate partial products. Therefore, an understanding of precision effects on the adaptive filter performance is necessary to a designer for intelligent selection of processor and components. Again, this is a topic that has been documented in many digital signal-processing texts treating hardware considerations; however, fundamentals are included here to motivate the special considerations due adaptive filtering applications.

Let us first consider the output calculation, an FIR filter with finite precision coefficients operating on finite precision data samples.* As noted, FIR convolution is best suited to multiply/accumulate operations, with each product summed into an internal accumulator. The reason for this is simple. A product of an n-bit data word and an m-bit filter coefficient is an $n + m$ bit result. The use of an internal accumulator means the entire product can be retained without prematurely truncating its precision and without resorting to extended-width I/O. For example, the Texas Instruments TMS320 family of signal-processing chips allows multiplication of 16-bit data and coefficients and an accumulation of the full 32-bit results. When the convolution is completed, i.e., N products are summed, the accumulator is normally rounded or truncated back to 16 bits for I/O.

Demonstration of the quantitative concepts behind this convolutional operation are instructive for the understanding of other precision concerns. For this reason, we digress briefly to present important concepts key to precision analysis: dynamic range, scaling, and performance measurement. The *dynamic range* of an n-bit integer sample can be represented on a dB scale, ranging from its maximum at 0 dB down to its 1-bit level at $20 \log 2^{-(n-1)} = -6(n-1)$ dB. Assuming that the signal is reasonably active, spanning several quantization levels, the quantization error will be uniformly distributed over the quantization step size. Therefore, the power level of the quantization noise floor is 1/12 or 11 dB below the 1-bit level. As an example, a 12-bit sample has a dynamic range of 66 dB, with quantization noise floor 77 dB below full scale.

Scaling of data and coefficients to best fit within the dynamic range requires some care. Usually some gain adjustment is introduced prior to the conversion operation, making certain that the input signal does not exceed the maximum voltage of the analog-to-digital converter, that is, its "clipping level." Likewise, each processing stage will call for some form of internal scaling. Clipping a

---

*While the use of floating-point or block floating-point representation often relieves a designer from headaches of scaling fixed precision arithmetic, it is usually unnecessary for the processing of real signals with limited dynamic range; the price of exercising less care in design is a substantial increase in processor complexity and cost. Consequently, we limit our discussion here to integer representation.

waveform at any point introduces a serious nonlinear distortion, which can be viewed as a bursty signal-dependent noise component that destroys the fidelity of the original signal during its peak intervals. We see that the effective noise level of a digitized sample is comprised of a clipping component as well as the basic quantization noise contribution. Design choices should be made that maximize the ratio of strength of the signal to that of the composite noise. Significant reduction of the input signal strength can essentially eliminate clipping effects, but then the strength of the signal is weakened relative to the basic quantization level. Thus, scaling becomes a matter of best fitting the signal between the upper bound (full-scale representation) and the lower bound (the half-bit quantization level). Selecting the signal gain requires an understanding of the signal character.

Consider three common signal examples. Shown in Figure 7.24 are probability densities for the magnitudes of (1) a sinusoid (whose properties are similar to those of data signals), (2) a Gaussian process, and (3) an actual long-term record of speech. In each case, the density is absolutely limited below by zero (i.e.

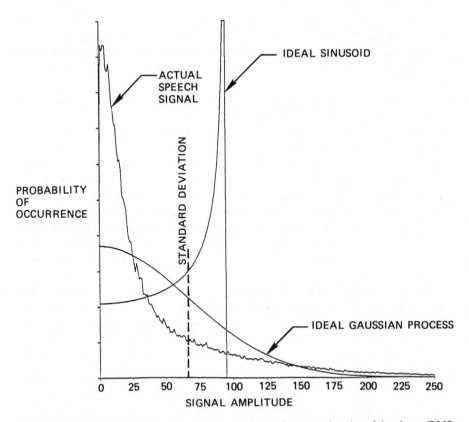

FIGURE 7.24. Probability of occurrence of the signal amplitude as a function of signal type (RMS amplitude of all three signals held constant).

the sample's magnitude), and after peaking, it effectively goes to zero for high values. The tonal case is absolutely limited above by the sinusoid's amplitude; because the waveform flattens only near its peak, the density is highest near this maximum amplitude. (Derivation of the density is a common exercise in most probability texts.) It is interesting to note that the Gaussian and speech processes are distributed similarly, with the notable exception of the long tail due to outlying speech transients associated with "voiced" and "stopped" consonant sounds. (Speech excursions can be modeled to a first order by a Gamma distribution, as discussed in [Paez and Glisson (1972)].) Such densities show how the signal samples are distributed for a given choice of scaling; the dynamic range must contain the bulk of the density. Any tail that extends beyond the upper limit of the dynamic range will be clipped to the maximum value.

Presented in this conventional form, the probability density is of little use for purposes of scaling. Shown in Figure 7.25 for the same processes are the *power density* functions, on logarithmic scales. (For our discussion this function is defined as $x^2p(x)$, that is, the power associated with a differential amplitude interval. The definite integral of this function gives the power of the samples contained between two levels.) Note that not only is the function value on a logarithmic scale, but the abscissa has been expressed in terms of dB, just as dynamic range. To maintain the integral property of the density, the function has been normalized to reflect the abscissa warping. In this form, the power of the clipped samples is the upper tail integral, while the power lost beneath the quantization level is the lower tail integral. All three processes shown in Figure 7.25 possess similar lower tails, reflecting flatness of the probability densities near the zero-magnitude point. The density of the tonal process still peaks at the sinusoid amplitude, and is zero beyond. Also note that relative to its peak location, the speech process exhibits an upper tail that extends several dB beyond that of the Gaussian process.

It is common engineering design to adjust the gain so that the average effect of the clipping noise contribution roughly equals that of the quantization noise; this must be done by means of the analog gain prior to the ADC, as well as for each subsequent digital processing stage. Any scaling choice, for example, the choice of filter coefficients, should reflect this basic rule. In terms of Figure 7.25, this means that the upper tail area should roughly equal the lower tail area, defined by clipping level above and the half-bit level below. The position of the signal within the dynamic range can be quantified in terms of the *crest factor* (CF), defined as the distance in dB between the upper limit (clipping level) and the total signal power. In practical terms, the crest factor is a measure of "headroom" available to accommodate signal excursions.

Thus, scaling is reduced to selection of a crest factor appropriate for the signal characteristics and the precision used in storage and arithmetic. Inspection of Figure 7.25a indicates that for the tonal case, the entire density is limited to 3 dB above the signal's power level. (That is, the tone's peak is $\sqrt{2}$ times its rms level.) Therefore, a tone or data signal can be scaled using a fairly low crest factor, e.g.,

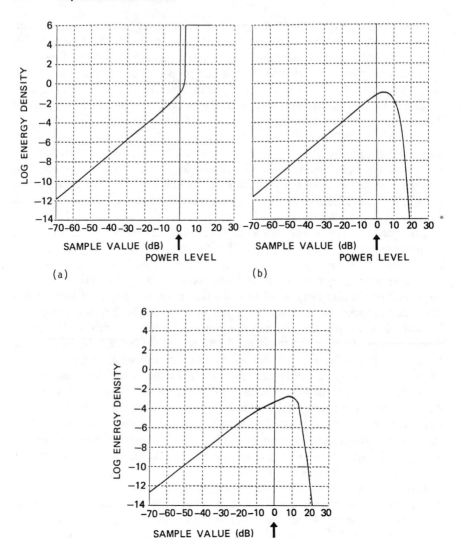

FIGURE 7.25. Log energy density functions of the unity power processes in Figure 7.24 (a) ideal tone (b) ideal gaussian signal (c) long term speech record.

6–8 dB, allowing a margin of safety. For samples represented with lower precision, this safety margin would normally be reduced.

For the Gaussian case, the problem is more complicated. The power density no longer cuts off for high values as in the tonal case, so the crest factor is chosen by balancing the upper and lower tails. In this case, this can be done numerically, to yield the "optimal" crest factor as a function of sample precision:

Number of Bits	CF (dB)
8	11.9
9	12.5
10	13.1
11	13.6
12	14.0
14	15
16	16
n	$\sim n/2 + 8$

Thus, for 12-bit representation, a purely Gaussian input should be adjusted to have its power 14 dB below the clipping level; after sampling, the standard deviation of the resulting samples will be about 407 for the full scale of 2047.

Lastly, consider the speech process. The Gaussian example allows calculation of optimal scaling simply because its mathematical form is known; it provides a baseline table for use with general inputs. Of course, actual signals are seldom characterized so completely, and crest factor selection must be done using rules of thumb. Deviations of a signal from the Gaussian character imply corresponding changes in crest factor. For example, the wider upper tail of the speech process is its principal variation from the Gaussian case. Consequently, we might call for a crest factor several dB above the optimal for a Gaussian signal; in this case, crest factor might be 16–22 dB depending on the arithmetic precision. As the signal becomes "more Gaussian," say, due to superposition of several indepndent sources, the crest factor can be correspondingly relaxed.

Measurement of performance or signal quality can be inferred from the ratio of power in the signal component to the power of all degradation. To diagram the relationship of the digitized signal to its corruption, the schematic representation shown in Figure 7.26 is used. Shown is the dynamic range in dB available for representation of a finite precision signal; full scale is shown at the 0-dB level, with the half-bit quantization noise "floor" shown at the bottom. When scaled using a given crest factor, the signal's rms level appears CF dB below the 0-dB level. Clipping distortion, not explicitly shown on the figure, is assumed to contribute no higher power than the quantization component, as would be the case for proper choice of crest factor. The signal-to-noise ratio is then just the dB difference between the signal level and noise floor. However, since both degrading components, quantization noise and clipping distortion, are signal-dependent, special consideration must be paid to evaluation of signal quality; the simple power ratio is of limited value.

A more meaningful measurement of signal-induced noise effects can be done by drawing on techniques developed for evaluating linearity and fidelity of analog devices. The noise power ratio (NPR), mentioned in Chapter 6, is a measurement of signal power to noise floor made with a specific input. A

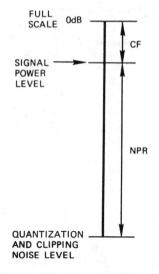

FIGURE 7.26. Input level diagram.

spectrally flat input, usually Gaussian noise which covers the band of interest, is passed through the circuit. At some frequency in the band, the input is notched with a very narrow and deep reject band. Nonideal effects in the circuit induce intermodulation terms that show as a noise floor. The depth of the notch that remains in the output is defined as the circuit NPR and provides a quantitative measurement of circuit fidelity.

In the same way, the input to a digital system can be notched and processed to determine the degree to which nonideal arithmetic effects degrade the noise floor. Often this is a more meaningful measurement of performance than SNR because it is more readily measured and concerns realistic broadband input excitation at the specified crest factor scaling. Figure 7.27 shows an example of an NPR test involving a 16-bit fixed-point implementation of a decimating lowpass filter. Figure 7.27a shows the spectrum of an infinite precision NPR sequence; Figure 7.27b shows the resulting output for the fixed-point lowpass filter. Note that the notch has been reduced from an "infinite" depth to about 70 dB due to nonideal noise contributions. Such testing is commonly used to characterize a digital system from end to end to verify that precision and scaling have not induced unacceptable levels of noise. Note that the NPR is very close to the SNR when the test signal is spread uniformly across the entire sampling bandwidth. For cases where the signal band is less than the sampling bandwidth, an adjustment factor relates the two.

Let us return to the arithmetic noise effects that specifically degrade a convolutional sum and determine the impact on NPR performance. The desired

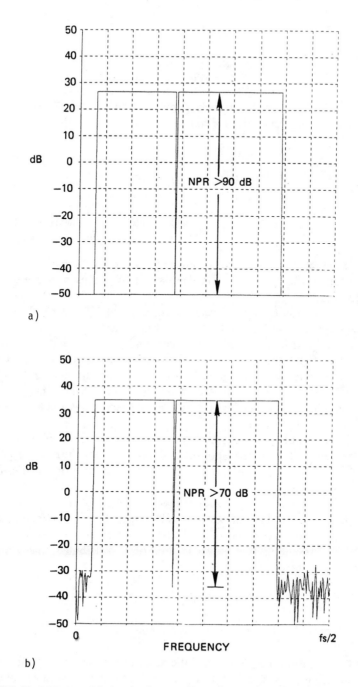

FIGURE 7.27. NPR test of decimating lowpass filter using 16-bit coefficients (a) input spectrum for NPR test (b) output spectrum for NPR test.

output for an FIR filter is expressible as

$$y(k) = \sum_{l=0}^{N-1} w(l)x(k-l). \tag{7.5.1}$$

A simple model assumes that each element of the sum is subject to some quantization error, denoted as $\delta w(i)$ and $\delta x(k-i)$, and that the output is actually

$$y(k) + \delta y(k) = \sum_{l=0}^{N-1} [w(l) + \delta w(l)][x(k-l) + \delta x(k-l)]$$

$$\cong \sum_{l=0}^{N-1} w(l)x(k-l) + \sum_{l=0}^{N-1} \delta x(k-l)w(l) + \sum_{l=0}^{N-1} \delta w(l)x(k-l), \tag{7.5.2}$$

neglecting the second-order error product. To this point we have assumed that full precision of the output sum is maintained by means of a double-length accumulator.

Actual calculation of signal level, noise floor, and NPR depend on specific filter characteristics. We can only make certain general statements that can be useful to a designer in assessment of an implementation. Signal power, i.e., the strength of the first term, is determined by the input signal power (i.e., $-CF$) and by the effective filter passband gain. Adjustment of the output crest factor, that is, output power level, is accomplished by filter gain setting. The level of the output noise floor may depend on the number of calculations and their precision as detailed in the following.

The second term in (7.5.2) is simply the input quantization noise shaped and scaled by the specified filter response. Its level relative to the signal power is nominally the same as that at the input, with minor improvements afforded by the filter selectivity.

The third term in (7.5.2) has a dependence on the number of coefficients and their accuracy; we can think of each tap as having an associated noise generator each with the same power, that of the half-bit quantization noise [Rabiner and Gold, 1975]. Assuming that noise is uniformly distributed, the composite strength seen at the output is

$$NP_s E\{\delta w^2\} = (N/12)P_s \tag{7.5.3}$$

or $10 \log(N) - CF - 11$ dB, where N is the number of taps and P_s is the signal power at the input.

Figure 7.28 shows the level schematic associated with an FIR filter output calculation using 16-bit coefficients and data. On the left, 7.28a shows the input level diagram with the crest factor of 16 dB and the ideal noise floor at $-6 \times 15 - 11$ dB $= -101$ dB. The net NPR at the input is then -16 dB $+ 101$ dB $= 85$ dB. Figures 7.28b and 7.28c show the corresponding

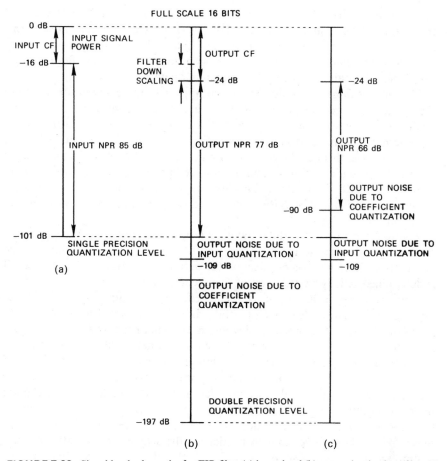

FIGURE 7.28. Signal level schematics for FIR filter (a) input level (b) output levels, short filter (c) output levels, long filter.

output level diagrams for a short filter (small *N*) and a long filter (large *N*), respectively. In both cases, the signal power appears at a level that has been reduced by 8 dB, reflecting a downscaling by the filter response for this example. For this hypothetical case, the output crest factor has been improved to 24 dB, most likely an excessive value. (In certain cases such increases may be necessary. For example, the filter may isolate an individual component that requires a high crest factor when viewed by itself.)

When the convolutional sum is calculated using a multiplier with an internal 32-bit accumulator, as found in the Texas Instruments TMS320 family, the internal quantization floor is far below the level of the input quantization noise, shown in Figures 7.28*b* and 7.28*c* at −197 dB. When transferred out of the accumulator, the result is rounded to 16 bits, and the inherent quantization floor is returned to the −101-dB level. At first glance it might seem that discarding the lower 16 bits has increased output noise level by 96 dB and thereby degraded the output NPR a similar amount. However, the self-generated noise terms associated with the convolutional sum (7.5.2) actually set the dominant noise floor far above this inherent double-precision quantization noise floor, so that very little if anything is lost by the rounding operation.

The first of these two noise terms that appear in the convolutional sum of (7.5.2), i.e. the filtered input noise, is shown at a reduced level at the output, and again reflects some degree of band isolation by the filter. Recall that the second noise term is dependent on the number of taps in the filter. In Figure 7.28b, the short filter case, this noise term is shown at a level below its companion noise component; rounding of the accumulator to 16 bits then raises the effective noise level to the -101-dB level, in this case roughly 8 dB above the actual noise present in the signal. The NPR is degraded to 77 dB due to the 8-dB loss in signal power.

Figure 7.28c, involving a filter having a large number of taps, shows the level of this second noise term at about -90 dB. In this case rounding of the accumulator to 16 bits has no effect on the output noise level; the NPR of 66 dB is determined strictly by the level of this second term.

Using this approach, we can see the net effect of truncating prematurely, i.e., prior to accumulation. Had each partial product term been truncated to single precision, the individual noise level would have been unchanged from the input; accumulation of N such terms would have meant a net roundoff degradation of $10 \log N$ dB worse than proper accumulation and truncation.

For fixed filters, these concepts are fairly well defined. However, for adaptive filters, the feedback path makes scaling a far more complex issue. The location of the filter output within its dynamic range is given by the yet-to-be-determined filter coefficients, which in turn depend on the level of the error feedback. In a sense, for the LMS algorithm in particular, the scaling of the entire process (output and filter gain) is set by the scaling of the "desired" signal. Reducing the power of $d(k)$ (increasing its crest factor) reduces the gain demand on the filter, thereby reducing the actual size of the tap values.

Overflow of filter coefficients is naturally something to be avoided, and it usually pays to maintain a fairly conservative approach to scaling. (Actually, saturated tap values clipped at the maximum magnitude degrade performance in a graceful manner; however, allowing taps to overflow and "wrap around" invariably proves catastrophic.) In many applications, the adaptive filter's job is to excise some offending additive component from its input, and as such it reduces the original incoming power level. Were it further expected to compensate the loss and restore the component of interest back to a favorable CF at the output, weights might well saturate. Therefore, the desired component should be applied to the adaptive filter scaled down from its optimal crest factor. At first glance, it might seem that NPR will necessarily be degraded by such an approach. While this may be true, improperly increased scaling can do nothing to enhance the NPR of the component of interest.

When updating weights according to (4.2.7), it is again necessary to make provision for ample precision in the accumulation. That is, the product $x(k)e(k)$

defines a double precision number, and further downscaling by μ extends the dynamic range further. This result must then be added to the original stored weight value to form the updated value.

We are faced by the same fundamental problem as in the convolution calculation. Insufficient precision provided in the weight register will be responsible for a performance degradation. Only in this case, it is not evident as an increased noise floor, but rather as a coefficient bias or stalling of convergence. Stalling of weights short of the optimal setting is then responsible for excess error.

Recall that, at startup, weights are normally far away from their optimal settings, thus generating a sizable error. For a given weight precision and step size μ the update term decreases and eventually achieves a near-zero mean. This does not mean that the weight has the optimal setting, but rather that it varies around the proper value. When weights jitter due to such ongoing adjustment, they degrade performance or error power by misadjustment noise, proportional to μ. Theory tells us to reduce step size μ to the point where a suitable trade-off of misadjustment noise and fast response time is reached [Widrow et al., 1976].

However, there is another factor entering into selection of step size. At some point the update term will "slide off" the bottom of the allotted dynamic range of the accumulator used for weight updating. In effect, $e(k)x(k)$ is truncated to zero and the weight stalls at a location dependent on its initial value. In reality, since $x(k)e(k)$ is a random quantity, stalling occurs gradually; fewer and fewer values exceed the quantization level and continue to prod the weight. As the weights improve and the error term decreases further, eventually the weight must stop changing almost completely.

A suboptimal weight setting may or may not result in a significant performance degradation. This largely depends on the processing configuration and bandwidths of signal components. Before considering an example, a discussion of reasonable design steps is in order.

Clearly, to minimize bias effects due to stalling, we simply need to make the weight accumulator longer, i.e., tolerant to small update contributions. Practicality, however, requires that a reasonable length be maintained to simplify hardware. (Extended precision accumulation often requires multiple memory accesses and objectionably wide adders.) It should be noted that block adaptation with its update accumulation accomplishes a degree of extended precision, thereby reducing the stalling effects. Thus, a designer must examine the application carefully and weigh the performance sensitivity to weight bias error against the misadjustment error seen with larger step size μ. This will be made clearer by an example.

For these reasons the weight storage precision may well exceed that of the input data. In such cases, the output convolution may use only the upper portion of the weight. Therefore, only when the accumulation of small update terms carries a change into this upper part will the output calculation use a new weight.

A simple example serves to illustrate most effects governing choice of step size. We consider the sinusoid cancellation problem, where effects are pronounced and easily characterized. Suppose the primary input $d(k)$ contains a dominant sinusoidal $A \cos(\omega_0 k)$ and the reference input to the filter $x(k)$ contains the same tone at a different amplitude, $a \cos(\omega_0 k)$. Primary crest factor is set to 9 dB, that is, A equals one-half of full scale. A single weight w_0 is adapted to minimize the error power

$$e(k) = (A - w_0 a)\cos(\omega_0 k)$$

$$w_0(k + 1) = w_0(k) + \mu e(k)a \cos(\omega_0 k)$$

$$\sim w_0(k) + \mu[A - w_0(k)a]a/2 \tag{7.5.4}$$

neglecting the double frequency product. In time, and with infinite precision, the weight converges to A/a and provides perfect cancellation.

However, suppose this were implemented in fixed-point arithmetic with n bits. At some point, update term $\mu e(k)x(k)$ drops below the quantization level, i.e.,

$$\mu(A - w_0 a)\frac{a}{2} \leqslant 2^{-n+1}, \tag{7.5.5}$$

at which point

$$w_0 \approx \frac{A}{a} - \frac{2^{-n+2}}{a^2 \mu} \approx \frac{A}{a} - \delta, \tag{7.5.6}$$

where the weight error $\delta = (4/a^2 \mu)2^{-n}$. The resulting error, that is, the amplitude of the residual sinusoid, is

$$a\delta = \frac{4 \cdot 2^{-n}}{a\mu}. \tag{7.5.7}$$

Cancellation, quantified as 20 log (amplitude before/amplitude after), is then

$$20 \log \frac{A}{4 \cdot 2^{-n}/a\mu} = 20 \log 2^{n-2} + 20 \log Aa\mu$$

$$= 6n - 12 + 20 \log Aa\mu \text{ dB.} \tag{7.5.8}$$

For 10-bit weights with $\mu = 2^{-4}$ and $A = a = 0.5$, this provides only 12 dB of cancellation, or 75 percent. Increasing μ to 2^{-3} improves the suppression by another 6 dB.

For this example the countering effects of larger μ can also be shown in closed form. Suppose that underlying the primary input with its dominant tone is a noiselike process of interest with power σ_s^2. This randomness is present in the error, and consequently serves to jitter the weight. This case has been examined in detail in [Glover, 1977; Treichler, 1979]. In infinite precision, the weight converges to a random variable with mean A/a and variance $\mu\sigma_s^2$; the amplitude of the residual sinusoidal component is likewise random with variance $a^2\mu\sigma_s^2$. This leads to a clear trade-off: Large step size generates an undesirable error due to weight jitter; and small step size is responsible for error due to weight stalling. In this simplified case, the effects can be quantified, but in general (longer filters and correlated signals) the exact expressions cannot be found in closed form. However, the relationships, and thus the compromise, with μ still hold, and an experienced designer can form fairly reliable choices for μ and n, dependent on application details. For example, depending on the strength of the error waveform expected at convergence, the length of the weight accumulator can be specified. In applications where larger errors may be acceptable it may need no more than the filter input precision; in other critical cases where residual error may be very small, weight storage may require in excess of the twice the data precision. Likewise, choice of step size should be such that the update term at convergence drops no closer than 2–3 bits to the accumulator's quantization level.

We conclude our example by demonstrating certain design choices that result.Certain key remaining parameters need to be specified. Primary input has a signal-to-interference ratio (SIR) of -14 dB with a CF of 9 dB ($\sigma_s^2 = 0.005$ and $A = 0.5$); we wish to eliminate essentially all of the tonal interferer so that we know the acceptable error power is $\sim\sigma_s^2$. A CF of 9 dB at the filter input means that the input level a is 0.5. In this example, cancellation must be done quite accurately for significant suppression to be achieved, so we might expect that a high precision accumulator will be required. If the input were provided with 12 bits of precision (66-dB dynamic range) cancellation will be limited to 57 dB (the original input level of -9 dB reduced to the quantization level) given infinite precision updating. Were $\mu = 2^{-8}$, i.e. an 8-bit shift, the misadjustment error would limit cancellation to only 47 dB; decreasing to $\mu = 2^{-10}$ allows 53 dB. However, this level can only be reached if the accumulator has adequate precision. In this case $n > 22.8$ bits, nearly twice the input data precision. For convenience in operations, this would be rounded to 24 bits, or double precision. Thus, weights would be maintained through the update process in double precision, but only the most significant half would enter into the output calculation.

This example is not meant to imply that extra weight precision is always required. However, it does show that instances exist for which an adaptive filter will be severely constrained by truncation of weight precision. For a further discussion on precision effects, see, for example, [Caraiscos and Liu (1984)].

7.6 ANALOG ALTERNATIVES

Recall that for purposes of discussion in this chapter a measure of complexity was defined in terms of net multiply rate, only because the size and power of digital circuitry is dominated by multipliers, augmented by memory control and multiplexing hardware. There are situations where this complexity measure maps to a device that is undesirable: (1) cases where performance requirements do not require the accuracy and flexibility of digital hardware, and (2) cases where bandwidth does not lend itself to acceptable forms of digital hardware. In such applications a designer may wish to consider analog or hybrid implementations.

The principle attraction of digital hardware is its accuracy; that is, it operates repeatably just as theory predicts. Analog devices are notorious for nonideal effects, e.g., trim adjustments, ripple, reflected energy, and frequency rolloff, which make them less than desirable for applications requiring accurate placement of resonances and notches. Thermal effects and aging further serve to complicate critical designs. However, in an adaptive mode, where there is an effective feedback closure around the filter, nonideal effects tend to be compensated, although leakage of analog weight storage is an unavoidable phenomenon. In a sense, an adaptive configuration optimizes performance of its components even when subjected to such handicaps. Assuming that reasonable care is taken in design to minimize the severity, a perfectly workable analog or hybrid system may be realizable with considerable lower power, size, and/or cost.

The basic analog building blocks include some form of signed product, e.g., a four-quadrant attenuator or mixer, summing amplifiers, and some form of "delay." This last element is loosely defined to be any component that provides some distinction between its input and output and involves some dynamic (and hopefully linear) time response. Its function is to generate a linearly independent set of observations of the input process that spans the space of the desired waveform.

Figure 7.29 shows a typical analog adaptive filter. Note that the "delay line" of the FIR structure may involve anything resembling a delay element with a suitably wide bandwidth. The weighting operation in the convolution is performed by analog mixers, whose outputs are summed. The error waveform then feeds back into other mixer components, and each is integrated to generate a voltage corresponding to a weight.

Recall that in any adaptive implementation the principal source of complexity is the output convolution; the update process can be performed at a pace permitted by the hardware. The analog implementation can be tailored to take advantage of this fact. Instead of a totally analog circuitry, the output calculation can be controlled by digital feedback. That is, the output (or error waveform) can be digitized, as can the delay line nodes, and the weights maintained and updated using the digital techniques discussed earlier; digital-to-analog converters provide the weighting voltage to the tap mixers. Block adaptation and/or sign-bit adaptation can make such adjustments suitably simple that their hardware contribution can be minimal. The advantage is that the accuracy afforded by even this fraction of digital hardware improves convergent behavior.

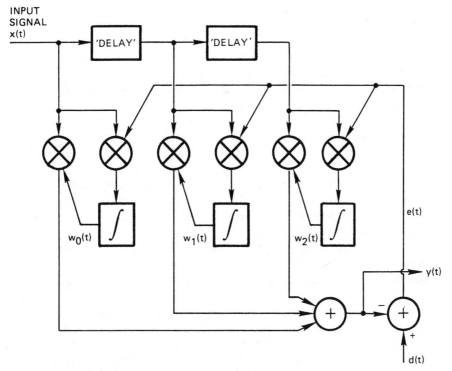

FIGURE 7.29. Block diagram of a typical analog adaptive filter.

Such techniques have their place for low-complexity, low-bandwidth applications, as in the hum cancellation problem. On the other hand, even while
development of digital hardware and associated software may be more costly,
the recurring cost of production is most likely considerably less due to simplified
testing. Consequently, a designer must also weigh volume of production.

For the other extreme, there may be no other practical alternative. For
equalization or cancellation of wideband signals, for example, RF PCM signals
in excess of 40-MHz bandwidth, a totally digital implementation might well
require racks of equipment, whereas an analog system could occupy several
boards. Such systems have found use in microwave systems for equalization,
sidelobe cancellation, and beam steering.

7.7 IIR CONSIDERATIONS

The attentive reader will note that this chapter has discussed only implementations involving the FIR filter structure. For the most part, this reflects
the fact that the overwhelming majority of current applications involve an FIR

implementation whereby the present competitive constraints on performance and complexity are best met. As discussed in earlier chapters, the FIR structure has the definite advantage of guaranteed bounded-input, bounded-output stability, which greatly simplifies specification of adaptation rules. Yet there are selected instances where an adaptive IIR filter appears to offer significant theoretical advantages; however, a designer must be aware of certain subtle practical costs associated with the IIR filter.

For example, suppose that the optimal processing filter in a given application were to achieve a narrowband resonance or passband; in such a case, an FIR implementation might require an extremely large number of taps to emulate a long impulse-response duration. Refer to the introductory paragraphs of Chapter 5. Such a structure not only would require a good deal of hardware for storage and calculation, but would no doubt be sluggish in its behavior. On the other hand, an IIR structure with the proper number of poles might require only a few adaptive parameters, and in a sense greatly reduce the basic complexity.

Overlooking the complications associated with adapting parameters which define pole locations (detailed in Chapter 5), there are other problems inherent in the feedback structure. First, accuracy of tap values becomes far more critical in cases where feedback is involved. In situations where a sharp resonance is of interest, the corresponding pole position will be near the unit circle. As tap values dither around their optimal values, this pole will tend to wobble in radius and may actually enter the region of instability. Proper algorithm behavior will prompt recovery from this momentary foray, but perhaps only after the associated transient is evident at the output. The basic sensitivity of pole positions to tap values becomes more pronounced for higher-order denominators and for poles near unity radius. From a hardware point of view, this would mean not only that storage of feedback taps require extra precision, but also that the output calculation would need to be done with a wider multiplier/accumulator. Thus, a single-chip signal processor using finite precision arithmetic might well be limited to low-order adaptive IIR implementations.

Secondly, the IIR structure tends to aggravate the self-generated computation noise associated with multiplier roundoff. When poles approach the unit circle, the effective noise seen at the filter's output exhibits spectral emphasis near the resonance [Rabiner and Gold, 1975]. Such noise properties tend to counterbalance much of the gain in hardware by costs in performance.

One interesting use of adaptive IIR filters occurs in conjunction with an FFT-based filter bank. [Bershad and Feintuch, 1980] used one-pole LMS-adapted IIR adaptive predictors at the outputs of the filter bank to improve the degrees of spectral resolution attainable. This technique was refined [Shynk and Gooch, 1985] considerably by using an accelerated algorithm to adapt the coefficients of second-order IIR filters.

7.8 SUMMARY

In this chapter we have presented certain practical aspects of adaptive filtering. Probably the most important points of this chapter can be summarized as shown in Table 7.1. This provides the designer with an overview of the implementation possibilities and their attributes. Each of the five implementations is compared in terms of speed, performance, and cost of development, production, and modification. We see that software approaches (the general-purpose CPU and the single-chip signal processor) have favorable performance and cost attributes and currently service low- to medium-bandwidth applications. For wider-bandwidth signals, discrete special-purpose digital designs can perform equally well, but at significantly higher cost figures. Higher speed and lower production cost can often be realized using custom integrated digital designs, but with extremely high development and modification costs. At the high extreme, we find the analog approach as a relatively high-cost and lower-performance means of implementation.

As digital technology improves in cost and speed in the next decade, we will find that the absolute definitions of bandwidth capability will expand. Nevertheless, the use of the general-purpose CPU will most likely remain limited and give way to the highly flexible, low-cost single-chip signal processor alternative.

TABLE 7.1. Comparison of the Attributes of Various Adaptive Filter Implementations

Implementation	Real-Time Bandwidth per Device	Incremental Cost	Initial Development Cost	Cost of Modification	Preservation of Signal Quality
Programmable Signal-Processing Chip	$\leqslant 50$ kHz	Low	Medium	Moderately low	Good
Discrete MSI/LSI/VLSI Components	$\leqslant 20$ MHz	Medium	High	High	Good
Custom VLSI Devices	$\leqslant 40$ MHz	Very low	Very high	Very high	Good
Analog Components	$\leqslant 500$ MHz	High	High	High	Poor to moderate
Computer Software	$\leqslant 1$ kHz	Medium	Low	Low	Good

Many of the applications now serviced by discrete special-purpose digital hardware will likewise be taken over by the signal-processing chip, due to cost and flexibility advantages. On the other hand, the more costly digital designs with their improved speed will also gain ground on applications now reserved for analog systems. The ever-increasing demands on bandwidth in communication systems will most likely provide the only applications requiring analog processing.

8

Design Examples

●*PRECIS   Representative applications spanning a wide range of bandwidth and complexity, first mentioned in Chapter 1, are developed in further detail. Design trade-offs and system considerations are discussed for each adaptive solution.*

8.1 INTRODUCTION

Recall that Chapter 1 presented several example applications for which adaptive processing has proven useful. Each case involved a phenomenon which was unknown and slowly time-varying and required a self-tuning filter to adjust to such changes. This chapter revisits these three cases and elaborates on considerations that enter into a practical design. While we do not attempt to weigh all implementational alternatives, these examples serve to demonstrate the evaluation of a design and the ultimate complexity required for favorable performance. The examples have been intentionally chosen as a set that spans the range of bandwidth and circuit complexity.

8.2 HUM REMOVAL FOR ELECTROCARDIOGRAM MONITOR

In Section 1.1 we introduced general concepts of adaptive cancellation of tonal interference. Here we concentrate on details associated with biomedical instrumentation and outline a specific functional design. In light of the computational requirements of the hum canceller, an assessment of the necessary hardware is given.

229

Measurement of biological activity by means of monitoring electrical discharge, as typified by the monitoring of heart patients, parallels the communication problem: A transmitter (electrical discharge) radiates energy through a propagation path (tissue) to a receiving antenna (electrode) positioned to maximize energy reception. Because the electrical discharge involves very low potentials, the received energy is very weak and requires care to prevent degradation of the signal content by added noises or filtering. Probably the strongest source of signal interference is 50/60–Hz pickup and its harmonics emanating from nearby electrical equipment such as lighting and instrument power supplies. Conventional means for dealing with such strong, concentrated interference is a fixed lowpass filter, which sacrifices waveform detail associated with spectral components above 50 Hz. Use of a notch filter suppressing energy in the appropriate narrow band represents an improvement; however, it still distorts the signal component of interest.

Chapter 1 described a means of removing the additive interference, not by filtering the signal path, but by coherently subtracting a replica of the interference waveform. This noise-cancelling approach to the problem requires a very accurate match of the replica to the actual interference to achieve adequate suppression. For 30 dB of suppression, the match in amplitude must be better than 3 percent, with a phase match better than 2°. To account for variations in frequency and amplitude of the stray interference, an adaptive filter can be used to adjust the phase and amplitude to maximize cancellation. The concept is shown in Figure 8.1.

Consider the development of a circuit for use with off-the-shelf medical instrumentation. It must provide significant rejection of the offending 50/60-Hz (nominally) component, say in excess of 30 dB, and provide the enhanced output in real time; the goal is to degrade the signal as little as possible. The processed signal is normally displayed using cathode-ray tube (CRT) deflection, so must be available in analog form. (A digital port would perhaps be attractive for further computer monitoring and analysis.)

There are two operational/economic concerns. First, setup should be as simple as possible, with no adjustments unfamiliar to the competent medical technician. Adequate internal self-testing is necessary to assure the user of proper operation. Secondly, as such signal enhancement might initially be viewed as a luxury, manufacturing costs, i.e., system complexity, should be kept as low as possible. These concerns are very important when developing instrumentation for a nonengineering field; potential for sales volume may exist, but only if the design is kept attractively simple.

Fortunately, this problem lends itself to a simple solution. First, its bandwidth is low, so that we have a wide selection of devices over which we can minimize costs and complexity. Arguments can be made that information of interest for the EKG waveform extends no higher than 100 Hz; a sampling rate of 512 Hz would be adequate, i.e., high enough to satisfy the Nyquist condition, and simplify specifications on the antialiasing filter yet low enough to keep processor complexity down. Secondly, the offending interference is made of up a dominant 50/60-Hz component, a second harmonic (from rectifying power

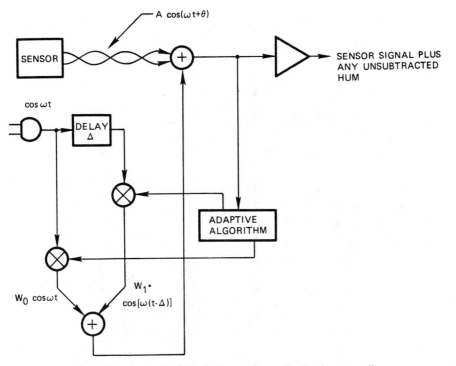

FIGURE 8.1. Simplified block diagram of an adaptive hum canceller.

supplies), a third harmonic (from nonlinearities in motors and transformers), and perhaps some low-energy higher harmonics. As implied by discussion in Chapter 3, the cancelling filter requires two degrees of freedom for each sinusoid (one for amplitude setting, one for phase alignment). Thus, six to eight active weights are probably adequate. Thirdly, the source of the reference waveform, serving as the filter input is conveniently at hand; the power supply of the adaptive canceller itself can be tapped to provide a 50/60-Hz source, at exactly the same frequency radiated by nearby equipment. Simple harmonic-generating distortion (e.g., a diode) can provide all spectral lines of interest to the filter. Lastly, the waveform of interest is always present and its amplitude and frequency vary slowly, so speed of adaptation is not a primary concern. Fast algorithms are unnecessary, and the simplicity of the LMS implementation is another place where significant savings can result.

To a first order the net complexity of this device is described by its multiply throughput. Eight weights at 512 Hz, each updated once per sample, lead to a multiply rate of under 10 kilomultiplies per second (k*/s). Since development and production costs are important and the bandwidth is reasonable, the single chip signal processor represents an ideal device around which to base such an instrument.

As pointed out in Chapter 7, the tone cancellation problem has been studied in detail and offers a number of closed-form design relations. First, it is necessary

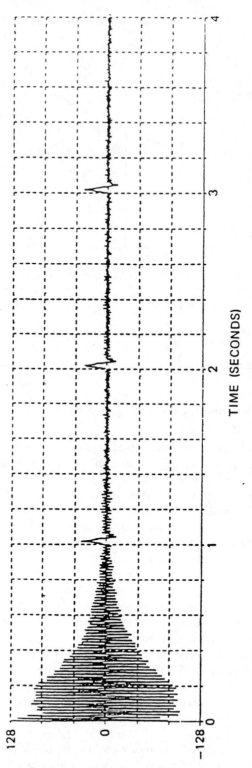

FIGURE 8.2. Output waveform of the adaptive hum canceller showing the error transient and the periodic signal of interest.

to specify scaling of the two analog to digital converter (ADC) inputs. Both are dominated by sinusoids, so it is safe to use a relatively low crest factor (CF), e.g., 6 dB. The primary input with its pulselike signal would require a larger crest factor were the 50/60-Hz component not dominant. Were the ADC to provide an 8-bit sample, with dynamic range of $6(n - 1) = 42$ dB, suppression could theoretically be as high as $42 - CF = 36$ dB. Accumulation should be done with a full 16 bits, so that roundoff noise in an 8-term convolution would not significantly degrade performance.

Effect of weight storage precision is the next concern. Clearly, 8-bit weights would enter into the convolution. A full 16 bits could conveniently be used for weight accumulation. Were μ set to a 6-bit shift (2^{-6}), the degradation due to weight stalling would still allow in excess of 36 dB of suppression.

Under normal circumstances this relatively large step size would tend to generate an unacceptable amount of weight jitter and residual tonal interference. Fortunately, there is a trick permitted by the nature of this application. The signal of interest is a low duty cycle pulse waveform. Once the interference has been reduced by 10–20 dB and the pulse integrity is evident, our processor can sense the onset of the pulse train, sync to it, and disable adaptation during the interval containing the pulse energy, i.e., "gate" the adaptation operation. Therefore, there will be little nonsinusoidal "noise" in the error process during adaptation that can cause weight jitter; the rather large step size should contribute little degradation to the cancellation.

Aside from front end analog circuitry, hardware requirements include two low-cost, low-speed 8-bit A/D converters, a processor chip with perhaps some meager external memory, and an 8-bit digital-to-analog converter (DAC) for output. Some form of analog automatic gain control (AGC) and lowpass filter condition the input to the ADC. A small digital circuit card, low-noise analog card, and a power supply should be adequate.

Figure 8.2 shows the behavior of a fixed-point simulation of this proposed hum canceller. Note the "heartbeat" spike is nearly totally masked initially, and within 2 s becomes essentially hum-free. The residual noise is due principally to additive measurement noise, near the 8-bit quantization level. Note that better fidelity of the underlying process of interest might well require higher-precision input sampling.

8.3 ADPCM IN TELEPHONY

Nearly the entire international telephone system has evolved around the 4-kHz-channel unit, since intelligible speech can be passed with a 4-kHz bandwidth. In long-distance service, these individual channels are normally frequency-division-multiplexed (FDM) into adjacent allocations to form various group-ings of voice circuits having sizable bandwidths for efficient transmission over cable or microwave. As discussed in Section 1.2, this type of analog service is giving way in recent years to digital formats, pulse code modulation (PCM), where individual channels are digitized at the Nyquist rate of 8 kHz, using a "companded" or nonlinear quantization rule (e.g., μ-law and A-law quantization), effectively mapping 13 bits of normal linear quantization into 8 information bits. Each voice circuit then generates a 64-kbps stream of data. Inverse processing at the receiving end reconstructs the original speech with "toll quality." Many such digital streams are time-division-multiplexed into higher bit rate streams and sent to modems for cable or microwave transmission. The entire digital portion of the telephone system is built on fundamental units of 64 kbps.

There are definite advantages associated with the digital form. First, pulsed signals possess a high level of noise immunity, so transmission quality is not degraded by effects other than those associated with original quantization. Secondly, there is ease of switching and routing; incoming bits are simply switched (digitally) to the appropriate output port. The FDM format requires isolation or demodulation of the appropriate channel, switching to the correct output carrier, and remultiplexing into a new channel grouping [Bellamy, 1982].

However, these advantages are counterbalanced by the significant increase in bandwidth. From an information point of view, speech contains far less than 64 kbps of information. Study has shown that a typical analog voice circuit can support on the order of 25 kbps of information, less than half of the rate allotted for PCM traffic [Lucky et al., 1968]. Therefore, this would indicate that more than one voice circuit could be multiplexed into each 64-kbps carrier. As demand for resources has increased, telephone administrations have felt sufficient economic pressure to define a standard means for reducing the bit rate for representing the speech waveform (as well as other traffic on the voice channel, such as modem and facsimile signals).

Means of bit-rate reduction were discussed in Section 1.2. There it was mentioned that this is not a simple matter of reducing the number of quantization levels. While 4-bit speech is intelligible, it suffers an excessive noise degradation due to quantization error. Rather, source encoding can reduce the redundancy or correlation of the speech samples, and generate a new, less correlated sequence of samples. Properly done, this new signal has lower excursions than speech and can be represented with more fidelity by 4-bit quantization. Thus, each incoming 64-kbps signal can be processed to become a 32-kbps stream. Two such coded signals can be multiplexed and sent in place of one uncoded channel, effectively doubling the number of trunks handled by a given facility. We might expect such hardware to first appear on terminations of high-cost circuits, i.e., satellite links and undersea cables. As the cost of the coding/decoding hardware permits, such devices might be used routinely on each digital carrier. There is potential for literally millions of such devices worldwide, and sufficient economic pressure to strive for "jellybean" costs for such devices. Sales potential for millions of encoding/decoding terminations even warrants investigation into custom VLSI development.

Since the early 1980s coding schemes have been proposed by several national telephone administrations. In principle, compatibility of coder and decoder need be important only within a geographic operating region; each administration would promote its own technique. However, in the interest of standardization for economy and international compatibility, the Consultative Committee on International Telephone and Telegraph (CCITT) has formulated a standard recommendation for international use of 32-kbps adaptive differential pulse code modulation (ADPCM). Manufacturers using the standard can be confident that their equipment can compete in the international market. Likewise, as international use of ADPCM spreads, the agreement will minimize the problems of interconnecting administrations.

Since the coding technique has been developed by committee and compromise, full description of its details, performance, and rationale is beyond the scope of this book. Suffice it to say that adaptive quantization and prediction underlies the reduction of bit rate. Figure 8.3 shows the basic structure of coder and decoder, using the notation of CCITT recommendation G.721. An adaptive predictor/filter forms an estimate $s_e(k)$ of the predictable portion of the incoming quantized speech sample $s_1(k)$. The error between the two is scaled by an "adaptive" gain (a sophisticated AGC), and the result is quantized to 4 bits. The magnitude of this 4-bit sample then enters into the adaptive gain adjustment, which tries to prevent saturation. There is a feedback loop that includes the inverse quantization/scaling operations to reconstruct a quantized version of the error $d_q(k)$, which is used by the predictor for the next iteration. The same quantizer and predictor hardware used in this feedback loop in the coder serves as the decoding system. Note that success of this system depends on the coder and decoder having nearly identical parameters (gain and filter coefficients) at each time sample; identical hardware design in coder and decoder is necessary to maintain tracking.

Without detracting from the importance of the adaptive gain function, the subject of primary interest to this book is the adaptive filter serving the predictor function. Unfortunately, the notation put forth by the CCITT in Recommendation G.721 departs from that used throughout this book and the adaptive filtering community. In the interest of clarity, the discussion of the predictor operation involves the following notational translations. The incoming quantized speech waveform shown as $s_1(k)$ in Figure 8.3 is the "desired" waveform $d(k)$ as originally discussed in Chapter 2. Similarly, the output of the predictor filter $s_e(k)$ is $y(k)$ using conventional notation. Lastly, the quantized error $d_q(k)$ entering the predictor is nothing more than the error sequence, $e(k) = d(k) - y(k)$, corrupted by an additive quantization component. Now, in terms of conventional notation, in principle the incoming speech signal $d(k)$ can be modeled as an uncorrelated "innovations" or noiselike process $w(k)$ passing through a linear correlating filter, as in (5.5.1), depicted in Figure 8.4. As discussed at the beginning of Chapter 5, speech may be generated by two such broadband processes, a true noise for "unvoiced" phonemes such as "s" and a periodic pulse train for "voiced" phonemes like "z." Due to the spectral shaping, the resulting signal's excursions may exhibit a wide range. A filter configured as

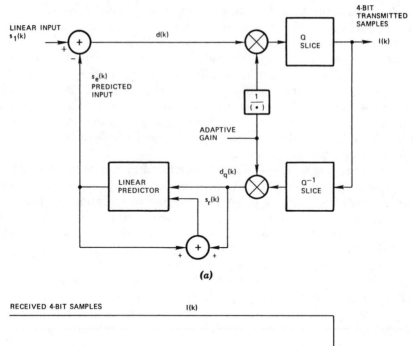

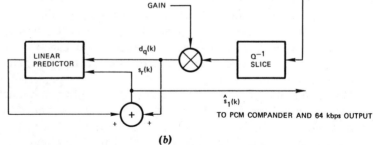

FIGURE 8.3. ADPCM (a) coder structure (b) decoder structure.

in Figure 5.11a can form an estimate of sample $d(k)$ from samples up to but not including $d(k)$. The more correlated or narrowband the signal $d(k)$, the better the estimate $y(k)$ will be. Upon convergence, the error, i.e., the part not predictable, is the driving uncorrelated innovations process; lower correlation implies smaller and more unpredictable excursions. At the receiver, the original speech waveform can be recovered using knowledge of the predictor filter, as shown in Figure 5.11b. To allow for gross level variations seen over short intervals, the adaptive gain network adjusts scaling of the innovations process prior to a quantizing lookup table.

The predictor filter specified for the CCITT ADPCM 32-kbps standard [Jayant and Noll, 1984] is a six-zero, two-pole filter (eight adaptive coefficients in all). It was determined by extensive testing that this filter order offered the best

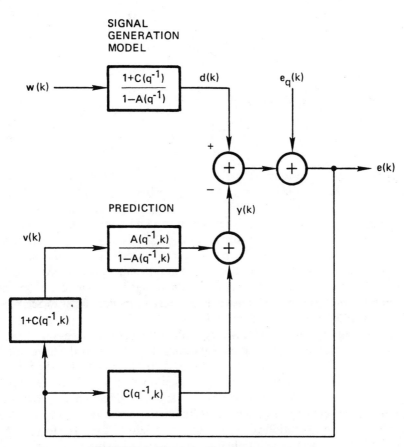

FIGURE 8.4. "Equivalent" prediction structure.

trade-off between recovered speech quality and hardware complexity. Due to the rapid time variations in the speech waveform, this filter must be adapted by a computationally simple algorithm to adequately track and predict the changing correlations with low implementation cost. Therefore, the eight parameters are adapted using a sign-sign type adaptive algorithm.

Feedforward coefficients:

$$c_s(k + 1) = (1 - 2^{-8})c_s(k) + 2^{-7}\text{sgn}[e(k)]\text{sgn}[e(k - s)], \qquad \text{for } s = 1, \ldots, 6 \tag{8.3.1}$$

Feedback coefficients:

$$a_1(k + 1) = (1 - 2^{-8})a_1(k) + 3 \cdot 2^{-8}\text{sgn}[v(k)]\,\text{sgn}[v(k - 1)] \tag{8.3.2}$$

$$a_2(k + 1) = (1 - 2^{-7})a_2(k) + 2^{-7}\{\text{sgn}[v(k)]\text{sgn}[v(k - 2)]$$

$$- g[a_1(k)]\text{sgn}[v(k)]\text{sgn}[v(k - 1)]\}, \tag{8.3.3}$$

where

$$g[a_1] = \begin{cases} 4a_1 & |a_1| \leq \frac{1}{2} \\ 2 \ \text{sgn}[a_1] & |a_1| > \frac{1}{2} \end{cases}$$

and the smoothed quantized prediction error v is

$$v(k) = e(k) + \sum_{s=1}^{6} c_s(k)e(k - s). \tag{8.3.4}$$

Note that these equations are driven by the data available to *both* coder and decoder (barring channel errors), i.e., the quantized prediction error $d(k) - y(k)$ sample, $e(k)$. As noted in Chapter 5, the leakage in (8.3.1)–(8.3.3) is added to encourage resynchronization of the transmitter and receiver parameterizations following a channel error. Step-size and weight precision have been specified by the CCITT to achieve acceptable reconstructed speech quality on a wide range of source waveforms. The CCITT standard also includes prespecified bounds on the magnitudes of the c_s and a_i. The bounds on the a_i are intended to maintain the stability of the receiver reconstruction transfer function. Refer to Sec. 6.5.3 of [Jayant and Noll, 1984] for further details.

A reasonable question at this stage might concern justification of (8.3.1)–(8.3.4). Upon close inspection these equations do not exhibit the form of either of the adaptive IIR Algorithms 9 and 12 suggested for the ADPCM problem in Section 5.5. Differences persist even if the sign operators and leakage factors of (8.3.1)–(8.3.3) are added to Algorithms 9 and 12 of Section 5.5. A principal distinction is that (8.3.1) uses neither the prediction error smoothing of (5.5.19)–(5.5.20) nor the information vector filtering of (5.5.9)–(5.5.10). The form of (8.3.2) and (8.3.3) bears even less resemblance to (5.5.7)–(5.5.8) or (5.5.18) and (5.5.20).

To understand the relationship of (8.3.1)–(8.3.4) to those algorithms discussed in Chapter 5 consider the predictor structure of Figure 5.11a has been redrawn in Figure 8.4. The transfer function from the transmitted signal $e(k)$ to predicted signal $y(k)$ reduces to

$$\frac{A(q^{-1}, k) + C(q^{-1}, k)}{1 - A(q^{-1}, k)},$$

assuming fixed filters. For the moment replace the IIR $A(q^{-1}, k)/[1 - A(q^{-1}, k)]$ operator in Figure 8.4 by a second-order finite-impulse-response (FIR) approximation $F(q^{-1}, k)$. With this replacement, the prediction y of Figure 8.4 can be written as

$$y(k) = \sum_{i=1}^{2} f_i(k)v(k - i) + \sum_{s=1}^{6} c_s(k)e(k - s), \tag{8.3.5}$$

where v is given by (8.3.4) and e is the quantized $d - y$. Using the sign–sign LMS update of (4.2.50) and the leakage of (4.2.51) yields (8.3.1). As noted earlier, these

c_s updates do not compensate for the underlying adaptive IIR filter form of the problem. However, the updates of the a_i do. The a_i updates are based on updates for f_i that do provide a time-varying prediction error smoothing compensation much like Algorithm 12 in Section 5.5, i.e.,

$$f_i(k + 1) = f_i(k) + \mu_i v(k - i)v(k). \tag{8.3.6}$$

For (8.3.5), we assumed that

$$F(q^{-1}) = \frac{A(q^{-1})}{1 - A(q^{-1})} = \frac{a_1 q^{-1} + a_2 q^{-2}}{1 - (a_1 q^{-1} + a_2 q^{-2})}$$

$$= a_1 q^{-1} + (a_2 + a_1^2)q^{-2} + \frac{a_1^3 q^{-3} + a_2(a_2 + a_1^2)q^{-4}}{1 - (a_1 q^{-1} + a_2 q^{-2})}. \tag{8.3.7}$$

This suggests the relationships $f_1 = a_1$ and $f_2 = a_2 + a_1^2$ or

$$a_1(k + 1) = a_1(k) + [f_1(k + 1) - f_1(k)] \tag{8.3.8}$$

and

$$a_2(k + 1) = a_2(k) - [f_1^2(k + 1) - f_1^2(k)] + [f_2(k + 1) - f_2(k)]. \tag{8.3.9}$$

Adding sign operators and a leakage factor to the combination of (8.3.6) and (8.3.8) yields (8.3.2). Similarly, adding sign operators and a leakage factor to the combination of (8.3.6) and (8.3.9) yields (8.3.3) with $g[a_1] = 2a_1$ after removing any terms scaled by (very small) μ_i^2. Therefore, in a sense the update of IIR coefficients has been determined by means of an FIR approximation.

Though the preceding discussion may seem to imply that the CCITT standard was engineered according to a curious mixture of mutations of algorithms developed in Chapter 5, this is not the case. Much of the theory of Chapter 5 was only first being developed after the CCITT standard had already begun to assume its particular form. It is curious that the "independent" CCITT standard can be reverse-engineered to Chapter 5. In fact, had the ideas of Chapter 5 been available, a stability maintenance scheme would have been suggested for the inverse of $[1 + C]$ in (8.3.4). The potential performance degradation with transmitted signal bursting due to temporary bouts of transmitter instability with the standard's omission of a $[1 + C]^{-1}$ stability maintenance scheme have been emphasized by [Macchi and Jaidane-Saidane, 1986]. This study was motivated by results stemming from the insights similar to those related in Chapter 5.

Though a curious mixture, the CCITT recommendation has performed adequately in a large variety of real-data tests. It becomes clear in examining its details that the CCITT went far beyond specifying the basic technique and formats for the coder. System equipment compatibility in this case requires that coder and decoder must exactly track in coefficients and gain. Thus, it was necessary that almost all details be specified, i.e., coefficient/internal precision, step sizes, quantization rules, gain attack/decay rates. Given this degree of specificity, the designer's task reduces to optimizing hardware necessary to perform the rather lengthy list of equations, branch tests, and lookups.

Based on the arithmetic requirements, the circuitry is quite simple. The most complex portion, the filter output calculation, involves only 8 multiplies at 8 kHz. Care was taken for the remainder of the circuit to use hard-wired binary shift operations, as well as adds and subtracts. Even though the adaptive quantizer involves a lengthy list of equations and branch tests, the operations by and large are relatively simple and within the capabilities of a single-chip signal processor. It is reasonable that one processor could serve as the coder/decoder for two such 64-kbps channels, generating two 32-kbps coded streams. A specially designed chip could as well serve multiple channels.

Were a larger capacity desired, for example, funneling two entire North American T1 carriers each with 24 channels into a single T1, a considerably different approach might be formulated. Using multiplexed high-speed LSI components, the circuitry serving 24 channels could well fit on one 8″ × 12″ board, requiring only one 10-MHz MAC chip multiplexed with an array of support and interface components. A device with two such cards and the necessary line interface cards could then serve as a two-into-one termination for T1 transmission.

It is interesting to note that variations on this basic coding scheme may be theoretically used to accommodate more than two speech circuits per 64 kbps link. On high-cost links serving many incoming trunks, intervals of high calling activity can dynamically trigger a reduction from the nominal 4-bit error representation to 3- or even 2-bit representation. This allows insertion of extra voice channels with minor degradation to toll quality.

8.4 MULTIPATH CORRECTION FOR TROPOSCATTER SIGNALS

8.4.1 The problem

Section 1.3 briefly introduced the problem of correcting or equalizing radio signals dispersed by propagating through the "tropospheric channel." In this section the tropospheric communication problem is described in more detail. An approach to solving the problem is also discussed in the next section, along with a description of the engineering steps needed to specify the resulting adaptive filter.

The basic troposcatter communications link is shown in Figure 1.6. A powerful transmitter sends a UHF or microwave signal in the direction of the receiving site. Even though the transmitter and receiver are obscured from each other by the curvature of the earth, enough energy is refracted by turbulence in the troposphere toward the receiver to demodulate the signal most of the time.

Troposcatter links use powerful transmitters, large antennas, and sensitive receivers, making them fairly expensive to build and operate. This is offset, however, by the fact that each "hop" can cover 300 to 800 km, making the cost per circuit mile competitive with communications satellites and undersea cables.

The principal shortcoming of tropo systems is circuit reliability, i.e., how much a given circuit is available to the overall telecommunications system. The objective for a typical end-to-end circuit is a maximum average outage rate of 0.01 percent, making the requirement for each hop of a circuit more stringent yet. In troposcatter systems, outages stem mainly from the deleterious effects of multipath propagation. As explained in Chapter 1, the multipath propagation filters the transmitted signal in a way that causes severe distortion at the receiver. This distortion, when it occurs, makes all circuits carried on the affected tropo unusable. This multipath distortion problem becomes worse as the bandwidth (and hence circuit-carrying capacity) is increased. Thus, the communications link designer must trade off between capacity (and revenue) and circuit reliability.

Link designers have traditionally used several approaches to improving link reliability to an adequate level. Straightforward approaches include increasing transmitter power, increasing receiver sensitivity, and increasing the sizes of the antennas at both sites. These methods serve only to increase the "link margin," the amount of signal strength which can be diluted by multipath-induced signal cancellation before the output quality falls to an intolerable level. They do not deal with so-called frequency-selective multipath, which leads to output distortion even when adequate power is seen at the receiver.

Both selective and nonselective multipath can be treated with "diversity reception." A diagram of four-way diversity is shown in Figure 8.5. Two transmitters are used, sending the same signal but at different frequencies.

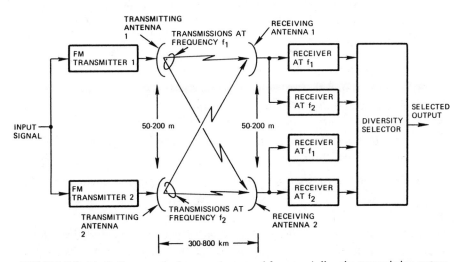

FIGURE 8.5. Block diagram of a four-way (space and frequency) diversity transmission system.

Four receivers are used at the receiving site, two with each of two antennas. Both transmitted signals are received at each of the two receiving antennas. The result is the reception of four versions of the transmitted signal. If the transmitting frequencies are sufficiently different and the receiving antennas far enough apart, the multipath fading behavior of each of the four propagation channels is essentially statistically independent. Therefore, the probability of all four signals having poor quality at the same instant is much less than any individual one. Capitalizing on this fact, a switch is used to connect the best of the four signals to the system output, thus producing the best end-to-end quality.

Diversity reception has proven a very powerful technique for improving the quality of troposcatter links, but this improvement comes at substantial cost. In particular, two transmitters are required instead of one, four receivers instead of one, and the number of large antennas is doubled. The large financial outlay for this diversity equipment has motivated the investigation of cheaper techniques that might attain the same performance level. In particular, we examine here the design of an adaptive filter which provides relief from frequency selective multipath, leaving the "flat-fading," i.e., nonselective, multipath as the only cause of channel outage. Correction of frequency selective fading can also allow the bandwidth and hence channel capacity of the link to be increased, lowering the cost per circuit.

For the purposes of this design example, the multipath correction filter is expected to operate in the following environment:

(a) The transmitted signal is frequency-modulated using an analog frequency-division multiplex (FDM) signal. The tropo system can be upgraded at any time to carry time-division multiplex (TDM) signals on a constant-envelope, bauded-modulation type, such as phase-shift keying.

(b) The bandwidth of the modulated signal may be as wide as 4 MHz, but perhaps as narrow as 1 MHz.

(c) The multipath propagation channel introduces a time spread of typically 0.5 μsec and at most 2 μsec. The Doppler spread is typically a few Hz and at most 20 Hz.

(d) The characteristics of the modulating signal are completely unknown to the receiver and may change at any time. (No special signal is added for the express purpose of training the adaptive filter.)

(e) The tropo system may already be equipped with two- or four-way diversity reception.

Given these conditions, we are ready to design the appropriate adaptive filter.

8.4.2 Design Steps

A prudent system design uses both adaptive equalization and diversity reception, since each accomplishes something the other cannot. Diversity reception neatly accommodates the problem of flat fading, that is, the condition when the signal at the input of one of the receivers is cancelled by antiphased paths. In this case, the diversity receiver simply picks the stronger signal and thereby preserves output quality. In the case of frequency-selective fading, however, only equalization of the propagation-induced filtering will restore quality. For this reason, we begin with the system design shown in Figure 8.6a. One transmitter, two receivers, and two receiving antennas are used. The receivers down-convert the input signals to a convenient intermediate frequency (IF). The correction filtering is applied at this point, and the resulting filter outputs are demodulated. A diversity switch then picks the best output. The correction filtering must be applied before the demodulator, since frequency modulation is a nonlinear process. Any filtering introduced by the channel after the transmitting modulator must be removed before the nonlinear receiver. The diversity switching is usually done after the demodulators for convenience and because it is usually necessary to demodulate both signals to decide which one of the two signals is the better.

Given the specifications listed, the following conclusions and observations can be made:

(a) *Digital implementation.* Digital implementations are usually chosen for adaptive filters if the bandwidth to be processed is not so high as to make the filter's construction prohibitively expensive. The principal reason for the inclination toward a digital implementation is that the filter and adaptive algorithm can be implemented with fewer errors and hence better performance.

Given a maximum bandwidth of 4 MHz, the input IF signal must be sampled at a rate of at least 5-MHz complex or 10-MHz real. For this example, we use the structure shown in Figure 8.7. The IF signal is centered at 2.5 MHz and digitized with 10-bit accuracy at a rate of 10 MHz. A digital Hilbert transformer [Gold and Rader, 1969] converts the real samples into quadrature pairs, i.e., complex samples, at a 5-MHz rate. These samples form the input to the correction filter itself.

The digital implementation is not inexpensive. On the other hand, it performs substantially better than known analog implementations. Its cost is amortized over many circuits, and it is substantially cheaper than the transmitters and antennas it is meant to replace.

(b) *Filter structure.* Since the multipath propagation channel has a finite impulse response, it is theoretically true that an infinite impulse response (IIR) filter offers the computationally cheapest form of correction filter. However the multipath channel cannot be guaranteed to have "minimum phase," that is, all its zeros inside the unit circle. When the channel has a zero outside of the unit circle, the IIR correction filter should move a pole outside of the unit circle to compensate for it. This makes the filter

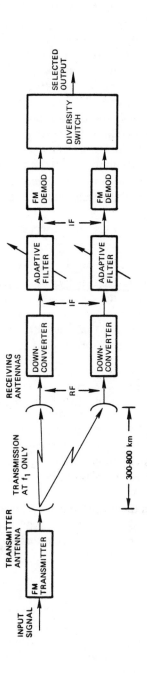

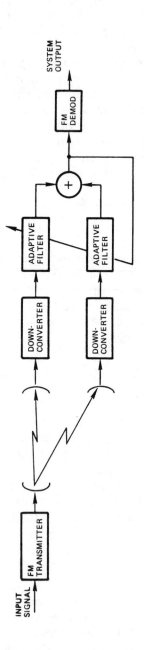

FIGURE 8.6. Two-way (space only) diversity receiving systems employing adaptive multipath correction in each path (a) diversity switching of the demodulator outputs (b)coherent combination of the received signals before demodulation.

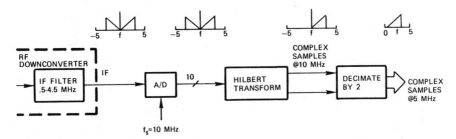

FIGURE 8.7. Processing steps and associated spectra used to generate digital input for the adaptive filter.

unstable and chaos quickly reigns. This and the fact that adaptive algorithms for IIR filters have not yet reached maturity make FIR filters the proper choice.

Given the choice of an FIR filter, we must now determine how many taps to use. The multipath channel itself can be modeled with 10 taps, since 10 taps at a 5-MHz sampling rate equals the maximum channel spread of 2 μsec. The correction filter, however, must be longer. Engineering experience has shown that the correction filter needs to be three to five times the length of the multipath spread to be effective. This can be rationalized by observing that the FIR correction filter is in some sense approximating the response of the theoretical IIR filter. An IIR filter with 10 taps would normally have an impulse response much longer than 10 time samples. Based on this analysis, we choose a 30-tap FIR filter as the basis for the adaptive correction filter. This filter is 12 times longer than the 0.5-μsec expected multipath delay and three times the maximum delay.

While many forms of FIR structure are available, including lattice, trans-multiplexer, and fast convolution, we will assume here the use of a conventional tapped-delay-line filter employing multiplier/accumulator chips, such as those discussed in Section 7.3. This type of filter is straightforward to design and test compared to the others and does not incur large penalties in size or performance.

(c) *Choice of adaptive algorithm.* For the problem of tropo multipath correction, the types of algorithms used are very restricted. Since the modulating signal structure is not known and no training signals are added at the transmitter, no desired signal $d(k)$ is available. This in turn implies that none of the algorithms discussed in Chapters 4 and 5 can be employed. A possible exception occurs in the case of "digital" or bauded signals in which so-called decision feedback can be used, but since this equalizer must work for both analog and digital modulation types, the sole use of decision feedback is inadequate.

The problem can be solved by observing that all of the modulation types of interest are of the constant envelope variety. This is not unusual in radio systems since constant-envelope signals have good resistance to the nonlinearity of powerful transmitters and have the optimum peak-to-average power ratio. The fact that the tropo systems use FM and PSK with constant envelopes means that property-restoral algorithms of the type explored in Chapter 6 can be employed. In particular, the constant-modulus algorithm (CMA) is an appropriate choice for this problem.

   (d) *Adaptation rate.* The adaptive filter must change its coefficients fast enough to "track" the changing impulse response of the multipath propagation channel. To achieve the desired high degree of multipath correction, the adaptive filter must track the channel very closely.

The degree of time variation in a propagation channel is indicated by the channel's Doppler spread. While Doppler spread has a well-defined technical meaning [Bello, 1963], as a practical matter it specifies the interval between significantly different realizations of the channel impulse response. In particular, a Doppler spread of 20 Hz indicates that a comparison between impulse responses taken 50 msec apart would show significant differences. Practical experience has shown that "tracking" this time variation means developing new adaptive solutions at least 10 times faster than the maximum Doppler spread. In this case, the adaptive logarithm driving the correction filter must be capable of essentially converging to new solutions in 5 msec.

Some theoretical work and extensive computer simulation [e.g., Treichler and Agee, 1983] have shown that the gradient descent version of the CMA will converge reliably within 500 μsec for a wide variety of multipath channels using either FM or phase shift keying. Thus, even the simplest type of adaptation algorithms converges more than fast enough to meet the 5-msec requirement. At this point the designer has (at least) two choices: (1) he can adapt the filter using only one-tenth the input data, leading to a reduction of 90 percent in the processing load needed to compute the filter coefficients, or (2) he can use all the data, allowing for either anomalously high Doppler spreads or, using a lower adaptation coefficient, providing smoother adaptation and lower "misadjustment." Adapting only one-tenth the time also reduces the finite word length requirements by 3 bits.

For the purpose of this design example, we assume that the coefficients are computed anew every 5 msec, allowing the coefficient update processor to be scaled down.

   (e) *Word length determination.* The sampled data entering the adaptive filter are assumed to be quantized to 10-bit accuracy. To provide for adequate dynamic range in the frequency response, the filter impulse response should be quantized no more coarsely than 12 bits. Assuming the use of 30 complex taps in the filter impulse response, each component of the filter output would specify 28 bits, even though only 10 or 11 are needed to describe the output waveform with sufficient quality.

The word lengths required to compute the error signal and adaptive updates depend strongly on the particular performance function chosen. Even within the limited class of gradient descent constant-modulus algorithms, several are available. We choose here the 1–1 algorithm, given by the performance function

$$J = \sum_{k=1}^{M} \left| (|y(k)| - A) \right|. \tag{8.3.1}$$

Both the signal modulus and its magnitude difference from A are raised only to the first power. The impact of this is that if $y(k)$ is adequately represented with N bits, then so are the elements of the instantaneous gradient estimate. This contrasts with the 2–2 version of CMA that requires $4N$ bits since the signal is squared twice.

The estimated gradient of J is given by

$$\hat{\nabla} J = \sum_{k=1}^{M} \mathbf{X}^*(k)\operatorname{sgn}(|y(k)| - 1)|\text{lim}\{y(k)\}, \tag{8.3.2}$$

where $\text{lim}\{\cdot\}$ denotes the unit length phasor in the direction of the argument. Since the last two terms in the product have unit length, each term of the gradient vector can be seen to specify $N + \log_2 M$ bits, where N is the input quantization and M is the number of terms in the sum. An averaging interval of 500 μsec and a sampling rate of 0.2 μsec imply that M is about 2500. If the input is quantized to 10 bits, then each term of the estimated gradient requires 22 bits to preserve full accuracy. If the adaptation constant μ is chosen to be in the range of 0.01 to 0.001 (usually in values equaling binary shifts), then the scaled gradient may require up to 32 bits.

By working through the requirements, we developed the following technical approach to the design of the adaptive correction filter. The input signal is down-converted to a center frequency of 2.5 MHz, sampled at 10 MHz with 10-bit accuracy, and converted to a 5-megasamples/sec complex format. The filter itself is a tapped-delay-line FIR structure with 30 complex taps. The 1–1 constant modulus algorithm is used to update the coefficients, but the channel variation is slow enough to allow a 90 percent reduction in the processing required by the adaptation hardware. The filter input would be presented by 10-bit samples, and the output computed to 28 bits, but actual output would be the most significant 12 bits. The coefficients must be updated with about 30 bits of accuracy but can be used with only the top 12 bits of accuracy in the filter output computation.

Even with the approach specified to this degree, there are still many implementation techniques open to the system designer. To continue the example, we make the arbitrary choice of using a design based on commercial multiplier/accumulator (MAC) devices such as the now-standard 1010J which can accomplish a 16-by-16-bit multiply and 35-bit accumulate operation in 155 nsec, comfortably less than the 200-nsec sampling interval. Since each MAC can perform only one cycle per input sample, some 120 of them are required to compute the 30-point complex dot product needed for each filter output. A convenient architecture for efficiently using all of these MACs is the reverse-form filter discussed in Section 7.3. Given the filter impulse response, the filter can be implemented with the MACs, a delay line for the impulse response, and a minor amount of control logic, as shown in Figure 7.10. The length of the filter and the complex coefficients force the filter to be subdivided into several circuit boards. Figure 8.8 shows one such board that can be used for this application. The board contains 20 MACs, organized as two 10-point filters with a common impulse-response delay line. The whole filter requires six such boards, three using the real part of the impulse response and three using the imaginary part. The four outputs of the two sections are appropriately subtracted and added to form the complex filter output.

The coefficient computation can also be accomplished using the reverse-form architecture of Section 7.3 and Figure 7.11. Because of the relaxed convergence rate requirements for this problem, the update computation circuit is shared over all 30 coefficients. Twelve MACs operating at the 5-MHz clock rate serve

FIGURE 8.8. 10-Tap FIR digital filtering circuit card used to process complex-valued data samples (courtesy of ARGOsystems, Inc.).

to update the coefficients. These MACs and the logic needed to compute the 1–1 CMA error function—a TI TMS32020, for example—can be placed on one card. The eighth card would contain the input analog-to-digital convertor and the Hilbert transformer needed to convert the real samples into the complex form used by the filter.

Given this proposed architecture, the designer must check that it satisfies all finite word length considerations. The 1010J MAC called above allows 16-bit input quantization and 16-bit coefficient quantization, provides a 35-bit accumulator, and can round the accumulated output. In short, it more than satisfies the requirements. A similar analysis shows that a 12-by-12-bit multiplier accumulator is marginal since its accumulator length is only 27 bits, where 30 might be needed. From this it appears that a 16-by-16-bit MAC should be used.

The final system consideration involves the form of the output. The filter's natural output is in the form of complex samples at a 5-megasample/sec rate. This signal must be processed in some fashion to make it compatible with the demodulator, FM or PSK, which follows. Any of three approaches might be used:

(a) Convert the complex samples into a 10-megasample/sec real format and apply those to a D/A convertor. The resulting analog signal is applied to a conventional analog demodulator.

(b) Instead of employing a complex adaptive filter, use a version with real coefficients. The real samples derived at 10 megasamples/sec from the input A/D are directly applied to the filter and the output sent directly to the D/A. A version of the constant modulus algorithm has been developed [Treichler and Larimore, 1985a] which allows adaptation of the real coefficients. The amount of filter computation is the same with the real filter, but the Hilbert transformer and output complex-to-real conversion are no longer necessary.

(c) Perform the demodulation in the digital domain as well. The complex filter output is a natural input to such a demodulator, and techniques have been developed which allow quality demodulation at the bandwidths required here [Treichler, 1981].

The overall system block diagram of the multipath correction filter outlined above is shown in Figure 8.9. Adaptive filters of this type have been actually built [Larimore and Goodman, 1985], and the degree of multipath correction and output signal quality match very well with the values predicted by the computer simulations presented in Section 6.3.4.

8.4.3 Incorporation of Diversity Combination

Part of the system concept for the tropo receiving processor was the use of diversity combination, i.e., receiving and processing two independent versions of the transmitted signal and then choosing the better of the two. This is usually

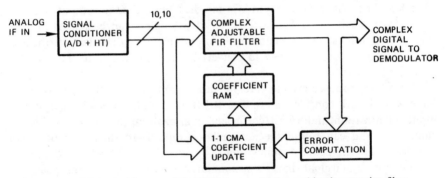

FIGURE 8.9. Architecture of the CMA-based adaptive multipath correction filter.

accomplished by measuring the signal quality at the output of the two demodulators and then choosing the better. Pre-detection combination, shown in Figure 8.6b, is used less frequently. In this case, the two IF signals are properly phased and then added coherently. At the cost of increased complexity, this technique can improve signal quality over that attained by switching alone since the added signal power is often more than either received signal separately.

The adaptive filter designed in the previous section can actually be extended to offer diversity combination as well. Consider the block diagram shown in Figure 8.10. The IF signals from the space-diversity receivers are both sampled

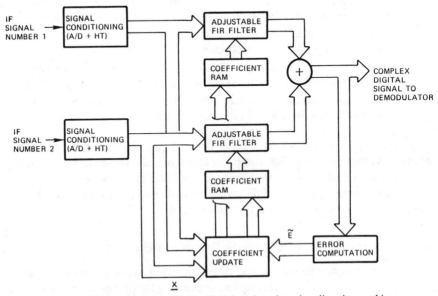

FIGURE 8.10. Block diagram of a CMA-based predetection diversity combiner.

and converted to a complex format. They are then applied to their own filters and the outputs added. The added signal forms both the input to the demodulator (by any of the three means discussed above) and the "output" to be used by the constant-modulus algorithm. The details are presented in [Treichler and Larimore, 1985]. In essence, CMA is used to adapt both filters simultaneously, introducing the spectral shaping, the gain, and the phase shifting required to coherently combine the two signals. The algorithm is no more complicated than before, a second demodulator is not needed, and no measurements of signal quality are required. The CMA adjusts both filters to make the sum signal as close to a constant-envelope signal as possible. In effect, this structure performs both the diversity combination needed to deal with non-frequency-selective multipath ("flat fading") and the spectral shaping needed to correct for frequency-selective fading.

The implementation of this adaptive processor requires two of the filters described. Even including the A/Ds and the Hilbert transformation circuitry, the whole processor can be packaged in one 24 in.-high rack-mounted chassis with no serious effort at size reduction. Note that this processor would execute approximately 1.32 gigamultiply/adds per second.

8.5 COMMENTS ON SOME OTHER APPLICATIONS

The design examples presented in this chapter range over very different application areas and over very different signal bandwidths. One could be implemented with a single digital signal-processing chip (e.g., the TI TMS320), while one could use a custom VLSI chip, and the last many high-speed dedicated multiplier/accumulator chips. Two examples used FIR filters, while one employed an IIR structure. The LMS algorithm was used in the first, a signed FIR-IIR mixture was used in the second, while a property-restoral algorithm (CMA) was used in the third. Even so, common threads exist between the implementations chosen. All, for various reasons, used digital implementations, and all, for other various reasons, found that approximate gradient descent algorithms, such as LMS, suffice rather than more complicated algorithms like recursive least squares (RLS). It is not reasonable to assume, however, that these common threads are rules in general. We briefly consider here several practical cases where other approaches are needed.

8.5.1 Adaptive Equalizers for High-Capacity PCM Signals

Multipath propagation effects like those discussed in Section 8.3 also degrade the quality of high-capacity PCM signals. These signals tend to use complicated modulation types, such as 16-level quadrature amplitude modulation (QAM), and occupy bandwidths of 20 to 50 MHz. Because of these high bandwidths, a fully digital implementation of the filter may be economically infeasible. As a result, such equalizers are usually based on an analog filter with some form of digital quality assessment and adaptation control.

8.5.2 Voiceband Data Signal Equalizers

The adaptive equalizers used in data modems to correct for the dispersion incurred in telephone voice channel have traditionally used the LMS or related approximate gradient search algorithms because they are robust, they require a relative minimum of hardware (or microcode cycles), and they converge rapidly enough for most traditional modem applications. Recently, however, there is considerable interest in equalizers that adapt very rapidly. One such application is the case where a central computer must poll many distant processors on a so-called multidrop circuit. The time required to equalize each modem (about half a second) can in fact be longer than the time required to transfer the needed data. Algorithms such as RLS and "fast" RLS can provide dramatically improved convergence speeds in such cases. Because of this improvement in efficiency, it may become cost-effective to use the more complicated RLS-type algorithms. Another application where RLS has been successfully employed is in HF data modems where the propagation channel changes too rapidly for a gradient-based algorithm to track the channel.

8.5.3 Frequency Domain Implementation of FIR Adaptive Filters

Although they are more complicated to design and control, transmultiplexer and fast convolution filters based on the fast Fourier transform (FFT) are the most efficient ways to implement FIR adaptive filters with long impulse responses. As a rule of thumb, any filter longer than 64 taps should be done in the frequency domain if computational efficiency is important to the system designer.

8.6 SUMMARY

As pointed out in the Preface, the purpose of this book was to deal with concerns of adaptive filtering important to research, development, and design:

(1) *Proper context for adaptive processing.* Chapters 1 and 8 were centered on real applications calling for adaptive processing. These chapters discussed the trade-offs and concerns surrounding the motivation for an adaptive solution, enabling the engineer to answer the following questions: When does addition of an adaptive filter make sense? What types of gains can be expected?

(2) *Means by which satisfactory parameter adjustment can be achieved.* Chapters 2 through 6 were devoted to development of many of the common adaptive techniques currently in use and under study.

(3) *Expected properties of such adaptive filters.* In Chapter 3, the concepts of optimality, convergence, and average behavior were discussed, along with the factors that determine convergence rate and performance.

(4) *Possible extensions.* By building on the "classical" least squares approaches of Chapters 3 and 4, Chapters 5 and 6 presented recently developed means of filter adjustment. In Chapter 5, the alternative to the finite-impulse-response (FIR) structures of Chapter 4 was the infinite-impulse-response (IIR) structure. A significant tool in this development is stability theory, in addition to the search and minimization procedures more common for adaptive FIR filters. However, connection was made between the statistical average behavior analysis of Chapters 3 and 4 and the stability theory-based averaging analysis of Chapter 5. Chapter 6 addressed the alternative performance function definition of property restoral. The formulations of Chapters 4 and 5, which relied on the presence of a desired signal for sample-by-sample prediction error generation, were shown to be only one special case of the more general property restoral concept. Certainly the development, analysis, and refinement of such extensions is a major component of the research effort in adaptive filter theory.

(5) *Practical concerns.* Given a filter structure and bandwidth requirement, there are a variety of implementation alternatives available. Chapter 7 discussed many of the practical aspects of digital designs, for example, precision and architecture, as well as providing a more general overview of implementation trade-offs.

Each of these points was considered in the applications cited in Chapter 1 and "resolved" for the same applications in Chapter 8. In each case, the practical concerns and objectives influenced the choices made in filter configuration, adaptive algorithm selection, tuning, and modification, and hardware realization. The background needed to perform these selections was provided by the intervening chapters of this book, which presented extendable fundamentals of the theory and design of adaptive filters.

A

FORTRAN Implementation of Selected Adaptation Algorithms

In this appendix listings of principal adaptive filtering algorithms are provided. In lieu of direct applications, these subroutines can serve as the framework for translation to other high-level languages or into assembly code for more efficiency. Specifically, the algorithms addressed are as follows:

1. LMS algorithm (Section 4.2.2)
2. Signed-error LMS algorithm (Section 4.2.6.1)
3. Normalized LMS algorithm (Section 4.2.5.2)
4. Griffiths's or P-vector algorithm (Section 4.2.5.5)
5. Recursive least squares algorithm (Section 4.3.2)
6. Constant-modulus algorithm (Section 6.3)
7. Simple hyperstable adaptive recursive filter algorithm (Section 5.2.1.1)

In each case, a subroutine has been written to perform one complete iteration of the algorithm, i.e., output convolution, error calculation, and weight update. Also given for each algorithm is an example calling program which includes a sequence of 20 noiselike input samples. In addition, a list of values (output and weights) that result from the test program is provided. In this way, a user can verify the integrity of any code transcription or translation.

Note that these specific implementations are simply for purposes of illustration, and no claim of optimality is made. As we have seen, many applications require careful tailoring of code in the interest of efficiency and may preclude use of the subroutine structure; rather, in-line code may well be preferable.

255

In any case, the calling example and tabulated results remain useful as a diagnostic tool for any modifications implemented by the user. The code itself is provided in the most fundamental form of FORTRAN to assure universal compatibility; however, the user should verify its validity, in particular I/O statements and the imbedded comment field.

A.1 THE LMS ALGORITHM

A.1.1 Adaptation Subroutine

```
      SUBROUTINE ADAPT(XIN,YOUT,D,RMU,W,NORD,XTDL,IPTR)
C
C     LMS ALGORITHM
C
C     Parameters:
C
C              XIN      — Real input sample at time k (Input)
C
C              YOUT     — Real output sample at time k (Output)
C
C              D        — "Desired" output sample at time k (Input)
C
C              RMU      — Real stepsize for weight updating (Input)
C
C              W        — Array containing weight vector at time k (Input)
C
C              NORD     — Number of taps in the filter (Input)
C
C              XTDL     — Array for storage of past data samples (Output)
C
C              IPTR     — Pointer for delay line array (Input)
C
C              This routine performs a single iteration of the LMS algorithm.
C     The input sample is placed into the "delay line" array, the output is
C     computed, and then the weights are updated.  Note that it is assumed
C     that initialization steps have been performed external to this routine,
C     i.e. delay line setup and initial weight setting definition.
C
      DIMENSION XTDL(1),W(1)

      XTDL(IPTR)=XIN                    !Write over oldest sample with incoming value

C
C     Do convolution to form the output
C
      ACCUM=0.                          !Clear the accumulator
      LPTR=IPTR                         !Initialize the data pointer to newest
      DO 10 L=1,NORD                    !Apply all weights to the data
      ACCUM=ACCUM+W(L)*XTDL(LPTR)       !Add in partial product
      LPTR=LPTR+1                       !Move data pointer to next oldest slot
10    IF(LPTR.GT.NORD) LPTR=1           !Check for wraparound of storage array

      YOUT=ACCUM                        !Send out accumulator

C
C     Do update of the weight vector
C
      ERROR=D-YOUT                      !LMS error term
      ERROR=RMU*ERROR                   !Scaling of error term
      LPTR=IPTR                         !Initialize the data pointer to newest

      DO 20 L=1,NORD                    !Update all weights
      W(L)=W(L)+ERROR*XTDL(LPTR)        !Actual update
      LPTR=LPTR+1                       !Move data pointer to next oldest slot
```

```
20    IF(LPTR.GT.NORD) LPTR=1              !Check for wraparound of storage array

C
C     Clean up by preparing for next data sample
C
      IPTR=IPTR-1                          !Point to oldest sample
      IF(IPTR.LE.0) IPTR=NORD              !Check for wraparound of storage array

      RETURN
      END
```

A.1.2 Example Calling Program and Results

```
C
C     860114 Test program for LMS algorithm routine.

C             This program generates a moving average sequence of real
C             samples, and uses the adaptive filter to form a direct model.
C             The generating plant has the impulse response 0,1,-.5,-1,0 ...;
C             the filter is allowed five weights.  The input sequence is a
C             pseudonoise process, with period of 20.  Stepsize is hardwired
C             to 0.005.
C

C
C     Generator arrays (input sequence, generator weights)
C
      DIMENSION XINPUT(20),GENER(5)

C
C     Arrays for adaptive filter (delay line, weights)
C
      DIMENSION XTDL(10),W(10)

C
C     Hardwired data for generator
C
      DATA GENER/0.,1.,-.5,-1.,0./
      DATA XINPUT/-0.125,0.010,-0.901,0.038,-0.355,-0.841,-2.891,
     &     0.989,-0.258,1.068,-1.469,1.489,-0.170,1.522,-1.022,
     &     1.233,0.930,-0.881,0.877,-1.275/

C
C     Setup for filter and generator
C
      IGPTR=1                              !Set up delay line pointer

      DO 10 L=1,10
   10 XTDL(L)=0.                           !Clear adaptive filter delay line
      IPTR=1                               !Set up filter delay line pointer

      ITER=0                               !Iteration counter
      NORD=5                               !Hardwired filter order
      RMU=.005                             !Hardwired stepsize
C
C     Execution loop
C
  100 XIN=XINPUT(IGPTR)                    !Get input sample from list
      ACCUM=0.                             !Clear accumulator

      LPTR=IGPTR                           !Point at current input value
      DO 110 L=1,5
      ACCUM=ACCUM+GENER(L)*XINPUT(LPTR)    !Accumulate partial products
      LPTR=LPTR+1                          !Point to next oldest input value
  110 IF(LPTR.GT.20) LPTR=1                !Check for wraparound

      IGPTR=IGPTR-1                        !Point to next sample
      IF(IGPTR.LE.0) IGPTR=20              !Check for wraparound
```

```
      D=ACCUM                        !Becomes "desired" sample for filter

      CALL ADAPT(XIN,YOUT,D,RMU,W,NORD,XTDL,IPTR)        !Call to LMS routine
      ITER=ITER+1

      WRITE(7,120) ITER,XIN,D,YOUT,(W(L),L=1,NORD)  !Write out evolving weights
  120 FORMAT(1X,I5,1X,10(F8.3,1X))

      GO TO 100
      END
```

------------------------------ Printed Results ------------------------------

Iter	Xinput	Desired	Youtput	W(1)	W(2)	W(3)	W(4)	W(5)
1	-0.125	0.423	0.000	0.000	0.000	0.000	0.000	0.000
2	-1.275	0.771	0.000	-0.005	0.000	0.000	0.000	0.000
3	0.877	-1.222	-0.004	-0.011	0.007	0.001	0.000	0.000
4	-0.881	1.640	0.015	-0.018	0.014	-0.010	-0.001	0.000
5	0.930	-0.044	-0.036	-0.018	0.014	-0.010	-0.001	0.000
6	1.233	0.493	-0.001	-0.015	0.017	-0.012	0.001	-0.003
7	-1.022	1.649	0.021	-0.023	0.027	-0.004	-0.006	0.004
8	1.522	-2.569	-0.077	-0.042	0.040	-0.020	-0.018	0.015
9	-0.170	0.800	0.080	-0.043	0.045	-0.023	-0.013	0.018
10	1.489	0.091	-0.070	-0.041	0.045	-0.022	-0.014	0.019
11	-1.469	0.052	0.090	-0.041	0.045	-0.022	-0.014	0.020
12	1.068	-2.043	-0.110	-0.051	0.059	-0.036	-0.013	0.005
13	-0.258	0.314	0.110	-0.052	0.060	-0.038	-0.011	0.005
14	0.989	0.677	-0.084	-0.048	0.059	-0.034	-0.017	0.010
15	-2.891	0.050	0.173	-0.046	0.058	-0.034	-0.017	0.011
16	-0.841	-3.128	-0.147	-0.034	0.101	-0.048	-0.013	-0.005
17	-0.355	-0.385	0.055	-0.033	0.103	-0.042	-0.016	-0.004
18	0.038	2.957	0.039	-0.032	0.098	-0.054	-0.058	0.010
19	-0.901	1.057	0.071	-0.037	0.098	-0.056	-0.062	-0.004
20	0.010	-0.565	-0.066	-0.037	0.100	-0.056	-0.061	-0.002
21	-0.125	0.423	0.055	-0.037	0.100	-0.058	-0.061	-0.003
22	-1.275	0.771	0.089	-0.041	0.100	-0.058	-0.064	-0.002
23	0.877	-1.222	-0.155	-0.046	0.107	-0.057	-0.064	0.002
24	-0.881	1.640	0.215	-0.052	0.113	-0.066	-0.065	0.002
25	0.930	-0.044	-0.124	-0.052	0.113	-0.066	-0.066	0.002
1000	0.010	-0.565	-0.567	-0.017	0.981	-0.493	-0.977	0.013
1001	-0.125	0.423	0.414	-0.017	0.981	-0.493	-0.977	0.013
1002	-1.275	0.771	0.775	-0.017	0.981	-0.493	-0.977	0.013
1003	0.877	-1.222	-1.226	-0.017	0.981	-0.493	-0.977	0.013
1004	-0.881	1.640	1.626	-0.017	0.981	-0.493	-0.977	0.013
1005	0.930	-0.044	-0.069	-0.017	0.981	-0.493	-0.977	0.013
1006	1.233	0.493	0.452	-0.017	0.981	-0.493	-0.977	0.013
1007	-1.022	1.649	1.641	-0.017	0.981	-0.493	-0.977	0.013
1008	1.522	-2.569	-2.556	-0.017	0.981	-0.493	-0.977	0.013
1009	-0.170	0.800	0.808	-0.017	0.981	-0.493	-0.977	0.013
1010	1.489	0.091	0.072	-0.017	0.981	-0.493	-0.977	0.013

A.2 THE SIGNED ERROR LMS ALGORITHM

A.2.1 Adaptation Subroutine

```
      SUBROUTINE ADAPT(XIN,YOUT,D,RMU,W,NORD,XTDL,IPTR)
C
C     Signed error LMS algorithm
```

```
C
C       Parameters:
C
C              XIN     — Real input sample at time k (Input)
C
C              YOUT    — Real output sample at time k (Output)
C
C              D       — "Desired" output sample at time k (Input)
C
C              RMU     — Real stepsize for weight updating (Input)
C
C              W       — Array containing weight vector at time k (Input)
C
C              NORD    — Number of taps in the filter (Input)
C
C              XTDL    — Array for storage of past data samples (Output)
C
C              IPTR    — Pointer for delay line array (Input)
C
C              This routine performs a single iteration of the signed-error LMS
C       algorithm. The input sample is placed into the "delay line" array, the
C       output is computed, and then the weights are updated. Note that it is
C       assumed that initialization steps have been performed external to this
C       routine, i.e. delay line setup and initial weight setting definition.
C

        DIMENSION XTDL(1),W(1)

        XTDL(IPTR)=XIN                  !Write over oldest sample with incoming value

C
C       Do convolution to form the output
C
        ACCUM=Ø.                        !Clear the accumulator
        LPTR=IPTR                       !Initialize the data pointer to newest
        DO 10 L=1,NORD                  !Apply all weights to the data
        ACCUM=ACCUM+W(L)*XTDL(LPTR)     !Add in partial product
        LPTR=LPTR+1                     !Move data pointer to next oldest slot
10      IF(LPTR.GT.NORD) LPTR=1         !Check for wraparound of storage array

        YOUT=ACCUM                      !Send out accumulator

C
C       Do update of the weight vector
C
        ERROR=D-YOUT                    !Output error term
        ERROR=RMU*SIGN(1.,ERROR)        !Scaling of sign of error
        LPTR=IPTR                       !Initialize the data pointer to newest

        DO 20 L=1,NORD                  !Update all weights
        W(L)=W(L)+ERROR*XTDL(LPTR)      !Actual update
        LPTR=LPTR+1                     !Move data pointer to next oldest slot
20      IF(LPTR.GT.NORD) LPTR=1         !Check for wraparound of storage array

C
C       Clean up by preparing for next data sample
C
        IPTR=IPTR-1                     !Point to oldest sample
        IF(IPTR.LE.Ø) IPTR=NORD         !Check for wraparound of storage array

        RETURN
        END
```

A.2.2 Example Calling Program and Results

```
C
C       860115 Test program for signed-error LMS algorithm routine.
C
C              This program generates a moving average sequence of real
C              samples, and uses the adaptive filter to form a direct model.
C              The generating plant has the impulse response 0,1,-.5,-1,0 ...;
C              the filter is allowed five weights! The input sequence is a
C              pseudonoise process, with period of 20. Stepsize is hardwired
C              to 0.005.
C


C
C       Generator arrays (input sequence, generator weights)
C
        DIMENSION XINPUT(20),GENER(5)


C
C       Arrays for adaptive filter (delay line, weights)
C
        DIMENSION XTDL(10),W(10)


C
C       Hardwired data for generator
C
        DATA GENER/0.,1.,-.5,-1.,0./
        DATA XINPUT/-0.125,0.010,-0.901,0.038,-0.355,-0.841,-2.891,
     &      0.989,-0.258,1.068,-1.469,1.489,-0.170,1.522,-1.022,
     &      1.233,0.930,-0.881,0.877,-1.275/


C
C       Setup for filter and generator
C
        IGPTR=1                             !Set up delay line pointer

        DO 10 L=1,10
 10     XTDL(L)=0.                          !Clear adaptive filter delay line
        IPTR=1                              !Set up filter delay line pointer

        ITER=0                              !Iteration counter
        NORD=5                              !Hardwired filter order
        RMU=.005                            !Hardwired stepsize

C
C       Execution loop
C
 100    XIN=XINPUT(IGPTR)                   !Get input sample from list
        ACCUM=0.                           !Clear accumulator

        LPTR=IGPTR                          !Point at current input value
        DO 110 L=1,5
        ACCUM=ACCUM+GENER(L)*XINPUT(LPTR)   !Accumulate partial products
        LPTR=LPTR+1                         !Point to next oldest input value
 110    IF(LPTR.GT.20) LPTR=1               !Check for wraparound

        IGPTR=IGPTR-1                       !Point to next sample
        IF(IGPTR.LE.0) IGPTR=20             !Check for wraparound

        D=ACCUM                             !Becomes "desired" sample for filter

        CALL ADAPT(XIN,YOUT,D,RMU,W,NORD,XTDL,IPTR)      !Call to LMS routine
        ITER=ITER+1

        WRITE(7,120) ITER,XIN,D,YOUT,(W(L),L=1,NORD)  !Write out evolving weights
 120    FORMAT(1X,I5,1X,10(F8.3,1X))

        GO TO 100
        END
```

------------------------------ Printed Results ------------------------------

Iter	Xinput	Desired	Youtput	W(1)	W(2)	W(3)	W(4)	W(5)
1	-0.125	0.423	0.000	-0.001	0.000	0.000	0.000	0.000
2	-1.275	0.771	0.001	-0.007	-0.001	0.000	0.000	0.000
3	0.877	-1.222	-0.005	-0.011	0.006	0.001	0.000	0.000
4	-0.881	1.640	0.014	-0.016	0.010	-0.006	-0.001	0.000
5	0.930	-0.044	-0.028	-0.020	0.015	-0.010	0.006	0.001
6	1.233	0.493	0.001	-0.014	0.019	-0.015	0.010	-0.006
7	-1.022	1.649	0.011	-0.019	0.025	-0.010	0.006	-0.001
8	1.522	-2.569	-0.061	-0.027	0.030	-0.016	0.001	0.003
9	-0.170	0.800	0.072	-0.028	0.038	-0.021	0.007	0.008
10	1.489	0.091	-0.078	-0.020	0.037	-0.014	0.002	0.014
11	-1.469	0.052	0.077	-0.013	0.030	-0.013	-0.005	0.019
12	1.068	-2.043	-0.047	-0.018	0.037	-0.020	-0.005	0.011
13	-0.258	0.314	0.065	-0.020	0.042	-0.027	0.003	0.011
14	0.989	0.677	-0.048	-0.015	0.041	-0.022	-0.005	0.018
15	-2.891	0.050	0.058	0.000	0.036	-0.021	-0.010	0.025
16	-0.841	-3.128	-0.096	0.004	0.051	-0.026	-0.009	0.020
17	-0.355	-0.385	0.017	0.006	0.055	-0.011	-0.014	0.021
18	0.038	2.957	0.050	0.006	0.053	-0.016	-0.028	0.026
19	-0.901	1.057	-0.050	0.001	0.053	-0.017	-0.032	0.012
20	0.010	-0.565	-0.047	0.001	0.058	-0.018	-0.030	0.016
21	-0.125	0.423	0.009	0.001	0.058	-0.022	-0.030	0.014
22	-1.275	0.771	0.019	-0.006	0.057	-0.022	-0.035	0.014
23	0.877	-1.222	-0.088	-0.010	0.064	-0.021	-0.035	0.019
24	-0.881	1.640	0.096	-0.014	0.068	-0.028	-0.035	0.019
25	0.930	-0.044	-0.055	-0.010	0.064	-0.023	-0.042	0.018
.								
.								
.								
1000	0.010	-0.565	-0.549	-0.016	0.989	-0.509	-0.994	0.000
1001	-0.125	0.423	0.433	-0.016	0.989	-0.505	-0.995	0.002
1002	-1.275	0.771	0.787	-0.009	0.989	-0.505	-0.990	0.002
1003	0.877	-1.222	-1.218	-0.014	0.996	-0.504	-0.990	0.006
1004	-0.881	1.640	1.652	-0.009	0.991	-0.498	-0.990	0.006
1005	0.930	-0.044	-0.058	-0.005	0.987	-0.494	-0.996	0.005
1006	1.233	0.493	0.467	0.002	0.992	-0.498	-0.992	-0.001
1007	-1.022	1.649	1.631	-0.003	0.998	-0.493	-0.996	0.003
1008	1.522	-2.569	-2.562	-0.011	1.003	-0.499	-1.001	0.008
1009	-0.170	0.800	0.812	-0.010	0.995	-0.494	-1.007	0.003
1010	1.489	0.091	0.096	-0.018	0.996	-0.502	-1.002	-0.003

A.3 THE NORMALIZED LMS ALGORITHM

A.3.1 Adaptation Subroutine

```
      SUBROUTINE ADAPT(XIN,YOUT,D,ALPHA,GAMMA,W,NORD,XTDL,IPTR,ENERGY)
C
C         Alpha-LMS algorithm
C
C         Parameters:
C
C                 XIN     — Real input sample at time k (Input)
C
C                 YOUT    — Real output sample at time k (Output)
C
C                 D       — "Desired" output sample at time k (Input)
C
C                 ALPHA   — Real normalized stepsize for weight updating (Input)
C
C                 GAMMA   — Real denominator bound for update scaling (Input)
C
C                 W       — Array containing weight vector at time k (Input)
```

```
C
C                 NORD   — Number of taps in the filter (Input)
C
C                 XTDL   — Array for storage of past data samples (Output)
C
C                 IPTR   — Pointer for delay line array (Input)
C
C                 ENERGY — Energy in the delay line at time k (Output)
C
C                 This routine performs a single iteration of the normalized-LMS
C        algorithm.  The input sample is placed into the "delay line" array, the
C        output is computed, and then the weights are updated.  Note that it is
C        assumed that initialization steps have been performed external to this
C        routine, i.e. delay line setup and initial weight setting definition.
C

         DIMENSION XTDL(1),W(1)

         ENERGY=ENERGY-XTDL(IPTR)**2+XIN*XIN    !Update the delay line energy
         XTDL(IPTR)=XIN                         !Write over oldest sample with incoming value

C
C        Do convolution to form the output
C
         ACCUM=0.                           !Clear the accumulator
         LPTR=IPTR                          !Initialize the data pointer to newest
         DO 10 L=1,NORD                     !Apply all weights to the data
         ACCUM=ACCUM+W(L)*XTDL(LPTR)        !Add in partial product
         LPTR=LPTR+1                        !Move data pointer to next oldest slot
10       IF(LPTR.GT.NORD) LPTR=1            !Check for wraparound of storage array

         YOUT=ACCUM                         !Send out accumulator

C
C        Do update of the weight vector
C
         ERROR=D-YOUT                       !LMS error term
         STEPSIZE=ALPHA/(GAMMA+ENERGY)      !Compute normalized stepsize for time k
         ERROR=STEPSIZE*ERROR               !Scaling of error term
         LPTR=IPTR                          !Initialize the data pointer to newest

         DO 20 L=1,NORD                     !Update all weights
         W(L)=W(L)+ERROR*XTDL(LPTR)         !Actual update
         LPTR=LPTR+1                        !Move data pointer to next oldest slot
20       IF(LPTR.GT.NORD) LPTR=1            !Check for wraparound of storage array

C
C        Clean up by preparing for next data sample
C
         IPTR=IPTR-1                        !Point to oldest sample
         IF(IPTR.LE.0) IPTR=NORD            !Check for wraparound of storage array

         RETURN
         END
```

A.3.2 Example Calling Program and Results

```
C
C        860115 Test program for alpha-LMS algorithm routine.
C
C                 This program generates a moving average sequence of real
C                 samples, and uses the adaptive filter to form a direct model.
C                 The generating plant has the impulse response 0,1,-.5,-1,0 ...;
C                 the filter is allowed five weights.  The input sequence is a
C                 pseudonoise process, with period of 20.  Alpha stepsize is
C                 hardwired to 0.1.  Normalizing floor GAMMA hardwired to .1.
```

```
C
C
C       Generator arrays (input sequence, generator weights)
C
        DIMENSION XINPUT(20),GENER(5)

C
C       Arrays for adaptive filter (delay line, weights)
C
        DIMENSION XTDL(10),W(10)

C
C       Hardwired data for generator
C
        DATA GENER/0.,1.,-.5,-1.,0./
        DATA XINPUT/-0.125,0.010,-0.901,0.038,-0.355,-0.841,-2.891,
     &       0.989,-0.258,1.068,-1.469,1.489,-0.170,1.522,-1.022,
     &       1.233,0.930,-0.881,0.877,-1.275/

C
C       Setup for filter and generator
C
        IGPTR=1                          !Set up delay line pointer

        DO 10 L=1,10
     10 XTDL(L)=0.                       !Clear adaptive filter delay line
        IPTR=1                           !Set up filter delay line pointer

        ITER=0                           !Iteration counter
        NORD=5                           !Hardwired filter order
        ALPHA=.1                         !Hardwired stepsize
        ENERGY=0.                        !Initial data energy
        GAMMA=.1                         !Energy floor protection

C
C       Execution loop
C
    100 XIN=XINPUT(IGPTR)                !Get input sample from list
        ACCUM=0.                         !Clear accumulator

        LPTR=IGPTR                       !Point at current input value
        DO 110 L=1,5
        ACCUM=ACCUM+GENER(L)*XINPUT(LPTR) !Accumulate partial products
        LPTR=LPTR+1                      !Point to next oldest input value
    110 IF(LPTR.GT.20) LPTR=1           !Check for wraparound

        IGPTR=IGPTR-1                    !Point to next sample
        IF(IGPTR.LE.0) IGPTR=20          !Check for wraparound

        D=ACCUM                          !Becomes "desired" sample for filter

        CALL ADAPT(XIN,YOUT,D,ALPHA,GAMMA,W,NORD,XTDL,IPTR,ENERGY)
        ITER=ITER+1                      !Call to adaptive filter routine

        WRITE(7,120) ITER,XIN,D,YOUT,(W(L),L=1,NORD)  !Write out evolving weights
    120 FORMAT(1X,I5,1X,10(F8.3,1X))

        GO TO 100
        END
```

--------------------------------- Printed Results ---------------------------------

Iter	Xinput	Desired	Youtput	W(1)	W(2)	W(3)	W(4)	W(5)
1	-0.125	0.423	0.000	-0.046	0.000	0.000	0.000	0.000
2	-1.275	0.771	0.058	-0.098	-0.005	0.000	0.000	0.000
3	0.877	-1.222	-0.079	-0.138	0.053	0.006	0.000	0.000
4	-0.881	1.640	0.161	-0.177	0.092	-0.052	-0.006	0.000

5	0.930	−0.044	−0.285	−0.172	0.087	−0.047	−0.013	−0.001
6	1.233	0.493	−0.100	−0.159	0.097	−0.056	−0.004	−0.014
7	−1.022	1.649	0.221	−0.188	0.132	−0.030	−0.029	0.011
8	1.522	−2.569	−0.493	−0.236	0.164	−0.068	−0.058	0.038
9	−0.170	0.800	0.324	−0.237	0.176	−0.077	−0.048	0.046
10	1.489	0.091	−0.394	−0.227	0.175	−0.066	−0.055	0.054
11	−1.469	0.052	0.467	−0.219	0.167	−0.066	−0.063	0.059
12	1.068	−2.043	−0.476	−0.240	0.196	−0.095	−0.059	0.029
13	−0.258	0.314	0.317	−0.240	0.196	−0.095	−0.059	0.029
14	0.989	0.677	−0.258	−0.226	0.192	−0.080	−0.080	0.050
15	−2.891	0.050	0.706	−0.212	0.187	−0.078	−0.085	0.058
16	−0.841	−3.128	−0.357	−0.191	0.258	−0.103	−0.079	0.032
17	−0.355	−0.385	0.061	−0.189	0.261	−0.090	−0.083	0.033
18	0.038	2.957	0.249	−0.188	0.252	−0.112	−0.160	0.059
19	−0.901	1.057	0.183	−0.196	0.252	−0.115	−0.167	0.034
20	0.010	−0.565	−0.203	−0.196	0.271	−0.116	−0.160	0.051
21	−0.125	0.423	0.108	−0.200	0.271	−0.143	−0.158	0.041
22	−1.275	0.771	0.364	−0.220	0.269	−0.143	−0.173	0.041
23	0.877	−1.222	−0.558	−0.238	0.295	−0.140	−0.173	0.059
24	−0.881	1.640	0.670	−0.264	0.321	−0.178	−0.177	0.060
25	0.930	−0.044	−0.467	−0.255	0.312	−0.169	−0.190	0.058

```
            .
            .
            .
```

500	0.010	−0.565	−0.565	−0.011	0.987	−0.499	−0.988	0.009
501	−0.125	0.423	0.420	−0.011	0.987	−0.499	−0.988	0.008
502	−1.275	0.771	0.776	−0.011	0.987	−0.499	−0.987	0.008
503	0.877	−1.222	−1.223	−0.011	0.987	−0.499	−0.987	0.008
504	−0.881	1.640	1.634	−0.011	0.987	−0.499	−0.987	0.008
505	0.930	−0.044	−0.059	−0.010	0.987	−0.499	−0.988	0.008
506	1.233	0.493	0.467	−0.010	0.987	−0.499	−0.987	0.008
507	−1.022	1.649	1.640	−0.010	0.987	−0.499	−0.988	0.008
508	1.522	−2.569	−2.565	−0.010	0.987	−0.499	−0.988	0.008
509	−0.170	0.800	0.804	−0.010	0.987	−0.499	−0.988	0.008
510	1.489	0.091	0.077	−0.010	0.987	−0.499	−0.988	0.008

A.4 THE GRIFFITHS ALGORITHM

A.4.1 Adaptation Subroutine

```
      SUBROUTINE ADAPT(XIN,YOUT,PVEC,RMU,W,NORD,XTDL,IPTR)
C
C     Griffiths LMS algorithm
C
C     Parameters:
C
C             XIN     — Real input sample at time k (Input)
C
C             YOUT    — Real output sample at time k (Output)
C
C             PVEC    — Crosscorrelation vector of input and desired (Input)
C
C             RMU     — Real stepsize for weight updating (Input)
C
C             W       — Array containing weight vector at time k (Input)
C
C             NORD    — Number of taps in the filter (Input)
C
C             XTDL    — Array for storage of past data samples (Output)
C
C             IPTR    — Pointer for delay line array (Input)
C
C             This routine performs a single iteration of the Griffiths LMS
C     algorithm.  The input sample is placed into the "delay line" array, the
C     output is computed, and then the weights are updated.  Note that it is
C     assumed that initialization steps have been performed external to this
C     routine, i.e. delay line setup and initial weight setting definition.
```

```
C

      DIMENSION XTDL(1),W(1),PVEC(1)

      XTDL(IPTR)=XIN                    !Write over oldest sample with incoming value

C
C     Do convolution to form the output
C
      ACCUM=0.                          !Clear the accumulator
      LPTR=IPTR                         !Initialize the data pointer to newest
      DO 10 L=1,NORD                    !Apply all weights to the data
      ACCUM=ACCUM+W(L)*XTDL(LPTR)       !Add in partial product
      LPTR=LPTR+1                       !Move data pointer to next oldest slot
10    IF(LPTR.GT.NORD) LPTR=1           !Check for wraparound of storage array

      YOUT=ACCUM                        !Send out accumulator

C
C     Do update of the weight vector
C
      LPTR=IPTR                         !Initialize the data pointer to newest

      DO 20 L=1,NORD                    !Update all weights
      UPDATE=RMU*(PVEC(L)-YOUT*XTDL(LPTR))  !P vector type update
      W(L)=W(L)+UPDATE                  !Actual update
      LPTR=LPTR+1                       !Move data pointer to next oldest slot
20    IF(LPTR.GT.NORD) LPTR=1           !Check for wraparound of storage array

C
C     Clean up by preparing for next data sample
C
      IPTR=IPTR-1                       !Point to oldest sample
      IF(IPTR.LE.0) IPTR=NORD           !Check for wraparound of storage array

      RETURN
      END
```

A.4.2 Example Calling Program and Results

```
C
C     860115 Test program for Griffiths algorithm routine.
C
C           This program uses an adaptive filter to form a direct model
C           of a generating plant with an impulse response 0,1,-.5,-1,0 ...;
C           the filter is allowed five weights. The input sequence is a
C           pseudonoise process, with period of 20. Adaptation is by means
C           of the Griffiths P-vector algorithm. Stepsize is
C           hardwired to 0.01.
C


C
C     Generator arrays (input sequence, generator weights)
C
      DIMENSION XINPUT(20),PVEC(5)

C
C     Arrays for adaptive filter (delay line, weights)
C
      DIMENSION XTDL(10),W(10)

C
C     Hardwired data for generator (measured P vector)
C
      DATA PVEC/-0.423676,1.070822,-0.638163,-0.610532,-0.011973/
      DATA XINPUT/-0.125,0.010,-0.901,0.038,-0.355,-0.841,-2.891,
     &    0.989,-0.258,1.068,-1.469,1.489,-0.170,1.522,-1.022,
     &    1.233,0.930,-0.881,0.877,-1.275/
```

```
C
C       Setup for filter and generator
C

        DO 10 L=1,10
   10   XTDL(L)=0.                       !Clear adaptive filter delay line
        IPTR=1                           !Set up filter delay line pointer

        ITER=0                           !Iteration counter
        NORD=5                           !Hardwired filter order
        RMU=.01                          !Hardwired stepsize

C
C       Execution loop
C
  100   XIN=XINPUT(IGPTR)                !Get input sample from list
        IGPTR=IGPTR-1                    !Point to next sample
        IF(IGPTR.LE.0) IGPTR=20          !Check for wraparound

        CALL ADAPT(XIN,YOUT,PVEC,RMU,W,NORD,XTDL,IPTR)
        ITER=ITER+1                      !Call to adaptive filter routine


        WRITE(7,120) ITER,XIN,YOUT,(W(L),L=1,NORD)  !Write out evolving weights
  120   FORMAT(1X,I5,1X,10(F8.3,1X))

        GO TO 100
        END
```

---------------------------------- Printed Results ----------------------------------

Iter	Xinput	Youtput	W(1)	W(2)	W(3)	W(4)	W(5)
1	0.000	0.000	-0.004	0.011	-0.006	-0.006	0.000
2	-1.275	0.005	-0.008	0.021	-0.013	-0.012	0.000
3	0.877	-0.035	-0.012	0.032	-0.019	-0.018	0.000
4	-0.881	0.063	-0.016	0.042	-0.025	-0.024	0.000
5	0.930	-0.042	-0.020	0.052	-0.031	-0.031	-0.001
6	1.233	0.025	-0.024	0.063	-0.037	-0.037	0.000
7	-1.022	0.100	-0.028	0.072	-0.044	-0.043	-0.001
8	1.522	-0.209	-0.029	0.081	-0.048	-0.047	-0.003
9	-0.170	0.116	-0.033	0.090	-0.053	-0.054	-0.005
10	1.489	-0.095	-0.036	0.100	-0.058	-0.061	-0.004
11	-1.469	0.121	-0.038	0.109	-0.064	-0.069	-0.002
12	1.068	-0.288	-0.039	0.116	-0.066	-0.076	0.002
13	-0.258	0.118	-0.043	0.125	-0.071	-0.084	0.002
14	0.989	-0.025	-0.047	0.136	-0.077	-0.090	0.002
15	-2.891	0.191	-0.046	0.144	-0.083	-0.098	0.005
16	-0.841	-0.431	-0.054	0.143	-0.085	-0.106	0.009
17	-0.355	0.038	-0.058	0.154	-0.090	-0.112	0.009
18	0.038	0.353	-0.062	0.166	-0.094	-0.108	0.006
19	-0.901	0.170	-0.065	0.176	-0.100	-0.113	0.011
20	0.010	-0.132	-0.069	0.186	-0.106	-0.119	0.009
21	-0.125	0.098	-0.073	0.197	-0.111	-0.125	0.010
22	-1.275	0.181	-0.075	0.208	-0.118	-0.130	0.009
23	0.877	-0.325	-0.076	0.214	-0.125	-0.136	0.006
24	-0.881	0.431	-0.077	0.221	-0.126	-0.142	0.006
25	0.930	-0.197	-0.079	0.230	-0.130	-0.150	0.006
.							
.							
.							
500	0.010	-0.568	-0.045	0.941	-0.475	-0.906	0.025
501	-0.125	0.400	-0.048	0.952	-0.478	-0.912	0.026
502	-1.275	0.761	-0.043	0.964	-0.484	-0.911	0.026
503	0.877	-1.238	-0.036	0.959	-0.492	-0.917	0.015
504	-0.881	1.615	-0.026	0.955	-0.478	-0.921	0.014
505	0.930	-0.112	-0.029	0.965	-0.483	-0.929	0.014

506	1.233	0.454	-0.039	0.971	-0.486	-0.939	0.020
507	-1.022	1.631	-0.027	0.962	-0.507	-0.931	0.005
508	1.522	-2.520	0.007	0.947	-0.483	-0.914	-0.017
509	-0.170	0.791	0.004	0.946	-0.481	-0.929	-0.024
510	1.489	0.034	0.000	0.956	-0.488	-0.935	-0.025

A.5 THE RECURSIVE LEAST SQUARES ALGORITHM

A.5.1 Adaptation Subroutine

```
      SUBROUTINE ADAPT(XIN,YOUT,D,RHO,W,NORD,XTDL,IPTR,RINV)
C
C         RLS ALGORITHM
C
C         Parameters:
C
C                 XIN     — Real input sample at time k (Input)
C
C                 YOUT    — Real output sample at time k (Output)
C
C                 D       — "Desired" output sample at time k (Input)
C
C                 RHO     — Real exponential "forgetting" factor (Input)
C
C                 W       — Array containing weight vector at time k (Input)
C
C                 NORD    — Number of taps in the filter (Input)
C
C                 XTDL    — Array for storage of past data samples (Output)
C
C                 IPTR    — Pointer for delay line array (Input)
C
C                 RINV    — Real array for storage of inverse correlation
C                           matrix ('NORD'x'NORD long)
C
C             This routine performs a single iteration of the RLS algorithm.
C         The input sample is placed into the "delay line" array, the output is
C         computed, and then the weights are updated. Note that it is assumed
C         that initialization steps have been performed external to this routine,
C         i.e. delay line setup, initial weight setting definition, and matrix
C         inverse setup.
C
      PARAMETER MAXORD=255           !MODIFY THIS IF FILTER LENGTH > 255
      DIMENSION Z(MAXORD)            !Local storage of information vector
      DIMENSION XTDL(1),W(1),RINV(1) !Dummy storage of delay line, weights, inverse
C
C         Execution starts here
C
      XTDL(IPTR)=XIN                 !Write over oldest sample with incoming value
C
C         Do convolution to form the apriori output
C
      ACCUM=0.                       !Clear the accumulator
      LPTR=IPTR                      !Initialize the data pointer to newest
      DO 10 L=1,NORD                 !Apply all weights to the data
      ACCUM=ACCUM+W(L)*XTDL(LPTR)    !Add in partial product
      LPTR=LPTR+1                    !Move data pointer to next oldest slot
10    IF(LPTR.GT.NORD) LPTR=1        !Check for wraparound of storage array

      YOUT=ACCUM                     !Send out accumulator

C
```

```
C          Compute the apriori error sample
C
       E0=D-YOUT

C
C          Compute filtered information vector Z
C
       DO 20 I=1,NORD                      !Matrix product of Rinv and Xtdl
       LPTR=IPTR                           !Initialize pointer to newest sample
       ACCUM=0.                            !Clear accumulator
       DO 18 J=1,NORD
       ACCUM=ACCUM+RINV((J-1)*NORD+I)*XTDL(LPTR)
       LPTR=LPTR+1                         !Point to next oldest sample
18     IF(LPTR.GT.NORD) LPTR=1             !Check for wraparound of storage
20     Z(I)=ACCUM                          !Information vector component

C
C          Form normalized error power
C
       ACCUM=0.                            !Clear accumulator for inner product
       LPTR=IPTR                           !Initialize pointer to newest sample
       DO 30 L=1,NORD
       ACCUM=ACCUM+XTDL(LPTR)*Z(L)         !Inner product of two vectors
       LPTR=LPTR+1                         !Point to next oldest sample
30     IF(LPTR.GT.NORD) LPTR=1             !Check for wraparound
       Q=ACCUM

C
C          Form gain constant for update
C
       V=1./(RHO+Q)

C
C          Do weight updating
C
       UPDATE=E0*V                         !Update stepsize
       DO 40 L=1,NORD
40     W(L)=W(L)+UPDATE*Z(L)               !Update each weight

C
C          Update inverse matrix with new information
C
       DO 50 I=1,NORD
       DO 50 J=I,NORD
       JPTR=(J-1)*NORD+I
       KPTR=(I-1)*NORD+J
       RINV(JPTR)=(RINV(JPTR)-Z(I)*Z(J)*V)/RHO
50     RINV(KPTR)=RINV(JPTR)               !Symmetric matrix

C
C          Clean up by preparing for next data sample
C
       IPTR=IPTR-1                         !Point to oldest sample
       IF(IPTR.LE.0) IPTR=NORD             !Check for wraparound of storage array

       RETURN
       END
```

A.5.2 Example Calling Program and Results

```
C
C          860120 Test program for RLS algorithm routine.
C
C                  This program generates a moving average sequence of real
C                  samples, and uses the adaptive filter to form a direct model.
C                  The generating plant has the impulse response 0,1,-.5,-1,0 ...;
C                  the filter is allowed five weights.  The input sequence is a
C                  pseudonoise process, with period of 20.  R inverse is initialized
C                  to 1000 of the identity matrix.  Forgetting factor RHO is
C                  hardwired to .9.
```

```
C
C
C      Generator arrays (input sequence, generator weights)
C
       DIMENSION XINPUT(20),GENER(5)
C
C      Arrays for adaptive filter (delay line, weights, inverse matrix)
C
       DIMENSION XTDL(10),W(10),RINV(25)
C
C      Hardwired data for generator
C
       DATA GENER/0.,1.,-.5,-1.,0./
       DATA XINPUT/-0.125,0.010,-0.901,0.038,-0.355,-0.841,-2.891,
     &    0.989,-0.258,1.068,-1.469,1.489,-0.170,1.522,-1.022,
     &    1.233,0.930,-0.881,0.877,-1.275/
C
C      Setup for filter and generator
C
       IGPTR=1                        !Set up delay line pointer

       DO 10 L=1,10
    10 XTDL(L)=0.                     !Clear adaptive filter delay line
       IPTR=1                         !Set up filter delay line pointer
       ITER=0                         !Iteration counter
       NORD=5                         !Hardwired filter order
       RHO=.9                         !Forgetting factor
C
C      Initial inverse matrix
C
       DO 20 I=1,NORD
    20 RINV((I-1)*NORD+I)=1000.       !Diagonal matrix
C
C      Execution loop
C
   100 XIN=XINPUT(IGPTR)              !Get input sample from list
       ACCUM=0.                       !Clear accumulator

       LPTR=IGPTR                     !Point at current input value
       DO 110 L=1,5
       ACCUM=ACCUM+GENER(L)*XINPUT(LPTR)  !Accumulate partial products
       LPTR=LPTR+1                    !Point to next oldest input value
   110 IF(LPTR.GT.20) LPTR=1          !Check for wraparound
       IGPTR=IGPTR-1                  !Point to next sample
       IF(IGPTR.LE.0) IGPTR=20        !Check for wraparound

       D=ACCUM                        !Becomes "desired" sample for filter

       CALL ADAPT(XIN,YOUT,D,RHO,W,NORD,XTDL,IPTR,RINV)
       ITER=ITER+1                    !Call to adaptive filter routine


       WRITE(7,120) ITER,XIN,D,YOUT,(W(L),L=1,NORD)  !Write out evolving weights
   120 FORMAT(1X,I5,1X,10(F8.3,1X))

       GO TO 100
       END
```

---------------------------- Printed Results ----------------------------

Iter	Xinput	Desired	Youtput	W(1)	W(2)	W(3)	W(4)	W(5)
1	-0.125	0.423	0.000	-3.196	0.000	0.000	0.000	0.000
2	-1.275	0.771	4.075	-1.010	3.934	0.000	0.000	0.000
3	0.877	-1.222	-5.902	-0.678	0.530	-0.368	0.000	0.000
4	-0.881	1.640	1.531	-0.679	0.537	-0.446	-0.007	0.000

5	0.930	-0.044	-1.487	-0.678	0.528	-0.350	-1.056	-0.097
6	1.233	0.493	-0.839	-0.678	0.529	-0.359	-0.966	-1.073
7	-1.022	1.649	0.922	-0.579	0.798	-0.240	-1.043	-0.844
8	1.522	-2.569	-2.220	-0.544	0.811	-0.367	-1.126	-0.742
9	-0.170	0.800	-0.377	-0.277	0.919	-0.440	-0.869	-0.172
10	1.489	0.091	-0.563	-0.117	0.958	-0.429	-0.940	-0.007
11	-1.469	0.052	0.249	-0.093	0.930	-0.473	-0.967	0.026
12	1.068	-2.043	-1.967	-0.089	0.943	-0.480	-0.984	0.007
13	-0.258	0.314	-0.269	-0.086	0.943	-0.486	-0.977	0.012
14	0.989	0.677	0.605	-0.078	0.954	-0.486	-0.986	0.019
15	-2.891	0.050	0.213	-0.047	0.957	-0.509	-0.994	0.028
16	-0.841	-3.128	-2.945	-0.024	0.997	-0.502	-1.009	0.007
17	-0.355	-0.385	-0.379	-0.024	0.997	-0.501	-1.009	0.007
18	0.038	2.957	2.991	-0.023	0.997	-0.497	-1.001	0.008
19	-0.901	1.057	1.055	-0.023	0.997	-0.497	-1.002	0.007
20	0.010	-0.565	-0.568	-0.023	0.997	-0.497	-1.002	0.007
21	-0.125	0.423	0.420	-0.023	0.997	-0.497	-1.002	0.007
22	-1.275	0.771	0.802	-0.018	0.997	-0.497	-0.999	0.006
23	0.877	-1.222	-1.240	-0.016	0.995	-0.498	-0.999	0.004
24	-0.881	1.640	1.646	-0.016	0.994	-0.497	-0.999	0.004
25	0.930	-0.044	-0.054	-0.015	0.994	-0.497	-1.000	0.003

A.6 THE CONSTANT MODULUS ALGORITHM

A.6.1 Adaptation Subroutine

```
      SUBROUTINE ADAPT(XIN,YOUT,RMU,W,NORD,XTDL,IPTR)
C
C        CMA ALGORITHM
C
C        Parameters:
C
C             XIN    — Complex input sample at time k (Input)
C
C             YOUT   — Complex output sample at time k (Output)
C
C             RMU    — Real stepsize for weight updating (Input)
C
C             W      — Complex array containing weight vector at time k (Input)
C
C             NORD   — Number of taps in the filter (Input)
C
C             XTDL   — Complex array to store past data samples (Output)
C
C             IPTR   — Pointer for delay line array (Input)
C
C             This routine performs a single iteration of the complex CMA
C        algorithm.  The input sample is placed into the "delay line" array, the
C        output is computed, and then the weights are updated.  Note that it is
C        assumed that initialization steps have been performed external to this
C        routine, i.e. delay line setup and initial non-zero weight setting
C        definition.
C

      COMPLEX XTDL(1),W(1),ERROR,YOUT,XIN,ACCUM

C
C        Begin execution here
C
      XTDL(IPTR)=XIN                        !Write over oldest sample with incoming value

C
C        Do convolution to form the output
```

```
C
        ACCUM=0.                          !Clear the accumulator
        LPTR=IPTR                         !Initialize the data pointer to newest
        DO 10 L=1,NORD                    !Apply all weights to the data
        ACCUM=ACCUM+W(L)*XTDL(LPTR)       !Add in partial product
        LPTR=LPTR+1                       !Move data pointer to next oldest slot
10      IF(LPTR.GT.NORD) LPTR=1           !Check for wraparound of storage array

        YOUT=ACCUM                        !Send out accumulator

C
C       Do update of the weight vector
C
        ERROR=(CABS(YOUT)-1.)*YOUT        !CMA error term
        ERROR=RMU*ERROR                   !Scaling of error term
        LPTR=IPTR                         !Initialize the data pointer to newest

        DO 20 L=1,NORD                    !Update all weights
        W(L)=W(L)-ERROR*CONJG(XTDL(LPTR)) !Actual update
        LPTR=LPTR+1                       !Move data pointer to next oldest slot
20      IF(LPTR.GT.NORD) LPTR=1           !Check for wraparound of storage array

C
C       Clean up by preparing for next data sample
C
        IPTR=IPTR-1                       !Point to oldest sample
        IF(IPTR.LE.0) IPTR=NORD           !Check for wraparound of storage array

        RETURN
        END
```

A.6.2 Example Calling Program and Results

Figure A.1 illustrates the behavior in the frequency domain for the given test signal. At convergence, the secondary sinusoid should exhibit significant rejection. This type of verification requires the user to supply appropriate spectral and display routines.

```
C
C       860121 Test program for CMA algorithm routine.
C
C               This program uses an input consisting of two complex tones and
C               low-level white noise. The adaptive filter is used to reject
C               the weaker tone. The filter is allowed five weights.
C

C
C       Input array
C
        COMPLEX XINPUT(20),YOUT
C
C       Arrays for adaptive filter (delay line, weights, inverse matrix)
C
        COMPLEX XTDL(10),W(10),XIN
C
C       Hardwired data for input
C
        DATA XINPUT/(1.496,-0.012),(-0.410,0.704),(-0.843,0.476),
     &   (0.154,-1.481),(0.591,0.300),(0.487,1.013),(-1.419,-0.296),
     &   (0.163,-0.500),(1.142,-0.459),(-0.409,1.291),(-0.493,0.009),
     &   (-0.408,-1.305),(1.152,0.467),(0.143,0.519),(-1.396,0.283),
     &   (0.506,-1.002),(0.615,-0.281),(0.158,1.485),(-0.847,-0.500),
     &   (-0.413,-0.706)/

C
C       Setup for filter
```

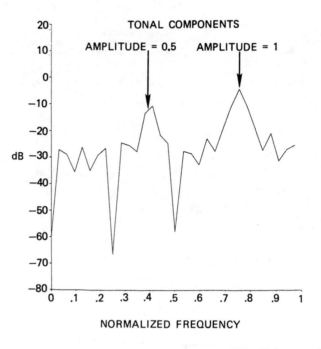

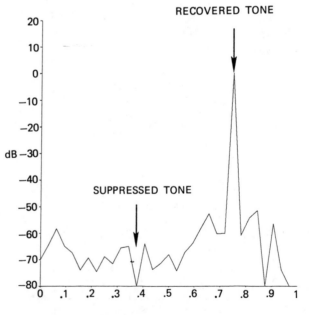

FIGURE A.1. Input and output spectral content for CMA test (a) input power spectral density (b) output power spectral density following convergence.

272

```
C
      W(1)=.001                        INontrivial initial weights
      IGPTR=1                          IPointer into input array
      DO 10 L=1,10
   10 XTDL(L)=0.                       IClear adaptive filter delay line
      IPTR=1                           ISet up filter delay line pointer
      ITER=0                           IIteration counter
      NORD=5                           IHardwired filter order
      RMU=.01                          IStepsize

C
C     Execution loop
C
  100 XIN=XINPUT(IGPTR)                IGet input sample from list

      CALL ADAPT(XIN,YOUT,RMU,W,NORD,XTDL,IPTR)
      ITER=ITER+1                      ICall to adaptive filter routine

  110 WRITE(7,120) ITER,XIN,YOUT,(CABS(W(L)),L=1,NORD)   IWrite out evolving weights
  120 FORMAT(1X,I5,1X,5(F8.3,1X),4(F6.3,1X))

      IGPTR=IGPTR-1                    IPoint to next input sample
      IF(IGPTR.LE.0) IGPTR=20          ICheck for wraparound

      GO TO 100
      END
```

---------------------------- Printed Results ----------------------------

						Weight Magnitudes			
Iter	Complex Xinput		Complex Youtput		W(1)	W(2)	W(3)	W(4)	W(5)	
1	1.496	-0.012	0.001	0.000	0.001	0.000	0.000	0.000	0.000	
2	-0.413	-0.706	0.000	-0.001	0.001	0.000	0.000	0.000	0.000	
3	-0.847	-0.500	-0.001	-0.001	0.001	0.000	0.000	0.000	0.000	
4	0.158	1.485	0.000	0.002	0.001	0.000	0.000	0.000	0.000	
5	0.615	-0.281	0.001	0.000	0.001	0.000	0.000	0.000	0.000	
6	0.506	-1.002	0.001	-0.001	0.001	0.000	0.000	0.000	0.000	
7	-1.396	0.283	-0.002	0.000	0.001	0.000	0.000	0.000	0.000	
8	0.143	0.519	0.000	0.001	0.001	0.000	0.000	0.000	0.000	
9	1.152	0.467	0.002	0.001	0.001	0.000	0.000	0.000	0.000	
10	-0.408	-1.305	0.000	-0.002	0.001	0.000	0.000	0.000	0.000	
11	-0.493	0.009	-0.001	0.000	0.001	0.000	0.000	0.000	0.000	
12	-0.409	1.291	0.000	0.002	0.001	0.000	0.000	0.000	0.000	
13	1.142	-0.459	0.002	-0.001	0.001	0.000	0.000	0.000	0.000	
14	0.163	-0.500	0.000	-0.001	0.001	0.000	0.000	0.000	0.000	
15	-1.419	-0.296	-0.002	0.000	0.001	0.000	0.000	0.000	0.000	
16	0.487	1.013	0.001	0.002	0.001	0.000	0.000	0.000	0.000	
17	0.591	0.300	0.001	0.000	0.001	0.000	0.000	0.000	0.000	
18	0.154	-1.481	0.000	-0.003	0.001	0.000	0.000	0.000	0.000	
19	-0.843	0.476	-0.002	0.001	0.001	0.000	0.000	0.000	0.000	
20	-0.410	0.704	0.000	0.002	0.001	0.000	0.000	0.000	0.000	
21	1.496	-0.012	0.003	0.000	0.001	0.000	0.000	0.000	0.000	
22	-0.413	-0.706	-0.001	-0.002	0.001	0.000	0.000	0.000	0.000	
23	-0.847	-0.500	-0.002	-0.001	0.001	0.000	0.000	0.000	0.000	
24	0.158	1.485	0.000	0.004	0.001	0.000	0.000	0.000	0.000	
25	0.615	-0.281	0.003	-0.001	0.002	0.000	0.000	0.000	0.000	
.										
.										
.										
100	-0.410	0.704	0.007	0.094	0.023	0.022	0.021	0.022	0.022	
101	1.496	-0.012	0.115	0.004	0.025	0.023	0.022	0.024	0.022	
102	-0.413	-0.706	-0.018	-0.120	0.026	0.024	0.022	0.025	0.024	
103	-0.847	-0.500	-0.107	0.005	0.026	0.025	0.024	0.026	0.025	
104	0.158	1.485	-0.002	0.134	0.028	0.026	0.024	0.027	0.026	
105	0.615	-0.281	0.133	-0.021	0.029	0.028	0.025	0.028	0.027	
106	0.506	-1.002	-0.001	-0.124	0.030	0.029	0.027	0.029	0.028	

```
107   -1.396    0.283   -0.156    0.000    0.032 0.030 0.028 0.031 0.029
108    0.143    0.519    0.023    0.147    0.032 0.032 0.029 0.032 0.031
109    1.152    0.467    0.143    0.003    0.034 0.032 0.031 0.033 0.032
110   -0.408   -1.305   -0.006   -0.177    0.036 0.034 0.031 0.035 0.033
         .
         .
         .
500   -0.410    0.704    0.057    0.988    0.198 0.189 0.232 0.189 0.197
501    1.496   -0.012    1.000   -0.048    0.198 0.189 0.232 0.189 0.197
502   -0.413   -0.706   -0.072   -1.003    0.198 0.189 0.232 0.189 0.197
```

A.7 THE SIMPLE HYPERSTABLE ADAPTIVE RECURSIVE FILTER ALGORITHM

A.7.1 Adaptation Subroutine

```
      SUBROUTINE ADAPT(XIN,YOUT,D,RMU,W,NA,NB,NC,XTDL,IPTR,JPTR,KPTR)
C
C       SHARF ALGORITHM
C
C       Parameters:
C
C           XIN    — Real input sample at time k (Input)
C
C           YOUT   — Real output sample at time k (Output)
C
C           D      — Real desired output sample at time k (Input)
C
C           RMU    — Real array, 2 stepsizes for weight updating
C                         RMU(1) for forward, RMU(2) for feedback (Input)
C
C           W      — Real array of weights, NA A's, NB B's, NC C's
C                         A's forward weights, B's feedback weights,
C                         C's error smoothing weights (Input)
C
C           NA     — Number of forward taps in the filter (Input)
C
C           NB     — Number of feedback taps in the filter (Input)
C
C           NC     — Number of error filter taps (Input)
C
C           XTDL   — Real array for storage of past data samples (Output)
C
C           IPTR   — Pointer for input delay line array (Input)
C
C           JPTR   — Pointer for output delay line array (Input)
C
C           KPTR   — Pointer for error delay line array (Input)
C
C
C           This routine performs a single iteration of the SHARF
C       algorithm.  The input sample is placed into the forward "delay line"
C       array, the output is computed, the error filtered, and the weights
C       updated.  Note that it is assumed that necessary initialization steps
C       have been performed external to this routine, i.e. delay line setup and
C       initialization of any non-zero weight settings.
C
      DIMENSION W(1),XTDL(1),RMU(1)


C
C       Begin execution here
C
      XTDL(IPTR)=XIN                          !Write over oldest sample with incoming value
```

```
C
C       Do forward convolution
C
        ACCUM=0.                            !Clear the accumulator
        LPTR=IPTR                           !Initialize the data pointer to newest
        DO 10 L=1,NA                        !Apply all weights to the data
        ACCUM=ACCUM+W(L)*XTDL(LPTR)         !Add in partial product
        LPTR=LPTR+1                         !Move data pointer to next oldest slot
10      IF(LPTR.GT.NA) LPTR=1               !Check for wraparound of storage array

        YOUT=ACCUM                          !Hold accumulator for now
C
C       Do feedback convolution
C
        ACCUM=0.                            !Clear the accumulator
        LPTR=JPTR+NA                        !Initialize the data pointer to newest
        DO 15 L=NA+1,NA+NB                  !Apply all weights to the data
        ACCUM=ACCUM+W(L)*XTDL(LPTR)         !Add in partial product
        LPTR=LPTR+1                         !Move data pointer to next oldest slot
15      IF(LPTR.GT.NA+NB) LPTR=NA+1         !Check for wraparound of storage array

        YOUT=YOUT+ACCUM                     !Hold accumulator for now


C
C       Filter error
C
        XTDL(KPTR+NA+NB)=D-YOUT             !Compute and save apriori error
        ACCUM=0.                            !Clear the accumulator
        LPTR=KPTR+NA+NB                     !Initialize the data pointer to newest
        DO 18 L=NA+NB+1,NA+NB+NC            !Apply all weights to the data
        ACCUM=ACCUM+W(L)*XTDL(LPTR)         !Add in partial product
        LPTR=LPTR+1                         !Move data pointer to next oldest slot
18      IF(LPTR.GT.NA+NB+NC) LPTR=NA+NB+1   !Check for wraparound of storage array

        V=ACCUM                             !Error for adaptation


C
C       Do update of the forward weights
C
        UPDATE=RMU(1)*V                     !SHARF error term
        LPTR=IPTR                           !Initialize the data pointer to newest

        DO 20 L=1,NA                        !Update forward weights
        W(L)=W(L)+UPDATE*XTDL(LPTR)         !Actual update
        LPTR=LPTR+1                         !Move data pointer to next oldest slot
20      IF(LPTR.GT.NA) LPTR=1               !Check for wraparound of storage array


C
C       Do update of the feedback weights
C
        UPDATE=RMU(2)*V                     !SHARF error term
        LPTR=JPTR+NA                        !Initialize the data pointer to newest

        DO 25 L=NA+1,NA+NB                  !Update feedback weights
        W(L)=W(L)+UPDATE*XTDL(LPTR)         !Actual update
        LPTR=LPTR+1                         !Move data pointer to next oldest slot
25      IF(LPTR.GT.NA+NB) LPTR=NA+1         !Check for wraparound of storage array



C
C       Clean up by preparing for next data sample
C
        IPTR=IPTR-1                         !Point to oldest sample
        IF(IPTR.LE.0) IPTR=NA               !Check for wraparound of storage array
        JPTR=JPTR-1
        IF(JPTR.LE.0) JPTR=NB
```

```
      XTDL(JPTR+NA)=YOUT                    !Save output for next pass

      KPTR=KPTR-1
      IF(KPTR.LE.0) KPTR=NC

      RETURN
      END
```

A.7.2 Example Calling Program and Results

```
C
C        860129 Test program for SHARF algorithm routine.
C
C              This program generates a sequence of an ARMA process of real
C              samples, and uses the adaptive filter to form a direct model.
C              The generating plant has forward taps of 0,1,-.5,0,..., and
C              feedback taps of -1.4 and .6.  The adaptive filter is allowed
C              4 forward and 2 backward weights.  Error smoothing is done with
C              1,-1.  The input sequence is a pseudonoise process, with period
C              of 20.  Stepsizes are hardwired to .1 and .05.
C
C
C
C        Generator arrays (input sequence, generator weights)
C
      DIMENSION XINPUT(20),AGEN(3),BGEN(2)
C
C        Arrays for adaptive filter (delay line, weights)
C
      DIMENSION XTDL(30),W(30),RMU(2)
C
C        Hardwired data for generator
C
      DATA AGEN/0.,1.,-.5/
      DATA BGEN/-1.4,.6/
      DATA XINPUT/-0.125,0.010,-0.901,0.038,-0.355,-0.841,-2.891,
     &     0.989,-0.258,1.068,-1.469,1.489,-0.170,1.522,-1.022,
     &     1.233,0.930,-0.881,0.877,-1.275/
C
C        Setup for filter and generator
C
      IGPTR=1                               !Set up delay line pointer

      DO 10 L=1,30
   10 XTDL(L)=0.

      IPTR=1                                !Set up filter delay line pointer
      JPTR=1                                ! for all three sections
      KPTR=1
      NA=3                                  !Number of forward filter coefficients
      NB=2                                  !Number of feedback filter coefficients
      NC=2                                  !Number of error smoothing coefficients
      W(6)=1.                               !Error smoothing coefficients
      W(7)=-1.

      RMU(1)=.1                             !Forward stepsize
      RMU(2)=.05                            !Feedback stepsize

      ITER=0                                !Iteration counter
C
C        Execution loop, generate desired signal, go to adaptation
C
  100 XIN=XINPUT(IGPTR)                     !Get input sample from list
      ACCUM=0.                              !Clear accumulator

      LPTR=IGPTR                            !Point at current input value
```

```
      DO 110 L=1,3
      ACCUM=ACCUM+AGEN(L)*XINPUT(LPTR)    !Accumulate partial products
      LPTR=LPTR+1                         !Point to next oldest input value
110   IF(LPTR.GT.20) LPTR=1               !Check for wraparound

      IGPTR=IGPTR-1                       !Point to next sample
      IF(IGPTR.LE.0) IGPTR=20             !Check for wraparound

      D=ACCUM-BGEN(1)*Y1-BGEN(2)*Y2       !Becomes "desired" sample for filter
      Y2=Y1
      Y1=D

      CALL ADAPT(XIN,YOUT,D,RMU,W,NA,NB,NC,XTDL,IPTR,JPTR,KPTR)
      ITER=ITER+1                         !Call to adaptive filter routine


      WRITE(7,120) ITER,XIN,D,YOUT,(W(L),L=1,NA+NB)  !Write out evolving weights
120   FORMAT(1X,I5,1X,10(F8.3,1X))

      GO TO 100
      END
```

---------------------------- Printed Results -----------------------------

Iter	Xinput	Desired	Youtput	A(1)	A(2)	A(3)	B(1)	B(2)
1	-0.125	0.461	0.000	-0.006	0.000	0.000	0.000	0.000
2	-1.275	0.515	0.007	-0.012	-0.001	0.000	0.000	0.000
3	0.877	-0.768	-0.010	-0.123	0.161	0.016	0.000	0.000
4	-0.881	0.130	0.229	-0.181	0.219	-0.068	-0.001	0.000
5	0.930	-0.676	-0.421	-0.195	0.232	-0.082	-0.003	0.000
6	1.233	0.346	0.049	-0.127	0.284	-0.131	-0.014	0.007
7	-1.022	1.658	0.355	-0.230	0.408	-0.037	-0.012	-0.015
8	1.522	0.475	-0.818	-0.232	0.409	-0.038	-0.012	-0.015
9	-0.170	1.703	0.705	-0.227	0.364	-0.008	0.000	-0.020
10	1.489	1.169	-0.396	-0.142	0.354	0.078	0.020	-0.043
11	-1.469	2.188	0.685	-0.133	0.345	0.079	0.021	-0.045
12	1.068	0.149	-0.499	-0.225	0.471	-0.048	-0.008	-0.028
13	-0.258	0.698	0.616	-0.210	0.410	0.035	0.006	-0.048
14	0.989	0.096	-0.249	-0.184	0.404	0.063	0.014	-0.054
15	-2.891	0.833	0.878	-0.072	0.365	0.073	0.019	-0.066
16	-0.841	-2.276	-0.890	0.041	0.753	-0.060	-0.040	-0.049
17	-0.355	-3.082	-0.483	0.084	0.855	0.291	0.014	-0.103
18	0.038	-2.884	-0.460	0.085	0.849	0.276	0.010	-0.110
19	-0.901	-1.973	-0.093	0.036	0.851	0.257	-0.003	-0.124
20	0.010	-1.951	-0.699	0.037	0.794	0.259	-0.006	-0.138
21	-0.125	-1.088	-0.213	0.032	0.795	0.225	-0.019	-0.140
22	-1.275	-0.482	-0.036	-0.023	0.789	0.226	-0.023	-0.155
23	0.877	-1.235	-1.021	-0.002	0.760	0.223	-0.024	-0.157
24	-0.881	0.075	0.414	0.009	0.749	0.239	-0.017	-0.157
25	0.930	-0.474	-0.289	0.023	0.735	0.252	-0.014	-0.165
	.							
	.							
	.							
1000	0.010	-1.943	-1.797	0.022	1.001	-0.415	1.325	-0.552
1001	-0.125	-1.079	-1.027	0.021	1.001	-0.424	1.317	-0.560
1002	-1.275	-0.475	-0.501	0.011	1.000	-0.424	1.313	-0.567
1003	0.877	-1.230	-1.288	0.014	0.996	-0.424	1.312	-0.569
1004	-0.881	0.078	-0.003	0.012	0.998	-0.427	1.310	-0.569
1005	0.930	-0.473	-0.513	0.008	1.001	-0.430	1.310	-0.567
1006	1.233	0.662	0.649	0.004	0.999	-0.428	1.311	-0.567
1007	-1.022	1.978	1.971	0.005	0.998	-0.428	1.311	-0.566
1008	1.522	0.734	0.675	0.013	0.993	-0.422	1.316	-0.565
1009	-0.170	1.874	1.715	0.011	1.008	-0.432	1.319	-0.555
1010	1.489	1.252	1.076	0.014	1.008	-0.429	1.321	-0.554
	.							
	.							
	.							
2000	0.010	-1.943	-1.941	0.000	1.000	-0.499	1.399	-0.599

B

Implementation of an Audio-Frequency Interference Canceller with a Texas Instruments TMS32020

B.1 BRIEF REVIEW OF THE PROBLEM

Section 8.2 developed a technical approach to solving the practical problem of removing or cancelling power line hum that inadvertantly appears in the output of medical instrumentation amplifiers such as those used for electrocardiogram (EKG) monitors. The "system design" was done by examining the expected signal amplitudes and spectra, and then engineering judgments were made that led to the general specification of a digitally based interference canceller that would employ an adaptive filter implemented with a single-chip signal processor such as the TI TMS320.

In this appendix we use that generic problem as the motivation for the specification of a simple signal processing system, including both a hardware and firmware segment. This has several uses—it demonstrates how a system-level description such as that in Section 8.2 can be reduced to an implementation, it provides a typical schematic of how a particular single-chip processor is interconnected with auxilliary components, it provides an example of assembly language algorithms for the LMS adaptive algorithm, and, for those interested in pursuing it, provides enough information to construct the circuit and gain practical experience with it. This example also contains various hardware and software modules that can be directly employed for other examples and applications.

B.2 GENERAL SYSTEM DESCRIPTION

A block diagram of the experimental system described here is shown in Figure B.1. This figure should be compared with Figures 8.1 and 1.3, which describe hum cancellers in various degrees of theoretical and practical detail. The basic hardware element of this system is a TMS32020 signal processor. It has two signal inputs and one signal output. The first input, often called the "primary", is the additive combination of potentially three signals, two of them externally generated, and the third, the "interference", is produced by a local function generator chip. These three signals are summed, digitized by a PCM coder–decoder chip (CODEC), and applied to the signal processor. The "interference" is sampled separately by another A/D and provided to the second input, often called the "reference input."

Interference cancellation is obtained by filtering the reference input with a 16-tap FIR LMS-directed adaptive filter and then subtracting the filter's output from the primary input. If the filter succeeds in adjusting the phase and amplitude of the reference signal properly, then it will accurately cancel the interference component of the primary input and the system output, the difference between the primary and filtered reference signals, will be devoid of the interference term. This system output is supplied to the D/A, the decoder portion of a PCM CODEC, and an analog output results.

A second system output is also provided via a parallel D/A. This output is used to display the weights being updated by the adaptive algorithm. Each of these coefficients is updated every 125 μsec, the input signal sampling rate, and, after each one is updated, it is sent out to the D/A. By connecting an oscilloscope to this output the evolution of the weights can be monitored in real time.

It should be noted that the input signal sampling rate used here, that is, 8 kHz, is substantially higher than the 500 Hz developed in Section 8.2 as a good

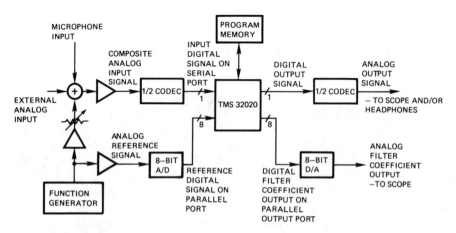

FIGURE B.1. System block diagram of the adaptive canceller.

value for the 50/60-Hz hum cancellation problem. Using the 8-kHz rate allows relatively inexpensive components, such as the PCM CODEC, to be employed and allows demonstration at audio frequencies, both desirable for experimental hardware. In the particular case of hum cancellation, however, the higher sampling rate leads to the use of more filter coefficients and a greater computation load than the lower sampling rate would require.

B.3 FLOW DIAGRAM OF THE REQUIRED ALGORITHMS

Given the basic system hardware environment just described (and in considerably more detail in Section B.5) the algorithms to be executed in the signal processing chip must be determined. Actual TMS32020 assembly language code is provided in Section B.6 but at this point it suffices to describe the algorithms to be implemented and their interconnection. A flow diagram is shown in Figure B.2. We begin assuming that the 8-kHz framing clock generated in the hardware segment has interrupted the TMS320. At this time the 8-bit digital sample is obtained from the PCM coder. Since the PCM coder introduces signal compression undesired for this application, it is removed by looking up the proper uncompressed value in a ROM-based table. Next, the output generated on the previous cycle, $e(k - 1)$, is processed. This difference signal is the basic output of the canceller, but it cannot be passed out to the PCM decoder in its linear form since the coder chip expects the signal to be compressed. As a result the basic output, the linear difference between the primary input and the filtered reference signal, must be compressed using a linear-to-μ-law conversion algorithm. The 8-bit result of this conversion is applied to the serial output port and sent to the PCM decoder for the generation of the analog system output.

The program then waits for the reference signal A/D to finish its conversion cycle. When this occurs, the new reference sample is read in. Using the new reference input sample and 15 past ones, the adaptive filter output is obtained by computing the dot product (convolutional sum) of the data and coefficient vector. After this dot product is complete, the result, a gain and phase corrected version of the reference input, is subtracted from the decompressed primary input to produce the system output $e(k)$.

After the system output is generated and stored, filter coefficients are updated. Each coefficient is updated and stored with 32-bit accuracy, but only the top 16 bits of each is used for the filter dot product calculation. The linear difference signal between the primary and filter output constitutes the error signal to be used in updating the filter coefficients. This error, a 16-bit variable, is multiplied by each of the 16 stored data values. Each such multiplication yields a 32-bit product. This product is then scaled by μ, the adaptation constant, which, for convenience, is assumed to be a negative power of two and can, therefore, be implemented using the barrel shifter. The resulting scaled product is then added to the 32-bit version of the respective coefficient. By repeating this cycle 16 times

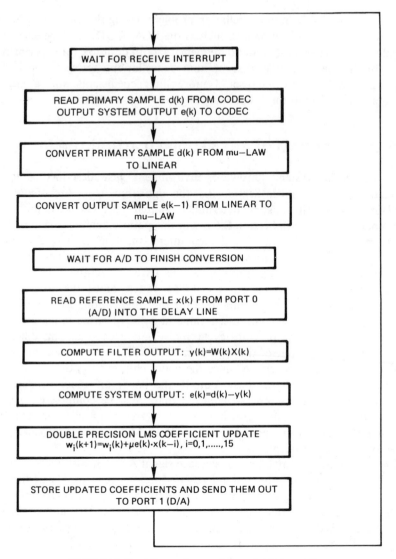

FIGURE B.2. Software flowgraph for adaptive canceller.

all of the filter's coefficients are updated. As each 32-bit coefficient is updated it is rounded to 16-bit accuracy and placed in storage for use in the dot product calculation. The top 10 bits are also sent out to the auxiliary output D/A convertor. It should be observed here that in many applications it is not necessary to update and maintain the filter coefficients in double precision arithmetic (see Section 7.5). Some applications do require it, however, and it is therefore included as a part of this example as a guide.

After these operations are complete, the processor enters a waiting mode until the next 125-μsec system frame pulse is received. The amount of waiting time can be determined on an oscilloscope by observing the time difference between the frame pulse (start) and the last coefficient output (end).

It is instructive at this point for the reader to review Section 7.2 and to recognize that the software execution cycle described here corresponds very closely to the general scenario for real-time operation described there.

B.4 EXPECTED BEHAVIOR

Chapters 3, 4, and 7 give us methods of predicting the behavior of this processor. Some aspects of this behavior are described below. This may be thought of as a review for the general reader but for the reader interested in building and experimenting with this system it provides both exercises and a basis to determine whether or not the system is working properly. The first observations discussed assume that the function generator is configured to generate sine waves. In this mode we expect the adaptive filter to adjust the gain and phase of the sinusoidal reference input so that it cancels the interference component in the primary input. From this we expect to see the following: (1) The coefficients displayed on the oscilloscope will usually grow from their zero initial condition into a sinusoidal formation. It can be confirmed analytically that a sinusoidal impulse response yields a bandpass frequency response centered at the frequency of the sinusoid. Thus the frequency of the interfering sinusoid can be inferred by examining the period of the weights. (2) As the amplitude of the function generator's contribution to the primary input is increased by turning the appropriate potentiometer, the weights will also grow in amplitude. This weight growth implies an increase in the filter gain to match the primary input amplitude. Note that the weights do not respond instantaneously but in fact take a fraction of a second to fully respond. This time lag is exactly the convergence time of the adaptive algorithm and is inversely proportional to both the power of the function generator signal when applied to the reference input and to the adaptation constant μ, a parameter in the TMS320 code. Changing either of these should change the convergence time appropriately. (3) When the frequency control potentiometer on the function generator is varied, the adaptive filter will respond by modifying its gain and phase for the new frequency. This change will be seen on the oscilloscope display of the weights. The response time of the system to this type of change is also determined by the algorithm convergence time and hence μ and the reference input power. (4) With no primary inputs other than the interference signal from the function generator, the weights should settle to steady values and the sinusoid should disappear from the system output. With other primary input components, however, the weights can be a bit "noisy" and not all of the interference will be suppressed. In the extreme case of an input sinusoid with frequency very close to that of the reference signal, the weights can be seen to evolve cyclically at a rate equal to the difference between the two signals. Glover [Glover, 1977] discusses this phenomenon in detail.

When the function generator chip is arranged for triangular or square wave outputs, the weights will usually evolve away from the sinusoidal form. It will usually take longer for the algorithm to converge and for the interference to be suppressed from the output. This can be understood by observing that a square or triangular wave can be decomposed into a set of harmonically related sinusoids. The adaptive filter will attempt to adjust the gain and phase for each one and the coefficients will therefore not appear to be a simple sinusoid. Convergence is slower because the convergence of each harmonic is inversely related to its own power (see Section 4.4 for an accurate description). Thus the high-order harmonics with their reduced amplitudes will drag out the convergence time. If spectrum analysis is performed on the system output, it will be seen that the fundamental component is suppressed much more quickly than the others.

B.5 DESCRIPTION OF THE HARDWARE SEGMENT

B.5.1 Analog Interface

An Intercil ICL8038 function generator (U1), shown in Figure B.3, is used to generate the interference signal. It can generate sinusoids, triangle waves, and square waves—the period of the wave is controlled by a potentiometer. The output of this function generator is used in two ways. First the interference reference signal is developed by scaling the generator output with a potentiometer and using it to drive a TL072 amplifier (U2), which in turn drives the reference signal A/D converter. The function generator also supplies a component of the primary input, also via a potentiometer. This scaled signal, an external input, and the output of an optional electret microphone are summed in another TL072 preamplifier (U2). Its output drives a TCM2916 combination filter and PCM coder–decoder chip (CODEC). The coder portion of the chip digitizes the primary signal, logarithmically compresses it in accordance with μ-law specifications, and transmits in an 8-bit serial format. The chip contains an antialiasing filter that allows the signal to be sampled at 8 kHz.

The master clock (MCLK) for the CODEC operates at 2.048 MHz. In the configuration shown, the master clock serves as the data clock for the CODEC. The framing pulse (FP) occurs every 125 μsec and signals the start of a new frame. Once receipt of this pulse, the codec transmits PCM data for the next eight clock pulses. This serial data is received by the serial port on the TMS32020 (U4).

B.5.2 Serial Interface

The reception operation in the TMS32020 is initiated by reception of the framing pulse (FP). After FP goes low, the TMS32020 starts clocking in serial data starting from the next falling edge of MCLK. The serial port on the

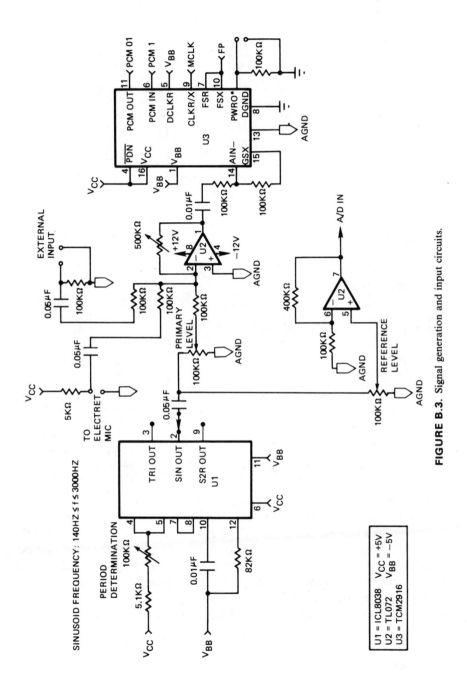

FIGURE B.3. Signal generation and input circuits.

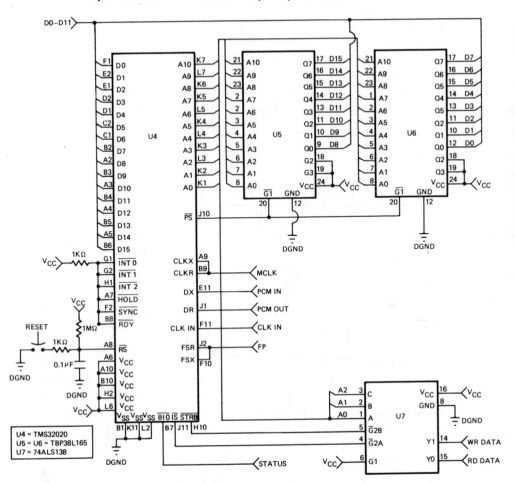

FIGURE B.4. TMS32020 circuit with associated memory.

TMS32020 is configured for 8-bit operation and after eight bits are received, the Data Receive Register (DRR) is filled and an internal reception interrupt is generated. This interrupt causes the TMS320 to read the DRR and store it in the input buffer. After the input is processed by the adaptive canceller, an output sample is produced and loaded into the Data Transmit Register (DXR). When the next frame pulse is received, transmission is initiated and continued until all eight bits are clocked into the decoder portion of the CODEC chip. There the 8-bit data is expanded into a linear scale, converted into analog format, and filtered.

B.5.3 Clock Circuit

A 20.48-MHz crystal is used with an SN74HC04 hex inverter (U8) to generate the input clock (CLKIN) for the TMS32020 (see Figure B.5). This clock is also divided by an SN74HC390 decade counter (U9) to produce the 2.048-MHz master clock (MCLK), which drives both the CODEC and the serial port of the TMS32020. The master clock (MCLK) is divided by a factor of 256 to produce the 8-kHz framing pulses (FP) used to control the sampling interval of the primary and reference inputs.

B.5.4 Parallel A/D and D/A Interfaces

The 12-bit reference signal A/D is strobed by the 8-kHz clock (CONV). Its output is read into an 12-bit external latch after the conversion process is complete, as indicated by the STATUS line. The TMS320 senses this completion signal and the 12-bit sample is read onto the TMS320 data bus. The TMS320 selects this latch for reading by selecting I/O port 1. Conversely, when the TMS320 is ready to send out a coefficient for display, it places the data on bits D6 through D15 of the data bus and selects I/O port 9. This selection has the effect of strobing the data into the latch (U15) which buffers the D/A input.

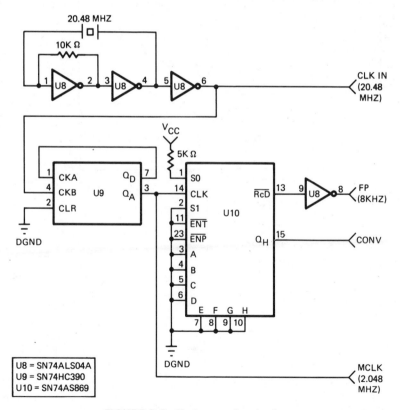

FIGURE B.5. Clock generation circuits.

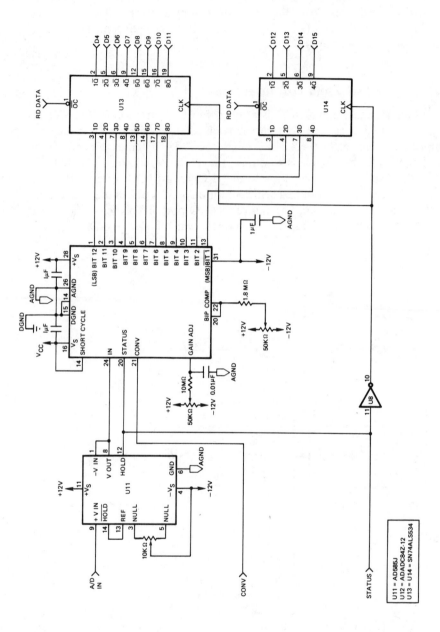

FIGURE B.6. Circuit for the reference A/D converter.

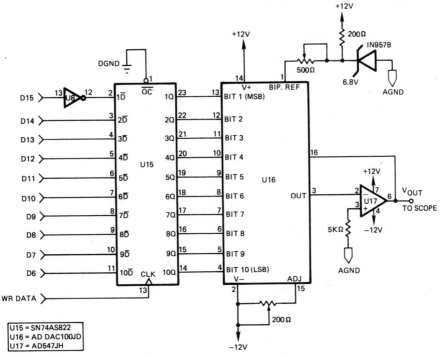

FIGURE B.7. Circuit for auxiliary D/A output.

B.5.5 Program Memory for the TMS32020

Two TBP38L165 2K by 8-bit PROMs (U5 and U6) are used to store the program for this processor. If other memory devices are used, they must be fast enough to meet the timing requirements for the TMS32020. The internal 544-word data RAM is more than sufficient for this system, and therefore no additional data memory is required. Table B.1 shows the expected processor and memory utilization for the firmware algorithms listed in Section B.6.

TABLE B.1. Processor Utilization

Routine	CPU Cycles at 200 nsec	RAM (Words)	ROM (Words)
μ-Law to linear conversion	11	1[a]	265
Linear to μ-law conversion	25	4	34
Computation of adaptive filter output	28	35[a]	12
Computation of system output	3	3[a]	3
Double precision filter coefficient update ($\mu = 2^{-9}$)	160	53[a]	20
TOTAL	244 ($\leqslant$625)	61 ($\leqslant$544)	423 ($\leqslant$2048)

[a]RAM locations indicated are used in more than one routine.

B.6 TMS32020 Assembly Language Listings

```
••••••••••••••••••••••••••••••••••••••••••••••••••••••••••••••••••••••••••••••
•••                                                                        •••
•••                    ADAPTIVE NOISE CANCELLER                            •••
•••                                                                        •••
•••                Written by: GEORGE TROULLINOS                           •••
•••                DIGITAL SIGNAL PROCESSING                               •••
•••                    TEXAS INSTRUMENTS                                   •••
•••                                                                        •••
••••••••••••••••••••••••••••••••••••••••••••••••••••••••••••••••••••••••••••••
```

*
*   The adaptation step size is $\mu = 2^{-N}$
*

N   EQU   9
*
*   Block B2 data memory allocation (DP=0 points to B2)
*

DRR	EQU	0	Serial Port Receive Register
DXR	EQU	1	Serial Port Transmit Register
IMR	EQU	4	Interrupt Mask Register
NEG7	EQU	96	NEG7 = −7
ONE	EQU	97	ONE = 1
BIAS	EQU	98	Bias for mu-law
MANT	EQU	99	Mantisa of the mu-law number
EXP	EQU	100	Exponent of the mu-law number
AX1	EQU	101	Address of $x(k-1)$
AX15	EQU	102	Address of $x(k-15)$
AW0H	EQU	103	Address of MSW of 1st coefficient
AW0L	EQU	104	Address of LSW of 1st coefficient

*
*   Block B0 data/program memory allocation
*   (DP=4,5 points to B0)
*

P4DM	EQU	512	Page 4 data memory address
W15H	EQU	0	MSW of 16th coefficient
W0H	EQU	15	MSW of 1st coefficient
W15L	EQU	16	LSW of 16th coefficient
W0L	EQU	31	LSW of 1st coefficient
W15PM	EQU	65280	Program memory address of B0

*
*   Block B1 data memory allocation (DP=6, 7 points to B1)
*

P6DM	EQU	768	Page 6 data memory address
DK	EQU	0	Primary input sample $d(k)$
YK	EQU	1	Adaptive filter output $y(k)$
EK	EQU	2	System output, error signal $e(k)$
TEMP	EQU	3	Temporary storage location
X0	EQU	4	Reference input sample $x(k)$
X15	EQU	19	Delayed ref. input sample $x(k-15)$

*
*

```
        AORG        0
        B           RESET           Power-up reset

        AORG        26
        B           INTRPT          Serial Port Receive Interrupt
```

```
* ..............................................................................
*                                                                              *
*    Serial Port Receive Interrupt Service Routine                             *
*                                                                              *
*    At every sampling interval (8 kHz sampling rate), a                       *
*    sample is read in and a sample is written out. The                        *
*    input sample is converted from mu-law to linear, and                      *
*    the output sample is converted from linear to mu-law.                     *
*                                                                              *
* ..............................................................................
*
```

```
            AORG        32
INTRPT      LDPK        0               Point to block B0 in data RAM
            LARP        1               Current AR = AR1
*
* ..............................................................................
*                                                                              *
*    Mu-Law to Linear Expansion Using a Table Look-Up                          *
*                                                                              *
* ..............................................................................
*
```

```
            LAC         DRR             Load Data Receive Register
            ANDK        >00FF           Mask off unwanted bits
            XORK        >00FF           Complement the input
            ADLK        TBLADD          Add the look-up table offset
            LDPK        6               Point to page 6 (Block B1)
            TBLR        DK              Read-in primary sample
*
* ..............................................................................
*                                                                              *
*    Linear to Mu-Law Compression                                             *
*                                                                              *
* ..............................................................................
*
```

```
            ZALH        EK              Output→ACC(31 − 16)
            LDPK        0               Point to page 0 (Block B2)
            BLZ         NEGAT           If ACC<0 go to NEGAT
            ADDH        BIAS            Add bias
            LAR         AR1,NEG7        AR1 = −7
            RPTK        6
            NORM                        Find MSB
            ANDK        >F000,14        Zero all bits except mantisa bits
            SACH        MANT            Store mantisa
            SAR         AR1,EXP         Store 2's complement of exponent
            ZALH        EXP
            ABS                         Get exponent
            ADD         MANT,2          Incorporate mantisa
```

```
              XORK      >FF00,4          Invert mantisa and exponent bits
              B         XMIT
NEGAT         ABS                        Get positive value
              ADDH      BIAS             Add bias
              LAR       AR1,NEG7         AR1 = −7
              RPTK      6
              NORM                       Find MSB
              ANDK      >F000,14         Zero all bits except mantisa bits
              SACH      MANT             Store mantisa
              SAR       AR1,EXP          Store 2's complement of exponent
              ZALH      EXP
              ABS                        Get exponent
              ADD       MANT,2           Incorporate mantisa
              XORK      >7F00,4          Invert all bits except sign bit
XMIT          SACH      DXR,4            Output mu-law sample
              EINT                       Enable interrupts
              RET
*
```

```
***********************************************************************************
*                                                                                *
*    Reset and Initialization                                                    *
*                                                                                *
***********************************************************************************
*
```

```
RESET         DINT                       Disable interrupts
              LDPK      0                Point to page 0 (Block B2)
              FORT      1                Set serial port for 8-bit mode
              LALK      >FFD0
              SACL      IMR              Enable receive port interrupts
              RTXM                       Set FSX pin to be an input
              CNFD                       Configure B0 as data RAM
              ROVM                       Remove overflow mode
              SSXM                       Set sign extention mode
              SPM       0                No shift of product register
              ZAC
              SACL      DRR              Initialize serial input
              SACL      DXR              Initialize serial output
              LALK      P6DM+X0+1
              SACL      AX1              Initialize pointer to x(k−1)
              LALK      P6DM+X15
              SACL      AX15             Initialize pointer to x(k−15)
              LALK      P4DM+W0H
              SACL      AW0H             Initialize pointer to W0H
              LALK      P4DM+W0L
              SACL      AW0L             Initialize pointer to W0L
*
*    Initialize page 4 (Block B0)
*
              ZAC
              LARP      AR1              Current AR = AR1
              LRLK      AR1,P4DM         AR1→B0
```

```
                RPTK        31
*               SACL        *+                  Zero filter coefficients
*       Initialize page 6 (Block B1)
*
                LRLK

                LRLK        AR1,P6DM
                RPTK        19
                SACL        *+                  Zero filter delay line
*
*       Initialize constants
*
                LACK        1
                SACL        ONE                 ONE=1
                LALK        -7
                SACL        NEG7                NEG7=-7
                LALK        132
                SACL        BIAS                BIAS=132 (for mu-law compression)
                EINT                            Enable interrupts
*

****************************************************************

*
*       Wait for Serial Port Receive Interrupt
*

****************************************************************

*
WCODEC          IDLE
*

****************************************************************

*
*       Wait for A/D to Finish Conversion
*

****************************************************************

*
WAIT            BIOZ        INPUT               If A/D done, input data
                B           WAIT                Else wait
INPUT           LDPK        6                   Point to page 6 (Block B1)
                IN          X0,PA0              Input reference sample
*

****************************************************************

*
*       Compute Filter Output
*

****************************************************************

*
                LDPK        0                   Point to page 0 (Block B2)
                LAR         AR1,AX15            AR1 points at x(k-15)
                LDPK        6                   Point to page 6 (Block B1)

                LAC         ONE,14              Round-off offset→ACC
                MPYK        0                   Clear the product register
                CNFP                            Configure B0 as program memory
                RPTK        15                  For n=15, . . . , 0 compute:
```

```
        MACD      W15PM,*-      x(k-n)w_n+ACC→ACC
        CNFD                    Reconfigure B0 as data memory
        APAC                    Accumulate last product
        SACH      YK,1          Filter output→y(k)
```

* *

*   *   Compute System Output

* *

```
        LAC       DK            Load primary sample
        SUB       YK            d(k)-y(k)→ACC
        SACL      EK            e(k)=d(k)-y(k)
```

* *

Least Mean Square Double Precision (32 bits) Update
of the Adaptive Filter Coefficients.

Filter coefficients are updated using:

$w_m(k+1)=w_m(k)+2^{-N}\{e(k)x(k-m)\}$

with $m=0, 1, 2, \ldots, 15$ and $N=7$

* *

Set up for update

```
        LRLK      AR0,15        AR0 is the loop counter
        LDPK      0             Point to page 0 (Block B2)
        LAR       AR1,AW0H      AR1 points at MSW of w_o
        LAR       AR2,AW0L      AR2 points at LSW of w_o
        LAR       AR3,AX1       AR3 points at x(0) (Shift introduced by
                                MACD is compensated)
        LDPK      6             Point to page 6 (Block B1)
        LARP      AR3           Current AR=AR3
        LAC       EK,16-N+1     Scale as specified by the
        SACH      TEMP          adaptation step size
        LT        TEMP          2^{(-N+1)}e(k)→T register
```

Update Coefficients, W0 (m=0) First and W15 (m=15) Last

```
LOOP    MPY       *+,AR2        Current AR points at XM (x(k-m))
        PAC                     2^{-N}\{e(k)x(k-m)\}→ACC
        ADDS      *,AR1         Current AR points at LSW of w_m
        ADDH      *             Current AR points at MSW of w_m
        SACH      *             Update the MSW of w_m
        OUT       *-,PA1,AR2    Output MSW of w_m to the D/A
        SACL      *-, 0,AR0     Update LSW of w_m
        BANZ      LOOP,*-,AR3   Repeat loop for 16 times
        B         WCODEC        Wait for serial receive interrupt
        PAGE
```

```
*••••••••••••••••••••••••••••••••••••••••••••••••••••••••••••••••••••••••••••••••••••••*
*                                                                                      *
*      ROM                                                                             *
*                                                                                      *
*      Mu-Law To Linear Conversion Table                                               *
*                                                                                      *
*••••••••••••••••••••••••••••••••••••••••••••••••••••••••••••••••••••••••••••••••••••••*
*

TBLADD       EQU          $
             DATA         0
             DATA         8/4
             DATA         16/4
             DATA         24/4
             DATA         32/4
             DATA         40/4
             DATA         48/4
             DATA         56/4
             DATA         64/4
             DATA         72/4
             DATA         80/4
             DATA         88/4
             DATA         96/4
             DATA         104/4
             DATA         112/4
             DATA         120/4
             DATA         132/4
             DATA         148/4
             DATA         164/4
             DATA         180/4
             DATA         196/4
             DATA         212/4
             DATA         228/4
             DATA         244/4
             DATA         260/4
             DATA         276/4
             DATA         292/4
             DATA         308/4
             DATA         324/4
             DATA         340/4
             DATA         356/4
             DATA         372/4
             DATA         396/4
             DATA         428/4
             DATA         460/4
             DATA         492/4
             DATA         524/4
             DATA         556/4
             DATA         588/4
             DATA         620/4
             DATA         652/4
             DATA         684/4
             DATA         716/4
             DATA         748/4
             DATA         780/4
```

```
DATA          812/4
DATA          844/4
DATA          876/4
DATA          924/4
DATA          988/4
DATA          1052/4
DATA          1116/4
DATA          1180/4
DATA          1244/4
DATA          1308/4
DATA          1372/4
DATA          1436/4
DATA          1500/4
DATA          1564/4
DATA          1628/4
DATA          1692/4
DATA          1756/4
DATA          1820/4
DATA          1884/4
DATA          1980/4
DATA          2108/4
DATA          2236/4
DATA          2364/4
DATA          2492/4
DATA          2620/4
DATA          2748/4
DATA          2876/4
DATA          3004/4
DATA          3132/4
DATA          3260/4
DATA          3388/4
DATA          3516/4
DATA          3644/4
DATA          3772/4
DATA          3900/4
DATA          4092/4
DATA          4348/4
DATA          4604/4
DATA          4860/4
DATA          5116/4
DATA          5372/4
DATA          5628/4
DATA          5884/4
DATA          6140/4
DATA          6396/4
DATA          6652/4
DATA          6908/4
DATA          7164/4
DATA          7420/4
DATA          7676/4
DATA          7932/4
DATA          8316/4
DATA          8828/4
```

```
DATA        9340/4
DATA        9852/4
DATA        10364/4
DATA        10876/4
DATA        11388/4
DATA        11900/4
DATA        12412/4
DATA        12924/4
DATA        13436/4
DATA        13948/4
DATA        14460/4
DATA        14972/4
DATA        15484/4
DATA        15996/4
DATA        16764/4
DATA        17788/4
DATA        18812/4
DATA        19836/4
DATA        20860/4
DATA        21884/4
DATA        22908/4
DATA        23932/4
DATA        24956/4
DATA        25980/4
DATA        27004/4
DATA        28028/4
DATA        29052/4
DATA        30076/4
DATA        31100/4
DATA        32124/4
DATA        0/4
DATA        −8/4
DATA        −16/4
DATA        −24/4
DATA        −32/4
DATA        −40/4
DATA        −48/4
DATA        −56/4
DATA        −64/4
DATA        −72/4
DATA        −80/4
DATA        −88/4
DATA        −96/4
DATA        −104/4
DATA        −112/4
DATA        −120/4
DATA        −132/4
DATA        −148/4
DATA        −164/4
DATA        −180/4
DATA        −196/4
DATA        −212/4
DATA        −228/4
```

```
DATA        -244/4
DATA        -260/4
DATA        -276/4
DATA        -292/4
DATA        -308/4
DATA        -324/4
DATA        -340/4
DATA        -356/4
DATA        -327/4
DATA        -396/4
DATA        -428/4
DATA        -460/4
DATA        -492/4
DATA        -524/4
DATA        -556/4
DATA        -588/4
DATA        -620/4
DATA        -652/4
DATA        -684/4
DATA        -716/4
DATA        -748/4
DATA        -784/4
DATA        -812/4
DATA        -844/4
DATA        -876/4
DATA        -924/4
DATA        -988/4
DATA        -1052/4
DATA        -1116/4
DATA        -1180/4
DATA        -1244/4
DATA        -1308/4
DATA        -1372/4
DATA        -1436/4
DATA        -1500/4
DATA        -1564/4
DATA        -1628/4
DATA        -1692/4
DATA        -1756/4
DATA        -1820/4
DATA        -1884/4
DATA        -1980/4
DATA        -2108/4
DATA        -2236/4
DATA        -2364/4
DATA        -2492/4
DATA        -2620/4
DATA        -2748/4
DATA        -2876/4
DATA        -3004/4
DATA        -3132/4
DATA        -3260/4
DATA        -3388/4
DATA        -3516/4
```

```
DATA        −3644/4
DATA        −3772/4
DATA        −3900/4
DATA        −4092/4
DATA        −4348/4
DATA        −4604/4
DATA        −4860/4
DATA        −5116/4
DATA        −5372/4
DATA        −5628/4
DATA        −5884/4
DATA        −6140/4
DATA        −6396/4
DATA        −6652/4
DATA        −6908/4
DATA        −7164/4
DATA        −7420/4
DATA        −7676/4
DATA        −7932/4
DATA        −8316/4
DATA        −8828/4
DATA        −9340/4
DATA        −9852/4
DATA        −10364/4
DATA        −10876/4
DATA        −11388/4
DATA        −11900/4
DATA        −12412/4
DATA        −12924/4
DATA        −13436/4
DATA        −13948/4
DATA        −14460/4
DATA        −14972/4
DATA        −15484/4
DATA        −15996/4
DATA        −16764/4
DATA        −17788/4
DATA        −18812/4
DATA        −19836/4
DATA        −20860/4
DATA        −21884/4
DATA        −22908/4
DATA        −23932/4
DATA        −24956/4
DATA        −25980/4
DATA        −27004/4
DATA        −28028/4
DATA        −29052/4
DATA        −30076/4
DATA        −31100/4
DATA        −32124/4


END
```

C

Authors' Preface

Appendix C presents the specific example of using the TI TMS32020 to efficiently implement a digital echo canceller for voice telephony. The appendix was prepared by Texas Instruments and the staff members of Teknekron Communications Systems, a leading company in the application of advanced digital signal-processing concepts to evolving communications requirements. While Appendix B is pedagogical in nature and directed toward the student, this appendix is aimed at the practicing engineer.

This appendix not only describes a practical problem and a good solution, but also presents a clever method of performing the update equation for the LMS adaptive algorithm. The particular architecture and instruction set of the TMS32020 are exploited with the use of block updating, producing an algorithm significantly more efficient in terms of computation time than the version of LMS used in Appendix B. This type of optimization is dependent on the processing architecture available and, in general, different designs will be required if other signal processing chips are used.

As with Appendix B, this appendix includes both a schematic diagram of the echo canceller circuit and a listing of the required assembly language code. An extended version of this appendix, as well as descriptions of other application examples, can be found in "Digital Signal Processing Applications with the TMS320 Family," a TI publication. This compilation of application notes will be updated by TI as new applications emerge and as new signal-processing devices are introduced.

Digital Voice Echo Canceller with a TMS32020

by Texas Instruments and Teknekron Communications Systems

C.1 INTRODUCTION

Echo cancellers using adaptive filtering techniques are now finding widespread practical applications to solve a variety of communications systems problems.[1] These applications are made possible by the recent advances in microelectronics, particularly in the area of Digital Signal Processors (DSPs). Cancelling echoes for long-distance telephone voice communications, full-duplex voiceband data modems, and high-performance "handsfree" audio-conferencing systems (including speakerphones) are a few examples of these applications.

This appendix describes the implementation of an integrated 128-tap (16-ms span) digital voice echo canceller on the Texas Instruments TMS32020[2] programmable signal processor. The implementation features a direct interface for standard PCM codecs (e.g., Texas Instruments TCM2913) and meets the requirements of the CCITT (International Telegraph and Telephone Consultive Committee) Recommendation G.165 for echo cancellers.[3] This report presents the requirements for echo cancellation in voice transmission and discusses the generic echo cancellation algorithms. The implementation considerations for a 128-tap echo canceller on the TMS32020 are then described in detail, as well as the software logic and flow for each program module. A hardware demonstration model of a 128-tap voice echo canceller using the TMS32020 has been constructed and tested.

C.2 ECHO CANCELLATION IN VOICE TRANSMISSION

C.2.1 Echoes in the Telephone Network

The source of echoes can be understood by considering a simplified connection between two subscribers, S1 and S2, as shown in Figure C.1. This connection is typical in that it contains two-wire segments on the ends, a four-wire connection

302

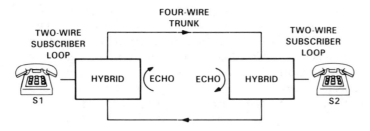

FIGURE C.1. A Simplified Telephone Connection.

in the center, and a hybrid at each end to convert from two-wire transmission to four-wire transmission. Each two-wire segment consists of the subscriber loop and possibly some portion of the local network. Over this segment, both directions of transmission are carried by the same wire pair, i.e., signals from speakers S1 and S2 are superimposed on this segment. On the four-wire section, the two directions of transmission are segregated. The speech from speaker S1 follows the upper transmission path, as indicated by the arrow, while speech originating from S2 follows the lower path. The segregation of the two signals is necessary where it is desired to insert carrier terminals, amplifiers, or digital switches.

The hybrid is a device that converts two-wire to four-wire transmission. The role of the hybrid on the right-hand side is to direct the signal energy arriving from S1 to the two-wire segment of S2 without allowing it to return to S1 via the lower four-wire transmission path. Because of impedance mismatches (unfortunately occurring in practice), some of this energy will be returned to speaker S1, who then hears a delayed version of his speech. This is the source of "talker echo."

The subjective effect of the talker echo depends on the delay around the loop. For short delays, the talker echo represents an insignificant impairment if the attenuation is reasonable (6 dB or more). This is because the talker echo is indistinguishable from the normal sidetone in the telephone. For satellite connections, the delay in each four-wire path is about 270 ms as a consequence of the high altitude of synchronous satellites. This means that the round-trip echo delay is approximately 540 ms, which makes it very disturbing to the talker, and can in fact make it quite difficult to carry on a conversation. When such is the case, it is essential to find ways of controlling or removing that echo. Since the subjective annoyance of echo increases with delay as well as echo level due to hybrid return energy, the measures for control depend on the circuit length.

For terrestrial circuits under 2,000 miles, the via net loss (VNL) plan,[4] which regulates loss as a function of transmission distance, is used to limit the maximum echo-to-signal ratio. On circuits over this length (e.g., intercontinental circuits), echo suppressors or cancellers are used. An echo suppressor is a voice-operated switch that attempts to open the path from listener to talker whenever the listener is silent. However, echo suppressors perform poorly since echo is not blocked during periods of doubletalk. They impart a choppiness to speech and background noise as the transmission path is opened and closed. Due to recent decreasing trends in DSP costs, digital echo cancellers are now viable as replacements for most of the circuits using echo suppressors.

For satellite circuits with full hop delays of 540 ms, echo suppressors are subjectively inadequate, and cancellers must be employed.

Digital Echo Cancellers in Voice Carrier Systems

The principle of the echo canceller for one direction of transmission is shown in Figure C.2. The portion of the four-wire connection near the two-wire interface is shown in this figure, with one direction of voice transmission between ports A and C, and the other direction between ports D and B. All signals shown are sampled data signals that would occur naturally at a digital transmission terminal or digital switch. The far-end talker signal is denoted $y(i)$, the undesired echo $r(i)$, and the near-end talker $x(i)$. The near-end talker is superimposed with the undesired echo on port D. The received signal from far-end talker $y(i)$ is available as a reference signal for the echo canceller and is used by the canceller to generate a replica of the echo called $\hat{r}(i)$. This replica is subtracted from the near-end talker plus echo to yield the transmitted near-end signal $u(i)$ where $u(i) = x(i) + r(i) - \hat{r}(i)$. Ideally, the residual echo error $e(i) = r(i) - \hat{r}(i)$ is very small after echo cancellation.

The echo canceller generates the echo replica by applying the reference signal to a transversal filter (tapped-delay line), as shown in Figure C.3. If the transfer function of the transversal filter is identical to that of the echo path, the echo replica will be identical to the echo, thus achieving total cancellation. Since the transfer function of the echo path from port C to port D is not normally known in advance, the canceller adapts the coefficients of the transversal filter. To reduce error, the adaptation algorithm infers from the cancellation error $e(i)$ (when no near-end signal is present) the appropriate correction to the transversal filter coefficients.

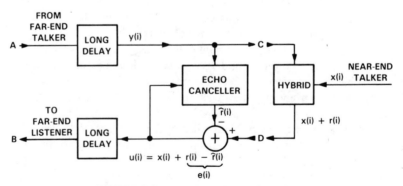

FIGURE C.2. Echo Canceller Configuration.

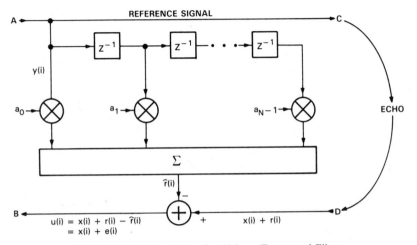

FIGURE C.3. Echo Estimation Using a Transversal Filter.

The number of taps in the transversal filter of Figure C.3 is determined by the duration of the impulse response of the echo path from port C to port D. The time span over which this impulse response is significant (i.e., nonzero) is typically 2 to 4 ms. This corresponds to 16 to 32 tap positions with 8-kHz sampling. However, because of the portion of the four-wire circuit between the location of the echo canceller and the hybrid, this response does not begin at zero, but is delayed. The number of taps N, must be large enough to accommodate that delay. With $N = 128$, delays of up to 16 ms (or about 1,200 miles of "tail" circuit) can be accommodated.

In practice, it is necessary to cancel the echoes in both directions of a trunk. For this purpose, two adaptive cancellers are used, as shown in Figure C.4, where one cancels the echo from each end of the connection. The near-end talker for one of the cancellers is the far-end talker for the other. In each case, the near-end talker is the "closest" talker, and the far-end talker is the talker generating

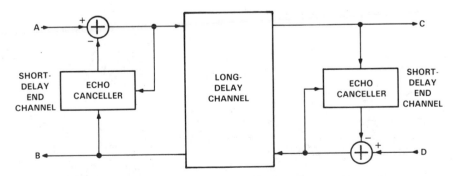

FIGURE C.4. Split-Type Echo Canceller for Two Directions of Transmission.

the echo being cancelled. It is desirable to position these two "halves" of the canceller in a split configuration, as shown in Figure C.4, where the bulk of the delay in the four-wire portion of the connection is in the middle. The reason is that the number of coefficients required in the echo-cancellation filter is directly related to the delay of the tail circuit between the location of the echo canceller and the hybrid that generates the echo. In the split configuration, the largest delay is not in the echo path of either half of the canceller. Therefore, the number of coefficients is minimized.

C.3 ECHO CANCELLATION ALGORITHMS

Generic algorithm requirements for each major signal processing function are discussed in this section. The signal processing flow for a single-channel digital voice echo canceller is shown in the block diagram of Figure C.5.

C.3.1 Adaptive Transversal Filter

The reflected echo signal r(i) at time i (see Figure C.2) can be written as the convolution of the far-end reference signal y(i) and the discrete representation h_k of the impulse response of the echo path between port C and D.

$$r(i) = \sum_{k=0}^{N-1} h_k y(i-k) \tag{1}$$

Linearity and a finite duration N of the echo-path response have been assumed. An echo canceller with N taps adapts the N coefficients a_k of its

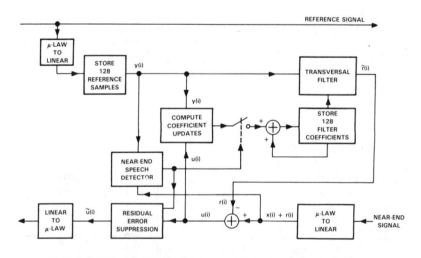

FIGURE C.5. Signal Processing for a Digital voice echo canceller.

transversal filter to produce a replica of the echo r(i) defined as follows:

$$\hat{r}(i) = \sum_{k=0}^{N-1} a_k y(i-k) \tag{2}$$

Clearly, if $a_k = h_k$ for $k = 0, \ldots, N-1$, then $\hat{r}(i) = r(i)$ for all time i and the echo is cancelled exactly.

Since, in general, the echo-path impulse response h_k is unknown and may vary slowly with time, a closed-loop coefficient adaptation algorithm is required to minimize the average or mean-squared error (MSE) between the echo and its replica. From Figure C.2 it can be seen that the near-end error signal u(i) is comprised of the echo-path error $r(i) - \hat{r}(i)$ and the near-end speech signal x(i), which is uncorrelated with the far-end signal y(i). This gives the equation

$$E(u^2(i)) = E(x^2(i)) + E(e^2(i)) \tag{3}$$

where E denotes the expectation operator. The echo term $E(e^2(i))$ will be minimized when the left-hand side of (3) is minimized. If there is no near-end speech (x(i) = 0), the minimum is achieved by adjusting the coefficients a_k along the direction of the negative gradient of $E(e^2(i))$ at each step with the update equation

$$a_k(i+1) = a_k(i) - \beta \frac{\partial E(e^2(i))}{\partial a_k(i)} \tag{4}$$

where β is the stepsize. Substituting (1) and (2) into (3) gives from (4) the update equation

$$a_k(i+1) = a_k(i) + 2\beta E\left[e(i)y(i-k)\right] \tag{5}$$

In practice, the expectation operator in the gradient term $2\beta E[e(i)y(i-k)]$ cannot be computed without a priori knowledge of the reference signal probability distribution. Common practice is to use an unbiased estimate of the gradient, which is based on time-averaged correlation error. Thus, replacing the expectation operator of (5) with a short-time average, gives

$$a_k(i+1) = a_k(i) + 2\beta \frac{1}{M} \sum_{m=0}^{M-1} e(i-m)y(i-m-k) \tag{6}$$

The special case of (6) for M = 1 is frequently called the least-mean-squares (LMS) algorithm or the stochastic gradient algorithm. Alternatively, the coefficients may be updated less frequently with a thinning ratio of up to M, as given in

$$a_k(i+M+1) = a_k(i) + 2\beta \sum_{m=0}^{M-1} e(i+M-m)y(i+M-m-k) \tag{7}$$

Computer simulations of this "block update" method show that it performs better than the standard LMS algorithm (i.e., M = 1 case) with noise or speech signals.[5] Many cancellers today avoid multiplication for the correlation function in (7), and instead use the signs of e(i) and y(i − k) to compute the coefficient updates. However, this "sign algorithm" approximation results in approximately a 50-percent decrease in convergence rate and an increase in degradation of residual echo due to interfering near-end speech.

The convergence properties of the algorithm are largely determined by the stepsize parameter β and the power of the far-end signal y(i). In general, making β larger speeds the convergence, while a smaller β reduces the asymptotic cancellation error.

It has been shown that the convergence time constant is inversely proportional to the power of y(i), and that the algorithm will converge very slowly for low-power signals.[6] To remedy that situation, the loop gain is usually normalized by an estimate of that power, i.e.,

$$2\beta = 2\beta(i) = \frac{\beta_1}{P_y(i)} \tag{8}$$

where β_1 is a compromise value of the stepsize constant and $P_y(i)$ is an estimate of the average power of y(i) at time i.

$$P_y(i) = (L_y(i))^2 \tag{9}$$

where $L_y(i)$ is given by

$$L_y(i + 1) = (1 - \rho)L_y(i) + \rho|y(i)| \tag{10}$$

The estimate $P_y(i)$ is used since the calculation of the exact average power is computationally expensive.

C.3.2 Near-End Speech Detector

When both near-end and far-end speakers are talking, the condition is termed "doubletalk." Since the error signal u(i) of Figure C.1 contains a component of the near-end talker x(i) in addition to the residual echo-cancellation error, it is necessary to freeze the canceller adaptation during doubletalk in order to avoid divergence. Doubletalk status can be detected by a near-end speech detector operating on the near-end and far-end signals y(i) and s(i), respectively.

A commonly used algorithm by A. A. Geigel[7] consists of declaring near-end speech whenever

$$|s(i)| = |x(i) + r(i)| \geqslant \frac{1}{2}\max\{|y(i)|, |y(i - 1)|, \ldots, |y(i - N)|\} \tag{11}$$

where N is the number of samples in the echo canceller transversal filter memory. It is necessary to compare s(i) with the recent past of the far-end signal rather than just y(i) because of the unknown delay in the echo path. The factor of one-half is based on the hypothesis that the echo-path loss through a hybrid is at least 6 dB. The algorithm in effect performs an instantaneous power comparison over a time window spanning the echo-path delay range.

A more robust version of this algorithm uses short-term power estimates, $\tilde{y}(i)$ and $\tilde{s}(i)$, for the power estimates of the recent past of the far-end receive signal $y(i)$ and the near-end hybrid signal $s(i)$, respectively. These estimates are computed recursively by the equations

$$\tilde{s}(i + 1) = (1 - \alpha)\tilde{s}(i) + \alpha|s(i)| \tag{12}$$

$$\tilde{y}(i + 1) = (1 - \alpha)\tilde{y}(i) + \alpha|y(i)| \tag{13}$$

where the filter gain $\alpha = 2^{-5}$. For this version of the algorithm, near-end speech is declared whenever

$$\tilde{s}(i) \geqslant \frac{1}{2} \max(\tilde{y}(i), \tilde{y}(i - 1), \ldots, \tilde{y}(i - N)) \tag{14}$$

Since the near-end speech detector algorithm detects short-term power peaks, it is desirable to continue declaring near-end speech for some hangover time after initial detection.

C.3.3 Residual Echo Suppressor

Nonlinearities in the echo path of the telephone circuit and uncorrelated near-end speech limit the amount of achievable suppression in the circuit from 30 to 35 dB. Thus, there is no merit in achieving more than a certain degree of cancellation.

The use of a residual echo suppressor algorithm has been found to be subjectively desirable.[6] During doubletalk, the residual suppressor must be disabled. A common suppression control algorithm is to detect when the return signal power falls below a threshold based on the receive reference signal power. If the return signal consists only of residual echo and the canceller has properly converged, then the residual echo level will be below the threshold and the transmitted return signal will be set to zero.

The return signal power is estimated by the equation

$$L_u(i + 1) = (1 - \rho)L_u(i) + \rho|u(i)| \tag{15}$$

The reference power estimate $L_y(i)$ is given by (10). Suppression is enabled on the transmitted signal $u(i)$ (i.e., $u(i) = 0$) whenever $L_u(i)/L_y(i) < 2^{-4}$. This corresponds to a suppression threshold of 24 dB.

C.4 IMPLEMENTATION OF A 128-TAP ECHO CANCELLER WITH THE TMS32020

The TMS32020 is ideally suited for the implementation of a single 128-tap digital voice echo canceller channel since it has the capability and features to implement all of the required functions with full precision. This section discusses an implementation approach that meets or exceeds the performance of currently available products and the requirements of the CCITT G.165 recommendations.[3]

C.4.1 Echo Canceller Performance Requirements

Echo cancellers have the following fundamental requirements:

1. Rapid convergence when speech is incident in a new connection
2. Low-returned echo level during singletalking (i.e., echo-return loss enhancement)
3. Slow divergence when there is no signal
4. Rapid return of the echo level to residual if the echo path is interrupted
5. Little divergence during doubletalking

The CCITT recommendation G.165 specifies echo canceller performance requirements with band-limited white-noise (300–3400 Hz) test signals at the near-end and far-end input signal ports. Digital voice echo canceller products are typically designed to accommodate circuits with tail delays of 16 ms or more and circuits with echo-return loss levels greater than 3 dB to 6 dB.

C.4.2 Implementation Approach

In the implementation of the generic echo-cancelling algorithms discussed above, the coefficient update process dominates the computational requirement and efficiency of DSP realizations. The DSP efficiency and speed, in turn, determines the maximum number of echo canceller taps that can be achieved with the processor.

The block update approach of (7) with $M = 16$ was chosen for the TMS32020 implementation because it takes advantage of the efficient multiply and accumulate capabilities of the processor. Using the block update approach, a full-performance 128-tap canceller can be realized with a small margin. During each sample period (125 μs), 8 out of 128 coefficients are updated using correlation of the 16 past error and signal values.

Computer simulation studies were undertaken to verify the performance of the block update algorithm ($M = 16$) in comparison with the stochastic gradient algorithm ($M = 1$), taking into account the finite-precision and word-length limitations of the TMS32020. Figures C.6 and C.7 show the simulation results

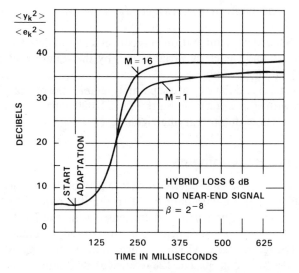

FIGURE C.6. Convergence performance of the block update algorithm and stochastic gradient algorithm.

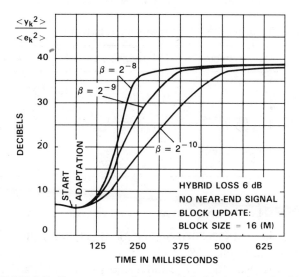

FIGURE C.7. Convergence performance of the block update algorithm.

for three values of the compromize stepsize constant β_1, defined in (8). The curves represent the average of 600 samples for single convergence runs from a zero initial condition with white-noise input. The block update algorithm performs better than the stochastic gradient algorithm for all three values. For values of β_1 larger than 2^{-8}, the algorithm can become unstable. Therefore, for both practical and performance reasons, the value $\beta_1 = 2^{-10}$ was chosen for implementation.

In the TMS32020 implementation, it is convenient and desirable to normalize both the stepsize and the error variables u(i) by the square root of the power estimate $P_y(i)$, i.e., $L_y(i)$ of (9).

Normalizing u(i) and the stepsize separately enables the product term of (7) to be computed with single precision on the TMS32020 without significant loss of precision or overflow due to varying signal level.

C.4.3 Memory Requirements

The echo canceller algorithm requires the storage of both reference samples and variable coefficients in on-chip data RAM so that the required FIR and block update convolution can be performed efficiently using the RPTK and MACD instructions. Therefore, the coefficients a_k are stored in block B0, which is configured as program memory. The 16 normalized error samples for coefficient updating are also stored in B0. The 128 reference signal samples y(i) are stored in data RAM along with an additional 16 reference samples $y(1 - 129), \ldots,$ $y(i - 143)$, which are used in the update of coefficients $a_{112}, \ldots, a_{127}$. The echo canceller data memory locations are summarized in Table C.1.

TABLE C.1. Echo canceller data memory locations

VARIABLE	SYMBOL	LOCATION	REMARK
$a_0, \ldots, a_{127}$	A0,....,A127	Block B0 767,766,....,640	A0 is in higher address
$y(k), \ldots, y(k-143)$	Y0,....,Y143	Block B1 768,769,....,911	Y128,....,Y143 required for block update
$un(k), \ldots, un(k-15)$	UN0,....,UN15	Block B0 512,....,527	

C.4.4 Software Logic and Flow

A flowchart of the TMS32020 program for the 128-tap digital voice echo canceller is shown in Figure C.8.

In Table C.2, the instruction cycle and memory requirements are listed for the various blocks of the program implementation. The blocks are listed in the order of execution.

The program loop is executed once per I/O data sample period of 125 μs. The program loop is interrupt-driven from the output data sample mark of a T1 frame. Depending on the near-end speech detector/hangover status, the coefficient update computation module may be skipped. An input data sample interrupt mark occurs during the program loop at a time dependent on the channel location within the T1 frame. In response to the interrupt, the main program execution is interrupted and saved until the new input samples have

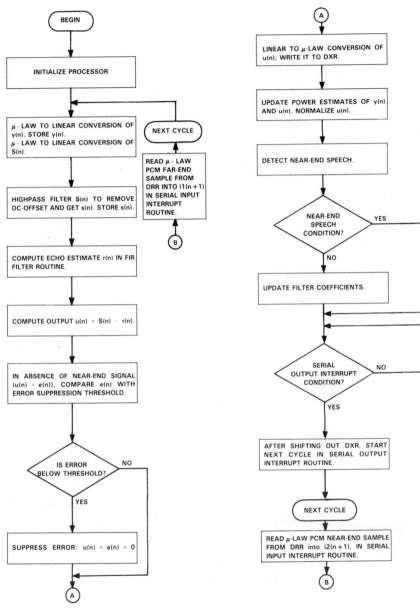

FIGURE C.8. Echo canceller program flowchart.

been read into memory. At the end of each program loop, the processor waits for the next output sample interrupt.

In the following subsections, the implementation of each major block is described in detail. Each variable used in an equation is referred to by its name in the program enclosed in parentheses.

TABLE C.2. Program module requirements

STEP	MODULE FUNCTION	DESCRIPTION	CPU CYCLES	PROGRAM MEMORY LOCATIONS	DATA* MEMORY LOCATIONS
1.	Cycle Start Routine	μ-law to linear conversions; take absolute value of inputs and high-pass filter s(i).	32	28	11
2.	Echo Estimation Routine	FIR convolution of reference samples and filter coefficients to get echo replica r(i).	156	14	258
3.	Compute Output	u(i) = s(i) − r(i) Store u(i).	6	6	2
4.	Residual Output Suppression Routine	If output power below threshold, set u(i) = 0.	12	15	4
5.	Linear to μ-law Compression Routine	Convert u(i) to μ-law.	26	35	4
6.	Power Estimation Routine	Estimate short-term power of u(i) and y(i).	28	14	6
7.	Output Normalization	Compute $u_n(i) = \dfrac{u(i)}{y(i)}$ and clip it.	28	25	19
8.	Near-end Speech Detection	Perform maximum test for near-end speech.	54	74	16
9.	Coefficient Increment Update Routine	If no near-end speech, compute increments for coefficient group.	183	63	26
10.	Coefficient Update Routine	Add increments to coefficient group.	43	43	2
11.	Cycle End Routine	Wait for interrupt.	1	3	0
12.	Receive Interrupt Service Routine	Save status and read input sample.	2 × 14	14	3
13.	Transmit Interrupt Service Routine	Branch to start.	2	2	0
14.	Interrupt Branches		12	6	0
15.	Processor Initialization**	Clear memory, initialize status and set parameters.	86**	86	0
16.	μ-law to Linear Conversion Table*		0	256	0
Total			614	676	351

*Locations are entered only for the routine that uses them first.

**Not in main cycle; CPU cycles not counted in total.

C.4.5 Cycle Start Routine

The voice echo canceller program has been implemented with either μ-law or A-law conversion routines as a program option.

The μ-law (or A-law) to linear input conversion routine is implemented by table lookup in order to minimize the number of instructions. The 256 14-bit two's-complement number corresponding to the 256 possible 8-bit μ-law numbers are stored in program memory. The 8 bits of the μ-law number specify the relative address of the corresponding linear number in the table, which is added to the first address in the table to form the absolute program memory address for the linear number. The TBLR instruction is then used to move the number from program to data memory.

In the cycle start routine, the μ-law input reference sample is read from memory location DRR2 and converted to its linear representation y(i) (Y0). Its absolute value is also stored in location ABSY0. The near-end input sample is then read and converted to a linear representation sdc(i) (SODC). The sample s(i) is next put through a highpass filter to remove any residual dc offset. The highpass filter is a first-order filter with a 3-dB frequency at 160 Hz. Its output s(i) (S0) is given by

$$s(i + 1) = (1 - \gamma)s(i) + \frac{1}{2}(1 - \gamma)(sdc(i) - sdc(i - 1)) \tag{16}$$

where $\gamma = 2^{-3}$.

Note that the filter implementation requires double-precision arithmetic, with S0 denoting the MSBs of s(i) and S0LSBS its LSBs.

C.4.6 Echo Estimation

The echo estimate $\hat{r}(i)$ (EEST) is formed by convolving the tap weight coefficients $a_0, \ldots, a_{127}$ (A0, $\ldots$, A127) with the 128 most recent reference samples $y(i) \ldots$, $y(i - 127)$ (Y0, $\ldots$, Y127).

$$\hat{r}(i) = \sum_{k=0}^{127} a_k y(i - k) \tag{17}$$

This operation is most efficiently implemented on the TMS32020 using the RPTK and MACD instruction. The samples $y(i), \ldots, y(i - 127)$ are stored in block B1 of data memory while $a_0, \ldots, a_{127}$ are stored in block B0 configured as program memory. Since the MACD instruction also performs a data move,

$$y(i - k + 1) \rightarrow y(i - k) \quad \text{for} \quad k = 1, \ldots, 128 \tag{18}$$

no data shifting is required for the computation of the next echo estimate.

The block update routine used for the coefficient adaptation requires the storage of $y(i - 128), \ldots, y(i - 143)$ (Y128, $\ldots$, Y143) in addition to the most recent 128 samples used in the convolution, they are updated using the RPTK and DMOV instructions.

$$y(i - k + 1) \rightarrow y(i - k) \quad \text{for} \quad k = 129, \ldots, 143 \tag{19}$$

The tap weight coefficients $a_0, \ldots, a_{127}$ are initially set to zero, and are adjusted by the algorithm to converge to the impulse response of the echo path $h_0, \ldots, h_{127}$.

$$a_k(i) \rightarrow h_k \qquad \text{for} \quad k = 0, \ldots, 127 \tag{20}$$

The $|h_k| < 1$, $\forall k$, because the power gain of the echo path is smaller than unity. The binary representation for the a_k's was chosen to be of the form (Q.15) with 15 bits after the binary points. This format represents a number between -1 and $(1 - 2^{-15})$. The reference samples and the echo estimate are represented as 16-bit two's-complement integers (no binary point). The 32-bit result of the convolution is therefore of the form (Q.15), and the 16 bits of the echo estimate are the MSB of accumulator low (ACCL) and the 15 LSBs of accumulator high (ACCH). One left shift of the accumulator is required before ACCH is stored in EEST.

C.4.7 Residual Error Suppression

The residual cancellation error is set to zero (or suppressed) whenever the ratio of a long-time average of the absolute value of the output (ABSOUT) to a long-time average of the absolute value of the reference signal (ABSY) is smaller than a fixed threshold. The two long-time averages are updated subsequently in the program as described below. The suppression is, of course, disabled when a near-end speech signal is present (HCNTR > 0). The suppression threshold is set at 1/16 or -24 dB.

C.4.8 Linear to μ-Law (A-Law) Conversion

The linear to μ-law (A-law) conversion routine is an efficient adaptation to the TMS32020 of the conversion routine written for the TMS32010 and described in the application report, "Companding Routines for the TMS32010."[8]

C.4.9 Signal and Output Power Estimation

An estimate of the long-time average of $|u(i)|$ is required by the residual error suppression routine. This estimate $L_u(i)$ (ABSOUT) is obtained by lowpass filtering $|u(i)|$ (ABSU0) using the following infinite impulse response (IIR) filter:

$$L_u(i + 1) = (1 - \alpha)L_u(i) + \alpha|u(i)| \tag{21}$$

where $\alpha = 2^{-7}$. In terms of the program variables, the IIR filter is given by

$$\text{ABSOUT} = 2^{-16}(2^{16} \times \text{ABSOUT} - 2^9 \times \text{ABSOUT} + 2^9 \times \text{ABSU0}) \tag{22}$$

Similarly, the estimate $L_y(i)$ (ABSY) of the long-term average of $y(i)$ (ABSY0) is the output of an IIR filter with the same α, but differs from the above filter by the addition of a cutoff term that prevents the estimate from taking values smaller than a desired level.

$$\text{ABSY} = 2^{-16}(2^{16} \times \text{ABSY} - 2^9 \times \text{ABSY} + 2^9 \times \text{ABSY0} + 2^9 \times \text{CUTOFF} \tag{23}$$

This insures that $\text{ABSY} \geqslant \text{CUTOFF}$ even if ABSY0 is zero for a long time.

Since $L_y(i)$ is used to normalize the algorithm stepsize, this feature is important in order to prevent excessively large stepsizes when the far-end talker is silent.

The stepsize is normalized according to

$$2\beta(i) = \frac{\beta_1}{L_y^2(i)} \tag{24}$$

In order to avoid double-precision arithmetic, this normalization is carried out in two stages (as described in the subsection on coefficient adaptation). Each of the stages requires a division by $L_y(i)$. It is more efficient to compute $L_y(i)^{-1}$ (IABSY) and replace the divisions by two multiplications.

Since ABSY is a positive integer, taking its inverse consists simply of repeating the SUBC instruction. IABSY is a positive fractional number of the form (Q.15), taking values between 0 and $1 - 2^{-15}$.

C.4.10 Output Normalization

The normalized output $u_n(i)$ (UN0) is defined as $u(i)/L_y(i)$ and replaces the actual error in the coefficient update routine for finite-precision considerations, described in the subsection on coefficient adaptation. In the absence of near-end speech, $u_n(i)$ is equal to a normalized cancellation error and is used in the coefficient update. In the presence of near-end speech, no coefficient update is carried out, and the normalized outputs are not used.

The block update approach requires the storage of the 16 most recent normalized outputs $u_n(i), \ldots, u_n(i - 15)$ (UN0, ..., UN15). In a given program cycle, only $u_n(i)$ is computed and stored, while $u_n(i - 1), \ldots, u_n(i - 15)$ computed in previous program cycles are only updated using the DMOV instruction

$$u_n(i - k + 1) = u_n(i - k) \quad \text{for} \quad k = 1, \ldots, 14 \tag{25}$$

In the absence of near-end speech, the normalized output should be a number smaller than one, which is represented as a (Q.15) fraction. To insure that the representation is adequate even in the presence of a near-end signal, the normalized output is clipped at $+1$ or -1, i.e.,

$$
\begin{aligned}
&\text{if } u_n(i) > 1.0, \text{ then } u_n(i) = 1.0 \\
&\text{if } u_n(i) < -1.0, \text{ then } u_n(i) = -1.0
\end{aligned}
\tag{26}
$$

C.4.11 Near-End Speech Detection

Near-end speech is declared if

$$
\tilde{s}(i) \geqslant \max(\tilde{y}(i),\, \tilde{y}(i-1),\, \ldots,\, \tilde{y}(i-127-h(i)))
\tag{27}
$$

where $\tilde{s}(i)$ (ABSSOF) is the output of a lowpass filter with input $2 \times |s(i)|$ (ABSSO). The variable $\tilde{y}(i)$ is a lowpass filtered version of $|y(i)|$, and $h(i)$ (H) a modulo-16 counter. The lowpass filters are IIR filters with short-time constants,

$$
s(i+1) = (1-\alpha)s(i) + 2\alpha|s(i)|
\tag{28}
$$

$$
y(i+1) = (1-\alpha)y(i) + \alpha|y(i)|
\tag{29}
$$

where $\alpha = 2^{-5}$.

The counter $h(i)$ is incremented by one for every input sample. The routines maintain nine partial maxima $m_0, m_1, \ldots, m_8$ (M0, M1, $\ldots$, M8), defined at time i by

$$
\begin{aligned}
m_0(i) &= \max(\tilde{y}(i),\, \ldots,\, \tilde{y}(i-h(i)+1)) \\
m_1(i) &= \max(\tilde{y}(i-h(i)),\, \ldots,\, \tilde{y}(i-h(i)-15)) \\
&\ \vdots \\
m_8(i) &= \max(\tilde{y}(i-h(i)-112),\, \ldots,\, \tilde{y}(i-h(i)-127))
\end{aligned}
\tag{30}
$$

where $i = 16k + h$, k is block index, and h denotes the position within the block.

Figure C.9 illustrates how the partial maxima are maintained. The condition for near-end speech declaration is then equivalent to

$$
\tilde{s}(i) \geqslant \max(m_0,\, \ldots,\, m_8)
\tag{31}
$$

The partial maxima are updated according to the following recursions:

$$
\begin{aligned}
&\text{if } h = 0, \text{ then } m_0(i) = \tilde{y}(i+1) \\
&\qquad \text{and } m_j(i) = m_{j-1}(i) \\
&\qquad\quad \text{where } j = 1, \ldots, 8
\end{aligned}
\tag{32}
$$

and

$$
\begin{aligned}
&\text{if } 0 < h \leqslant 15, \text{ then } m_0(i+1) = \max(m_0(i),\, \tilde{y}(i+1)) \\
&\qquad \text{and } m_j(i+1) = m_j(i) \\
&\qquad\quad \text{where } j = 1, \ldots, 8
\end{aligned}
$$

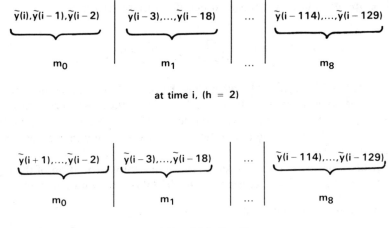

at time i, (h = 2)

at time i + 1, (h = 3)

FIGURE C.9. Partial maxima for near-end speech detection.

If near-end speech is declared, a hangover counter (HCNTR) is set equal to a hangover time (HANGT), which was chosen to be 600 samples or 75 ms. If no near-end speech is declared, then the hangover counter is decremented by one, unless it is zero. If the hangover counter is larger than zero, then the coefficient update routine is skipped. Moreover, if the reference signal power estimate $L_y(i)$ is smaller or equal to the cutoff value of -48 dB, then adaptation is also disabled to avoid divergence during long silences of the far-end talker.

C.4.12 Coefficient Adaptation

The 128 coefficients of the transversal filter are divided into 16 groups of 8 coefficients each, as shown in Table C.3.

The coefficients in only one of the groups are updated in a given program cycle, while the other coefficients are not modified. A modulo-16 counter h(i) (H) points to the index of the group to be updated, and is incremented by one during every program cycle.

TABLE C.3. The coefficient groups

GROUP	COEFFICIENTS
0	$a_0, a_{16}, a_{32}, \ldots, a_{112}$
1	$a_1, a_{17}, a_{33}, \ldots, a_{113}$
.	.
.	.
.	.
15	$a_{15}, a_{31}, a_{47}, \ldots, a_{127}$

The update equation is repeated here for ease of reference.

$$a_k(i + 1) = a_k(i) + \frac{\beta_1}{(L_y(i))^2} \sum_{m=0}^{15} e(i - m)y(i - k - m) \tag{34}$$

for $k = h, h + 16, \ldots, h + 112$, where h is the value of the counter and goes from 0 to 15. The error terms $e(i - m)(m = 0, \ldots, 15)$ are the most recent cancellation errors. In this case, the errors are equal to the 15 most recent canceller outputs $u(i), \ldots, u(i - 15)$ since the adaptation is carried out only in the absence of a near-end signal.

For finite-precision considerations, the actual implementation of the update equation by the routine is carried out in the following two main steps:

1. Compute eight partial updates:

$$\gamma_k(i) = \sum_{m=k}^{k+15} \frac{u(i - m)}{L_y(i)} y(i - k - m) \tag{35}$$

where $k = h, h + 16, \ldots, h + 112$.

The normalized outputs $u_n(i), \ldots, u_n(i - 15)$ have already been computed and stored.

2. Update the coefficients:

$$a_k(i + 1) = a_k(i) + (2^4 \times (L_y(i)^{-1} \times 2^G) \times \gamma_k(i))2^{-16} \tag{36}$$

where G (GAIN) is a program parameter that determines the stepsize of the algorithm and has the value $0, 1, 3, \ldots, 15$.

The partial updates $\gamma_k(i)$ are computed using the MAC instruction in repeat mode. The result is rounded and stored in temporary locations INC0, ..., INC0 + 7 in block B1.

For the second step of the update, $L_y(i)^{-1}$ (IABSY) is first loaded in the T register with a left shift of G (GAIN). It is then multiplied by each of the $\gamma_k(i)$'s. SPM is set to 2 to implement the 2^4 multiplication by shifting the P register four positions to the right before adding it to the accumulator (APAC).

C.4.13 Interrupt Service Routines

At the end of the cycle, the program becomes idle until a receive interrupt occurs followed by a transmit interrupt that sends it back to the beginning of the cycle. The transmit interrupt routine simply enables interrupts and branches back to the start. The receive interrupt must store the status register ST0 and the accumulator, then read the received sample from DRR, zero its eight most significant bits, and store it in DRR1. It restores the accumulator and status register ST0 before returning to the main program.

C.4.14 External Processor Hardware Requirements

Very little external hardware is required to implement a complete single-channel 128-tap echo canceller with the TMS32020. In addition to the processor, only two external 1K × 8 PROMs and some system-dependent interface logic are required. A typical interface circuit for the demonstration system is shown in Figure C.10.

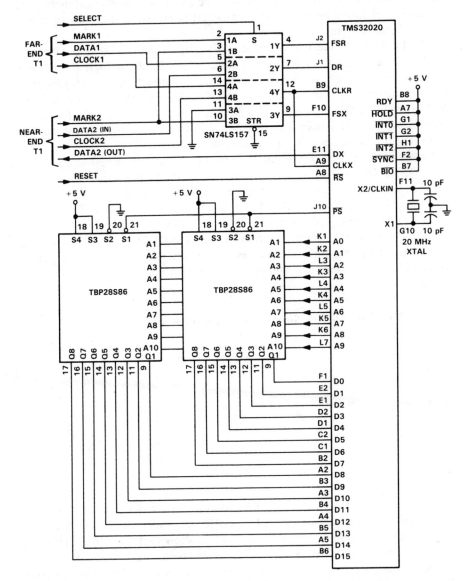

FIGURE C.10. Hardware schematic of the single-channel demonstration processor.

The TMS32020 serial I/O ports allow direct interfacing of the echo canceller to a digital T1 carrier data stream.

Three I/O functions must be performed during each T1 frame (125 μs). The far-end and the near-end signals must be read in, and the processed near-end signal must be written out. To perform these functions, a timing circuit must extract the T1 clock and the T1 frame marks for each direction of transmission. The timing circuit uses the frame mark to generate a channel mark that selects the desired channel out of the 24 present in the T1 frame. The channel mark goes to a high level during the clock cycle, immediately preceding the eight serial bits of the desired sample.

The T1 clock, channel mark, and serial data signals are directly input into the TMS32020 serial clock (CLKR), serial input control (FSR), and serial input port (DR), respectively. Because data is read in from two directions of transmission, a triple two-to-one multiplexer (e.g., SN74LS157) is required to select one of the two sets of T1 signals to be input into the TMS32020. During each T1 frame, the multiplexer alternates once between each direction of transmission, under the control of the timing circuit.

Since data is written out in only one direction, the TMS32020 serial output port (DX) is directly tied to the outgoing T1 data line. The serial output clock (CLKX) and the serial output control (FSX) signals are the same as the near-end direction-of-transmission CLKR and FSR signals. If the far-end T1 channel-frame location overlaps the near-end T1 channel location in time, it is necessary to delay each far-end sample external to the TMS32020 to permit to be read following the sample from the near-end direction. This requires an eight-bit serial shift register and some additional timing circuits.

TABLE C.4. TMS32020 echo canceller performance

TEST DESCRIPTION	CCITT G.165 PERFORMANCE REQUIREMENT	TMS32020 ECHO-CANCELLER PERFORMANCE
1. Final echo return loss after convergence; singletalk mode	-40 dbm0	< -48 dbm0
2. Convergence rate, singletalk mode	≥ 27 dB	> 38 dB
3. Leak rate	≤ 10 dB	≈ 0 dB
4. Infinite return loss convergence	-40 dbm0	< -48 dbm0

The performance of the TMS32020 echo canceller was measured for white-noise input, as suggested in the CCITT G.165 recommendation. The measurement results are summarized in Table C.4 and show that the TMS32020 echo canceller performance exceeds the CCITT requirements in all the tests described. The subjective performance on speech was also found to be very good in both singletalk and doubletalk modes, with no audible distortion of the signal.

C.5 CONCLUSION

The development of novel variations of the generic least-mean-squared (LMS) echo cancelling algorithm and the near-end speech and residual suppression control algorithms has resulted in the implementation of a complete 128-tap single-channel echo canceller on a single TMS32020 programmable Digital Signal Processor. The echo canceller performance exceeds all requirements of the CCITT G.165 recommendations and the performance of similar currently available products. The only external hardware required are two program PROMs and a serial data multiplexer. A direct T1-rate serial interface is available to minimize component count in four-wire VF and T1 carrier configurations.

The single-channel TMS32020 echo canceller program provides a high-performance building block for low-cost systems, which can be tailored to a wide variety of system applications. Programmability offers the flexibility to implement custom requirements, such as cascaded sections for longer tail delay range, short-range multichannel versions, or other special-purpose functions.

The echo canceller application illustrates the power and versatility of the TMS32020 single-chip programmable signal processor. Applications of this technology can be expected to benefit many other complex signal processing tasks in communications products, including voiceband data modems, voice codecs, digital subscriber transceivers, and TDM/FDM transmultiplexers.

REFERENCES

1. M. L. Honig and D. G. Messerschmitt, *Adaptive Filters,* Kluwer Academic Publishers (1984).
2. TMS32020 User's Guide, Texas Instruments.
3. "Recommendation G.165, Echo Cancellers," *CCITT,* Geneva (1980).
4. H. R. Huntley, "Transmission Design of Intertoll Telephone Trunks," *Bell System Technical Journal,* Vol 32, No. 5, 1019–36 (September 1953).
5. M. J. Gingell, B. G. Hay, and L. D. Humphrey, "A Block Mode Update Echo Canceller Using Custom LSI," *GLOBECOM Conference Record,* Vol 3, 1394–97 (November 1983).
6. D. G. Messerschmitt, "Echo Cancellation in Speech and Data Transmission," *IEEE Journal on Selected Topics in Communications,* SAC-2, No. 2, 283–303 (March 1984).
7. D. L. Duttweiler, "A Twelve-Channel Digital Voice Echo Canceller," *IEEE Transactions on Communications,* COM-26, No. 5, 647-53 (May 1978).
8. L. Pagnucco and C. Ershine, "Companding Routines for the TMS32010/TMS32020," *Digital Signal Processing Applications with the TMS320 Family,* Texas Instruments (1986).

```
0001
0002
0003                         128-TAP ECHO-CANCELLER PROGRAM
0004                         (C) COPYRIGHT TEXAS INSTRUMENTS INC., 1985
0005
0006
0007
0008 0000            IDT 'EC128'
0009

          . . .

0384
0385
0386
0387                         ECHO ESTIMATION ROUTINE
0388
0389 0099
0390 0099 C800  EESTR  LDPK  0
0391 009A
0392                   * MOVE Y128,Y129,....Y142 TO NEXT HIGHER MEMORY LOCATION
0393
0394
0395 009A 5589         LARP  AR1                 * 1 -> AR POINTER
0396 009A
0397 009B 316F         LAR   AR1,ADY142          * ADY142 -> AR1
0398 009B
0399 009C CB0E         RPTK  14                  * K = 142,141,....128
0400 009C
0401 009D
0402 009D 5690         DMOV  *-                  * Y(K) -> Y(K+1)
0403 009E
0404
0405                   * CONVOLVE REFERENCE SAMPLES WITH FIR COEFFICIENTS
0406
0407 009E
0408 009E 2E6E  FIR    LAC   ONE,14              * ROUND-OFF OFFSET -> ACC
0409 009F
0410 009F A000         MPYK  0                   * 0 -> P
0411 00A0
0412 00A0 CE05         CNFP                      * ARI STILL POINTS AT Y127
0413 00A1
0414 00A1 CB7F         RPTK  127                 * K = 127,126,....0
0415 00A2
0416 00A2 5C90         MACD  A127PM,*-           * Y(K) * A(I-K) + ACC -> ACC
     00A3 FF80
0417 00A4
0418 00A4 CE04         CNFD
0419 00A5
0420 00A5 CE15         APAC                      * P + ACC -> ACC
0421 00A6
0422 00A6 696B         SACH  EEST,1              * 2 * HIGH ACC -> EEST
0423 00A7

0424
0425
0426                         COMPUTE THE OUTPUT
0427 00A7
0428 00A7 2070         LAC   S0                  * S0 -> ACC
0429 00A8
0430 00A8 106B         SUB   EEST               * ACC - EEST -> ACC
0431 00A9
0432 00A9 606C         SACL  OUTPUT             * ACC -> OUTPUT
0433 00AA
0434 00AA C807         LDPK  7
0435 00AB
0436 00AB CE1B         ABS
0437 00AC
0438 00AC 6065         SACL  ABSE0              * ACC -> ABSE0 ON PAGE 7
0439
0440
0441                         RESIDUAL OUTPUT SUPPRESSION ROUTINE
0442
0443
0444
0445 00AD
0446 00AD
0447 00AD 3C6B  SPRS   LT    IABSY              * IABSY -> T REG
0448 00AE
0449 00AE 3866         MPY   ABSOUT             * ABSOUT * IABSY -> P REG
0450 00AF
0451 00AF 2062         LAC   HCNTR              * NEAR END SPEECH FLAG -> ACC
0452 00B0
0453 00B0 C800         LDPK  0
0454 00B1
0455 00B1 F180         BGZ   WOUT               * IF N.E. SPEECH NO SPRS
     00B2 00BB
0456 00B3
0457 00B3 CE14         PAC                       * P REG -> ACC
0458 00B4
0459 00B4 106D         SUB   THRES              * ACC - THRES -> ACC
0460 00B5
0461 00B5 F180         BGZ   WOUT               * IF THRES EXCEEDED SKIP SPRS
     00B6 00BB
0462 00B7 FA80         BIOZ  WOUT               * IF BIO PIN LOW SKIP SPRS
     00B8 00BB
0463 00B8
0464 00B9 CA00         ZAC                       * 0 -> ACC
0465 00BA
0466 00BA 606C         SACL  OUTPUT             * ACC -> OUTPUT
0467 00BB
0468 00BB 3C6C  WOUT   LT    OUTPUT             * OUTPUT -> r REG  (FOR UN0)
0469 00BB
```

```
              .............
              *
              *  POWER ESTIMATION ROUTINE
              *
              .............
0536
0537
0538
0539
0540
0541 00DF          *  T REG STILL CONTAINS OUTPUT
0542 00DF C807 NORM  LDPK  7
0543 00E0
0544              *
0545              * UPDATE LONG TAU OUTPUT POWER ESTIMATE (ABSOUT)
0546              *
0547 00E0          ZALH  ABSOUT        * ABSOUT -> HIGH ACC
0548 00E0 4066
0549 00E1          ADDS  AELSBS        * AELSBS -> LOW ACC
0550 00E1 4967
0551 00E2          SUB   ABSOUT,LTAU   * ACC - ABSOUT * 2**LTAU -> ACC
0552 00E2 1966
0553 00E3          ADD   ABSE0,LTAU    * ACC + ABSE0 * 2**LTAU -> ACC
0554 00E3 0965
0555 00E4          SACH  ABSOUT        * HIGH ACC -> ABSOUT
0556 00E4 6866
0557 00E5          SACL  AELSBS        * LOW ACC -> AELSBS
0558 00E5 6067
0559 00E6
0560              *
0561              * UPDATE LONG TAU REFERENCE POWER ESTIMATE (ABSY)
0562              *
0563 00E6          ZALH  ABSY          * ABSY -> HIGH ACC
0564 00E6 4069
0565 00E7          ADDS  AYLSBS        * AYLSBS -> LOW ACC
0566 00E7 496A
0567 00E8          SUB   ABSY,LTAU     * ACC - ABSY * 2**LTAU -> ACC
0568 00E8 1969
0569 00E9          ADD   ABSY0,LTAU    * ACC + ABSY0 * 2**LTAU -> ACC
0570 00E9 0968
0571
0572 00EA          ADD   CUTOFF,LTAU   * ACC + CUTOFF * 2**LTAU -> ACC
0573 00EA 096C
0574 00EB          SACH  ABSY          * HIGH ACC -> ABSY
0575 00EB 6869
0576 00EC          SACL  AYLSBS        * LOW ACC -> AYLSBS
0577 00EC 606A
0578 00ED
0579              ...
0580              * COMPUTE 1/ABSY (DIVIDE 1 BY ABSY)
0581
0582 00ED 4012     ZALH  AONE
0583 00EE          RPTK  14
0584 00EE CB0E
0585 00EF          SUBC  ABSY
0586 00EF 4769
0587 00F0          SACL  1ABSY
0588 00F0 606B
```

```
              .............
              *
              *  NEAR-END SPEECH DETECTION ROUTINE
              *
              .............
0646
0647
0648
0649
0650
0651 0109     NESP  LDPK  7
0652 0109 C807
0653 010A
0654              *
0655              * UPDATE SHORT TAU REFERENCE POWER ESTIMATE (ABSYOF)
0656              *
0657 010A          ZALH  ABSYOF        * ABSYOF * 2**16 -> ACC
0658 010A 4060
0659 010B          SUB   ABSYOF,STAU   * ACC - ABSYOF * 2**STAU -> ACC
0660 010B 1B60
0661 010C          ADD   ABSY0,STAU    * ACC + ABSY0 * 2**STAU -> ACC
0662 010C 0B68
0663 010D          SACH  ABSYOF        * HIGH ACC -> ABSYOF
0664 010D 6860
0665 010E
0666              *
0667              * UPDATE SHORT TAU NEAR END POWER ESTIMATE (ABSSOF)
0668              *
0669 010E          ZALH  ABSSOF        * ABSSOF * 2**16 -> ACC
0670 010E 4064
0671 010F          SUB   ABSSOF,STAU   * ACC - ABSSOF * 2**STAU -> ACC
0672 010F 1B64
0673 0110          ADD   ABSS0,STAU+NER * ACC + ABSS0*2**STAU+NER -> ACC
0674 0110 0C63
0675 0111          SACH  ABSSOF        * HIGH ACC -> ABSSOF
0676 0111 6864
0677 0112
0678              *
0679              * UPDATE MODULO 16 COUNTER (H)
0680              *
0681 0112          LAC   H             * H -> ACC
0682 0112 2060
0683 0113          ADD   AONE          * ACC + 1 -> ACC
0684 0113 0012
0685 0114          ANDK  >000F         * IF ACC = 16 THEN 0 -> ACC
0686 0114 D004
0687 0116          SACL  H             * ACC -> H
0688 0116 6060
0689 0117          BGZ   NESP1         * IF H > 0 THEN GO TO NESP1
0690 0117 F180
     0118 011F
0691              ...
0692              * MOVE M0,M1,....M7 TO NEXT HIGHER MEMORY LOCATION
0693              *
0694
0695 0119          LAR   AR1,ADM7      * ADM7 -> AR1
0696 0119 3117
0697 011A          RPTK  7             * K=7,6,....0
0698 011A CB07
0699 011B          DMOV  *-            * M(K) -> M(K+1)
0700 011B 5690
```

```
0701  011C
0702  011C 566D          DMOV   ABSYOF     * ABSYOF -> M0
0703  011D FF80          B      NESP3      * ON MEMORY MOVES SKIP DETECTION
0704  011D 0149
0705  011E
0706  011E                                 UPDATE MOST RECENT LOCAL MAXIMA (M0)
0707                        • • •
0708  011F
0709  011F 206D   NESP1   LAC    ABSYOF     * ABSYOF -> ACC
0710  0120 106E           SUB    M0         * ACC - M0 -> ACC
0711  0120
0712  0121 F280           BLEZ   NESP2      * IF M0 > ABSYOF THEN NO UPDATE
0713  0121 0124
0714  0122 566D           DMOV   ABSYOF     * ABSYOF -> M0
0715  0123
0716  0123                                  COMPARE REFERENCE POWER TO NEAR-END POWER
0717  0124
0718  0124
0719                        • • •
0720  0124
0721  0124 2064   NESP2   LAC    ABSSOF     * ABSSOF -> ACC
0722  0125 106E           SUB    M0         * ACC - M0 -> ACC
0723  0125
0724  0126 F280           BLEZ   NESP3      * NO N.E. SPEECH IF M0 > ABSSOF
0725  0126 0149
0726  0126
0727  0128
0728  0128 2064           LAC    ABSSOF     * ABSSOF -> ACC
0729  0129 106F           SUB    M0+1       * ACC - M1 -> ACC
0730  0129
0731  012A F280           BLEZ   NESP3      * NO N.E. SPEECH IF M1 > ABSSOF
0732  012A 0149
0733  012A
0734  012C
0735  012C 2064           LAC    ABSSOF     * ABSSOF -> ACC
0736  012C 1070           SUB    M0+2       * ACC - M2 -> ACC
0737  012D
0738  012E F280           BLEZ   NESP3      * NO N.E. SPEECH IF M2 > ABSSOF
0739  012E 0149
0740  012F
0741  0130
0742  0130 2064           LAC    ABSSOF     * ABSSOF -> ACC
0743  0130
0744  0131 1071           SUB    M0+3       * ACC - M3 -> ACC
0745  0131
0746  0132 F280           BLEZ   NESP3      * NO N.E. SPEECH IF M3 > ABSSOF
0747  0132 0149
0748  0134
0749  0134
0750  0134 2064           LAC    ABSSOF     * ABSSOF -> ACC
0751  0135

0752  0135 1072           SUB    M0+4       * ACC - M4 -> ACC
0753  0136
0754  0136 F280           BLEZ   NESP3      * NO N.E. SPEECH IF M4 > ABSSOF
0755  0137 0149
0756  0138
0757  0138
0758  0138 2064           LAC    ABSSOF     * ABSSOF -> ACC
0759  0139
0760  0139 1073           SUB    M0+5       * ACC - M5 -> ACC
0761  013A F280           BLEZ   NESP3      * NO N.E. SPEECH IF M5 > ABSSOF
0762  013A 0149
0763  013C
0764  013C
0765  013C 2064           LAC    ABSSOF     * ABSSOF -> ACC
0766  013D
0767  013D 1074           SUB    M0+6       * ACC - M6 -> ACC
0768  013E F280           BLEZ   NESP3      * NO N.E. SPEECH IF M6 > ABSSOF
0769  013E 0149
0770  0140
0771  0140
0772  0140 2064           LAC    ABSSOF     * ABSSOF -> ACC
0773  0141
0774  0141 1075           SUB    M0+7       * ACC - M7 -> ACC
0775  0142 F280           BLEZ   NESP3      * NO N.E. SPEECH IF M7 > ABSSOF
0776  0142 0149
0777  0144
0778  0144
0779  0144 2064           LAC    ABSSOF     * ABSSOF -> ACC
0780  0145
0781  0145 1076           SUB    M0+8       * ACC - M8 -> ACC
0782  0146 F280           BLEZ   NESP3      * NO N.E. SPEECH IF M8 > ABSSOF
0783  0146 0149
0784  0148
0785  0148
0786  0148                                  NEAR-END SPEECH DETECTED   SET HANGOVER COUNTER (HCNTR)
0787  0148 5661           DMOV   HANGT      * HANGT -> HCNTR
0788  0148 5661
0789  0149 0149
0790  0149                                  CHECK AND UPDATE HANGOVER COUNTER
0791                        • • •
0792
0793  0149         NESP3
0794  0149 2062           LAC    HCNTR      * HCNTR -> ACC
0795  014A F680           BZ     NESP4      * IF HCNTR = 0 THEN GO TO NESP4
0796  014A 0150
0797  014B
0798  014C 1012           SUB    AONE       * ACC - 1 -> ACC
0799  014C
0800  014D 6062           SACL   HCNTR      * ACC -> HCNTR
0801  014D
0802  014E FF80           B      LOOP       * GO TO CYCLE END
      014E 0149
```

```
0803  014F  01BE
0804  0150
0805
0806               *   CHECK IF LTAU REFERENCE POWER ESTIMATE IS BELOW CUTOFF
0807
0808  0150  2069  NESP4  LAC   ABSY        * ABSY -> ACC
0809  0151  106C         SUB   CUTOFF      * ACC - CUTOFF -> ACC
0810
0811  0152  F280  P.EZ   BZ    LOOP        * IF ABSY < CUTOFF THEN LOOP
0812  0153  01BE


                 * * *
                 * * *   COEFFICIENT INCREMENT UPDATE ROUTINE
                 * * *


0814
0815
0816
0817
0818
0819  0154
0820  0154  2015  UPINC  LAC   ADY1        * ADY1 -> ACC   (Y0 IS NOW IN Y1)
0821  0155
0822  0155  0060         ADD   H           * ACC + H -> ACC
0823  0156
0824  0156  6010         SACL  TEMP3
0825  0157
0826  0157  3110         LAR   AR1,TEMP3   * ADY1 + H -> AR1
0827  0158
0828  0158  CE05         CNFP
0829  0159
0830  0159  3C11         LT    CUN0        * UN0 -> T REG
0831  015A
0832  015A
0833  015A  2F12         LAC   AONE,15     * ROUND-OFF OFFSET -> ACC
0834  015B
0835  015B  38A0         MPY   *+          * UN(0) * Y(0+H) -> P REG
0836  015B
0837  015C  CB0E         RPTK  14          * K = 1,2,....15
0838  015D
0839  015D  5DA0         MAC   UN0PM+1,*+  * UN(K) * Y(K+H) + ACC -> ACC    UN0 -> T
0840  015E  FF01
0841  015F
0842  015F  3D11         LTA   CUN0        * P REG + ACC -> ACC    UN0 -> T
0843  0160
0844  0160  6878         SACH  INC0        * HIGH ACC -> INC(0)
0845  0161
0846  0161
0847  0161  2F12         LAC   AONE,15     * ROUND-OFF OFFSET -> ACC
0848  0162
0849  0162  38A0         MPY   *+          * UN(0) * Y(16+H) -> P REG
0850  0163
0851  0163  CB0E         RPTK  14          * K = 1,2,....15
0852  0164  5DA0         MAC   UN0PM+1,*+  * UN(K) * Y(K+16+H) + ACC -> ACC    UN0 -> T
0853  0165  FF01
0854  0166
0855  0166  3D11         LTA   CUN0        * P REG + ACC -> ACC    UN0 -> T
0856  0167
0857  0167  6879         SACH  INC0+1      * HIGH ACC -> INC(1)
0858  0168
0859  0168
0860  0168  2F12         LAC   AONE,15     * ROUND-OFF OFFSET -> ACC
0861  0169
0862  0169  38A0         MPY   *+          * UN(0) * Y(32+H) -> P REG
0863  016A
0864  016A  CB0E         RPTK  14          * K = 1,2,....15
0865  016B  5DA0         MAC   UN0PM+1,*+  * UN(K) * Y(K+32+H) + ACC -> ACC    UN0 -> T
0866  016C  FF01
0867  016D  3D11         LTA   CUN0        * P REG + ACC -> ACC    UN0 -> T
```

```
0868 016E
0869 016E 687A      SACH  INC0+2
0870 016F
0871 016F
0872 016F 2F12      LAC   AONE,15
0873 0170
0874 0170 38A0      MPY   *+
0875 0171
0876 0171 CB0E      RPTK  14
0877 0172
0878 0172 5DA0      MAC   UN0PM+1,*+
     0173 FF01
0879 0174
0880 0174 3D11      LTA   CUN0
0881 0175
0882 0175 687B      SACH  INC0+3
0883 0176
0884 0176
0885 0176 2F12      LAC   AONE,15
0886 0177
0887 0177 38A0      MPY   *+
0888 0178
0889 0178 CB0E      RPTK  14
0890 0179
0891 0179 5DA0      MAC   UN0PM+1,*+
     017A FF01
0892 017B
0893 017B 3D11      LTA   CUN0
0894 017C
0895 017C 687C      SACH  INC0+4
0896 017D
0897 017D
0898 017D 2F12      LAC   AONE,15
0899 017E
0900 017E 38A0      MPY   *+
0901 017F
0902 017F CB0E      RPTK  14
0903 0180
0904 0180 5DA0      MAC   UN0PM+1,*+
     0181 FF01
0905 0182
0906 0182 3D11      LTA   CUN0
0907 0183
0908 0183 687D      SACH  INC0+5
0909 0184
0910 0184
0911 0184 2F12      LAC   AONE,15
0912 0185
0913 0185 38A0      MPY   *+
0914 0186
0915 0186 CB0E      RPTK  14
0916 0187
0917 0187 5DA0      MAC   UN0PM+1,*+
     0188 FF01
0918 0189
0919 0189 3D11      LTA   CUN0
0920 018A

* HIGH ACC -> INC(2)

0921 018A 687E      SACH  INC0+6
0922 018B
0923 018B
0924 018B 2F12      LAC   AONE,15
0925 018C
0926 018C 38A0      MPY   *+
0927 018D
0928 018D CB0E      RPTK  14
0929 018E
0930 018E 5DA0      MAC   UN0PM+1,*+
     018F FF01
0931 0190
0932 0190 3D11      LTA   CUN0
0933 0191
0934 0191 687F      SACH  INC0+7
0935 0192
0936 0192
0937 0192 CE04      CNFD
```

```
0939          *
0940          *
0941          *   COEFFICIENT UPDATE ROUTINE
0942          *
0943          *
0944  0193
0945  0193 C010    LARK  AR0,16          * 16 -> AR0    (AR2 INCREMENT)
0946  0194
0947  0194 3116    LAR   AR1,ADINC0      * ADINC0 -> AR1
0948  0195
0949  0195 2014    LAC   ADA0            * ADA0 -> ACC
0950  0196
0951  0196 1060    SUB   H               * ACC - H -> ACC
0952  0197
0953  0197 6010    SACL  TEMP3
0954  0198
0955  0198 3210    LAR   AR2,TEMP3       * ADA0 - H -> AR2
0956  0199
0957  0199 CE0A    SPM   2               * SET 4 BIT LEFT SHIFT OF P REG
0958  019A
0959  019A 236B    LAC   |ABSY,GAIN      * |ABSY * 2**GAIN -> ACC
0960  019B
0961  019B 6010    SACL  TEMP3           * ACC -> TEMP3
0962  019C
0963  019C 3C10    LT    TEMP3           * TEMP3 -> T REG
0964  019D
0965  019D
0966  019D 38AA    MPY   *+,AR2          * INC(0)  T REG -> P REG
0967  019E
0968  019E 4080    ZALH  *               * A(H) . 2**16 -> ACC
0969  019F
0970  019F CE15    APAC                  * P REG + ACC -> ACC
0971  01A0
0972  01A0 6BD9    SACH  *0-,AR1         * HIGH ACC -> A(H)
0973  01A1
0974  01A1
0975  01A1 38AA    MPY   *+,AR2          * INC(1)  T REG -> P REG
0976  01A2
0977  01A2 4080    ZALH  *               * A(16+H) . 2**16 -> ACC
0978  01A3
0979  01A3 CE15    APAC                  * P REG + ACC -> ACC
0980  01A4
0981  01A4 6BD9    SACH  *0-,AR1         * HIGH ACC -> A(16+H)
0982  01A5
0983  01A5
0984  01A5 38AA    MPY   *+,AR2          * INC(2)  T REG -> P REG
0985  01A6
0986  01A6 4080    ZALH  *               * A(32+H) . 2**16 -> ACC
0987  01A7
0988  01A7 CE15    APAC                  * P REG + ACC -> ACC
0989  01A8
0990  01A8 6BD9    SACH  *0-,AR1         * HIGH ACC -> A(32+H)
0991  01A9
0992  01A9
0993  01A9 38AA    MPY   *+,AR2
0994  01AA
0995  01AA 4080    ZALH  *
0996  01AB
0997  01AB CE15    APAC
0998  01AC
0999  01AC 6BD9    SACH  *0-,AR1
1000  01AD
1001  01AD
1002  01AD 38AA    MPY   *+,AR2
1003  01AE
1004  01AE 4080    ZALH  *
1005  01AF
1006  01AF CE15    APAC
1007  01B0
1008  01B0 6BD9    SACH  *0-,AR1
1009  01B1
1010  01B1
1011  01B1 38AA    MPY   *+,AR2
1012  01B2
1013  01B2 4080    ZALH  *
1014  01B3
1015  01B3 CE15    APAC
1016  01B4
1017  01B4 6BD9    SACH  *0-,AR1
1018  01B5
1019  01B5
1020  01B5 38AA    MPY   *+,AR2
1021  01B6
1022  01B6 4080    ZALH  *
1023  01B7
1024  01B7 CE15    APAC
1025  01B8
1026  01B8 6BD9    SACH  *0-,AR1
1027  01B9
1028  01B9
1029  01B9 38AA    MPY   *+,AR2
1030  01BA
1031  01BA 4080    ZALH  *
1032  01BB
1033  01BB CE15    APAC
1034  01BC
1035  01BC 6BD9    SACH  *0-,AR1
1036  01BD
1037  01BD
1038  01BD CE08    SPM   0               * SET NO SHIFT OF P REG
```

References

[Albert and Gardner, 1967]
Albert, A. E., and L. A. Gardner, *Stochastic Approximation and Non-Linear Regression*, MIT Press, Cambridge, Mass., 1967.

[Alexander, 1986]
Alexander, S. T., *Adaptive Signal Processing, Theory and Applications*, Springer-Verlag, New York, 1986.

[Altay, 1984]
Altay, B. K., "Elimination of Real Positivity Condition and Error Filtering in Parallel MRAS," *IEEE Trans. on Automat. Control*, vol. AC-29, pp. 1017–1019, November, 1984.

[Anderson et al., 1986]
Anderson, B. D. O., R. R. Bitmead, C. R. Johnson, Jr., P. V. Kokotovic, R. L. Kosut, I. M. Y. Mareels, L. Praly, and B. D. Riedle, *Stability of Adaptive Systems: Passivity and Averaging Analysis*, MIT Press, Cambridge, Mass., 1986.

[Anderson and Johnson, 1982a]
Anderson, B. D. O., and C. R. Johnson, Jr., "Exponential Convergence of Adaptive Identification and Control Algorithms," *Automatica*, vol. 18, pp. 1–13, January, 1982.

[Anderson and Johnson, 1982b]
Anderson, B. D. O., and C. R. Johnson, Jr., "On Reduced-Order Adaptive Output Error Identification and Adaptive IIR Filtering," *IEEE Trans. on Automat. Control*, vol. AC-27, pp. 927–933, August, 1982.

[Anderson and Johnstone, 1983]
Anderson, B. D. O., and R. M. Johnstone, "Adaptive Systems and Time-Varying Plants," *Int. J. Control*, vol. 37, pp. 367–377, 1983.

[Bellamy, 1982]

Bellamy, J., *Digital Telephony*, Wiley, New York, chap. 2, 1982.

[Bello, 1963]

Bello, P. A., "Characterization of Randomly Time-Invariant Linear Channels," *IEEE Trans. on Communication Syst.*, vol. CS-11, pp. 360–393, December, 1963.

[Bershad and Feintuch, 1980]

Bershad, N. J., and P. L. Feintuch, "The Recursive LMS Filter: A Line Enhancer Application and Analytical Model for Mean Weight Behavior," *IEEE Trans. on Acoustics, Speech, and Signal Processing.*, vol. ASSP-28, pp. 652–660, December,1980

[Burrus and Parks, 1985]

Burrus, C. S., and T. W. Parks, *DFT/FFT and Convolution Algorithms*, Wiley–Interscience, New York, 1985.

[Cadzow, 1976]

Cadzow, J. A., "Recursive Digital Filter Synthesis via Gradient Based Algorithms," *IEEE Trans. on Acoustics, Speech, and Signal Processing*, vol. ASSP-24, pp. 349–355, October, 1976.

[Caraiscos and Liu, 1984]

Caraiscos, C. and B. Liu, "A Roundoff Error Analysis of the LMS Adaptive Algorithm," *IEEE Trans. on Acoustics, Speech, and Signal Processing*, vol. ASSP-32, pp. 34–41, February, 1984.

[Carlson, 1975]

Carlson, A. B., *Communication Systems*, McGraw-Hill, New York, 1975.

[Cioffi and Kailath, 1984]

Cioffi, J. M., and T. Kailath, "Fast, Recursive-Least-Squares Transversal Filters for Adaptive Filtering," *IEEE Trans. on Acoustics, Speech, and Signal Processing*, vol. ASSP-32, pp. 304–337, April, 1984.

[Cohn and Melsa, 1975]

Cohn, D. L., and J. L. Melsa, "The Residual Encoder: An Improved ADPCM System for Speech Digitization," *IEEE Trans. on Communications*, vol. COM-23, pp. 935–941, September, 1975.

[Dasgupta and Johnson, 1986]

Dasgupta, S., and C. R. Johnson, Jr., "Some Comments on the Behavior of Sign–Sign Adaptive Identifiers," *Systems and Control Letters*, vol. 7, pp. 75–82, April, 1986.

[Dentino et al., 1978]

Dentino, M., J. McCool, and B. Widrow, "Adaptive Filtering in the Frequency Domain," *Proc. IEEE*, vol. 66, pp. 1658–1659, December, 1978.

[Desoer and Vidyasagar, 1975]

Desoer, C. A., and M. Vidyasagar, *Feedback Systems: Input–Output Properties*, Academic Press, New York, 1975.

[Dugard and Goodwin, 1985]

Dugard, L., and G. C. Goodwin, "Global Convergence of Landau's 'Output Error with Adjustable Compensator' Adaptive Algorithm," *IEEE Trans. on Automat. Control*, vol. AC-30, pp. 593–595, June, 1985.

[Feintuch, 1976]
Feintuch, P. L., "An Adaptive Recursive LMS Filter," *Proc. IEEE*, vol. 64, pp. 1622–1624, November, 1976.

[Ferrara, 1980]
Ferrara, E. R., "Fast Implementation of the LMS Adaptive Filter," *IEEE Trans. on Acoustics, Speech, and Signal Processing*, vol. ASSP-28, pp. 474–475, August, 1981.

[Ferrara, 1981]
Ferrara, E. R., "The Time-Sequenced Adaptive Filter," *IEEE Trans. on Acoustics, Speech, and Signal Processing*, vol. ASSP-29, pp. 679–683, June, 1981.

[Freeney, 1980]
Freeney, S. L., "TDM/FDM Translation as an Application of Digital Signal Processing," *IEEE Communications Magazine*, vol. 18, pp. 5–15, January, 1980.

[Friedlander, 1982a]
Friedlander, B., "A Modified Prefilter for Some Recursive Parameter Estimation Algorithms," *IEEE Trans. on Automat. Control*, vol. AC-27, pp. 232–235, February, 1982.

[Friedlander, 1982b]
Friedlander, B., "System Identification Techniques for Adaptive Signal Processing," *IEEE Trans. on Acoustics, Speech, and Signal Processing*, vol. ASSP-30, pp. 240–246, April, 1982.

[Gersho, 1969]
Gersho, A., "Adaptive Equalization of Highly Dispersive Channels for Data Transmission," *Bell System Tech. J.*, vol. 48, pp. 55–70, January, 1969.

[Gersho, 1984]
Gersho, A., "Adaptive Filtering with Binary Reinforcement," *IEEE Trans. on Info. Thy.*, vol. IT-30, pp. 191–199, March, 1984.

[Gibson, 1980]
Gibson, J. D., "Adaptive Prediction in Speech Differential Encoding Systems," *Proc. IEEE*, vol. 68, pp. 488–525, April, 1980.

[Gitlin et al., 1982]
Gitlin, R. D., H. C. Meadors, and S. B. Weinstein, "The Tap-Leakage Algorithm: An Algorithm for the Stable Operation of a Digitally Implemented, Fractionally Spaced, Adaptive Equalizer," *Bell Syst. Tech. J.*, vol. 61, pp. 1817–1840, October, 1982.

[Glover, 1977]
Glover, J. R., "Adaptive Noise Cancelling Applied to Sinusoidal Interference," *IEEE Trans. on Acoustics, Speech, and Signal Processing*, vol. ASSP-25, pp. 484–491, December, 1977.

[Godard, 1980]
Godard, D. N., "Self-Recovering Equalization and Carrier Tracking in Two-Dimensional Data Communication Systems," *IEEE Trans. on Communications*, vol. COM-28, pp. 1867–1875, November, 1980.

[Gold and Rader, 1969]
Gold, B., and C. M. Rader, *Digital Processing of Signals*, McGraw-Hill, New York, chap. 8, 1969.

[Goodwin and Sin, 1984]
Goodwin, G. C., and K. S. Sin, *Adaptive Filtering, Prediction, and Control*, Prentice-Hall, Englewood Cliffs, N.J., 1984.

[Gray, 1972]
Gray, R. M., "On the Asymptotic Eigenvalue Distribution of Toeplitz Matrices," *IEEE Trans. on Info. Thy.*, vol. IT-18, pp. 725–730, November, 1972.

[Gray and Markel, 1973]
Gray, A. H., and J. D. Markel, "Digital Lattice and Ladder Filter Synthesis," *IEEE Trans. on Audio and Electroacoustics*, vol. AU-21, pp. 491–500, December, 1973.

[Griffiths, 1967]
Griffiths, L. J., "A Simple Adaptive Algorithm for Real Time Processing in Antenna Arrays," *Proc. IEEE*, vol. 57, pp. 1696–1704, October, 1967.

[Honig and Messerschmitt, 1984]
Honig, M. L., and D. G. Messerschmitt, *Adaptive Filters: Structures, Algorithms, and Applications*, Kluwer Academic, Hingham, MA., 1984.

[Horvath, 1980]
Horvath, S., Jr., "A New Adaptive Recursive LMS Filter," in *Digital Signal Processing* (ed. V. Cappellini and A. G. Constantinides), Academic Press, New York, pp. 21–26, 1980.

[Jayant and Noll, 1984]
Jayant, N. S., and P. Noll, *Digital Coding of Waveforms: Principles and Applications to Speech and Video*, Prentice-Hall, Englewood Cliffs, N.J., 1984.

[Johnson, 1979]
Johnson, C. R., Jr., "A Convergence Proof for a Hyperstable Adaptive Recursive Filter," *IEEE Trans. on Info. Thy.*, vol. IT-25, pp. 745–749, November, 1979.

[Johnson and Larimore, 1977]
Johnson, C. R., Jr. and M. G. Larimore, "Comments on and Additions to 'An Adaptive Recursive LMS Filter,'" *Proc. IEEE*, vol. 65, pp. 1399–1401, September, 1977.

[Johnson et al., 1981]
Johnson, C. R., Jr., M. G. Larimore, J. R. Treichler, and B. D. O. Anderson, "SHARF Convergence Properties," *IEEE Trans. on Acoustics, Speech, and Signal Processing*, vol. ASSP-29, pp. 659–670, June, 1981.

[Johnson and Taylor, 1980]
Johnson, C. R., Jr. and T. Taylor, "Failure of a Parallel Adaptive Identifier with Adaptive Error Filtering," *IEEE Trans. on Automat. Control*, vol. AC-25, pp. 1248–1250, December, 1980.

[Kaczmerz, 1937]
Kaczmerz, M. S., "Angenäherte Auslösung von Systemen Linearer Gleichungen," *Academie Polonaise des Sciences et des Lettres*, bull. A, vol. 3, pp. 355–357, 1937.

[Kailath, 1980]
Kailath, T., *Linear Systems*, Prentice-Hall, Englewood Cliffs, N.J., 1980.

[Knopp, 1956]
Knopp, K., *Infinite Sequences and Series*, Dover, New York, 1956.

[Landau, 1976]
Landau, I. D., "Unbiased Recursive Identification Using Model Reference Adaptive Techniques," *IEEE Trans. on Automat. Control*, vol. AC-21, pp. 194–202, April, 1976.

[Landau, 1978]
Landau, I. D., "Elimination of the Real Positivity Condition in the Design of Parallel MRAS," *IEEE Trans. on Automat. Control*, vol. AC-23, pp. 1015–1020, December, 1978.

[Larimore and Goodman, 1985]
Larimore, M. G., and M. J. Goodman, "Implementation of the Constant Modulus Algorithm at RF Bandwidths," *Proc. 19th Asilomar Conf. on Circuits, Sys., and Computers*, Pacific Grove, Calif., November, 1985.

[Larimore et al., 1980]
Larimore, M. G., J. R. Treichler, and C. R. Johnson, Jr., "SHARF: An Algorithm for Adapting IIR Digital Filters," *IEEE Trans. on Acoustics, Speech, and Signal Processing*, vol. ASSP-28, pp. 428–440, August, 1980.

[Lawrence and Johnson, 1986]
Lawrence, D. A., and C. R. Johnson, Jr., "Recursive Parameter Identification Algorithm Stability Analysis Via Pi-Sharing," *IEEE Trans. on Automat. Control*, vol. AC-31, pp. 16–24, January, 1986.

[Ljung, 1977]
Ljung, L., "On Positive Real Functions and the Convergence of Some Recursive Schemes," *IEEE Trans. on Automat. Control*, vol. AC-22, pp. 539–551, August, 1977.

[Ljung et al., 1978]
Ljung, L., M. Morf, and D. Falconer, "Fast Calculation of G in Matrices for Recursive Estimation Schemes," *Int. J. Control*, vol. 27, pp. 1–19, 1978.

[Ljung and Söderström, 1983]
Ljung L., and T. Söderström, *Theory and Practice of Recursive Identification*, MIT Press, Cambridge, Mass., 1983.

[Lucky et al., 1968]
Lucky, R. W., J. Salz, and E. J. Weldon, Jr., *Principles of Data Communications*, McGraw-Hill, New York, chap. 3, 1968.

[Macchi and Jaidane-Saidane, 1986]
Macchi, O., and M. Jaidane-Saidane, "Stability of an Adaptive ARMA Predictor in Presence of Narrow-Band Input Signals," *Proc. 2nd IFAC Workshop on Adaptive Systems, in Control and Signal Processing*, Lund Sweden, pp. 417–422, July, 1986.

[Melsa and Sage, 1973]
Melsa, J. L., and A. P. Sage, *An Introduction to Probability and Stochastic Processes*, Prentice-Hall, Englewood Cliffs, N.J., 1973.

[Mendel, 1973]
Mendel, J. M., *Discrete Techniques of Parameter Estimation: The Equation Error Formulation*, Marcel Dekker, New York, 1973.

[Monson, 1971]
Monson, P., "Feedback Equalization for Fading Dispersive Channels," *IEEE Trans. on Info. Thy.*, vol. IT-17, pp. 56–64, January, 1971.

[Narasimha and Peterson, 1979]
Narasimha, M. J., and A. M. Peterson, "Design of a 24-Channel Transmultiplexer," *IEEE Trans. on Acoustics, Speech, and Signal Processing*, vol. ASSP-27, pp. 752–762, December, 1979.

[Noble and Daniel, 1977]
Noble, B., and J. W. Daniel, *Applied Linear Algebra*, Prentice-Hall, Englewood Cliffs, N.J., 1977.

[Owen, 1982]
Owen, F. F. E., *PCM and Digital Transmission Systems*, McGraw-Hill, New York, 1982.

[Paez and Glisson, 1972]
M. D. Paez and T. H. Glisson, "Minimum Mean Squared-Error Quantization in Speech," *IEEE Trans. on Communications*, vol. COM-20, pp. 225–230, April, 1972.

[Parikh et al., 1980]
D. Parikh, N. Ahmed, and S. Stearns, "An Adaptive Lattice Algorithm for Recursive Filters," *IEEE Trans. on Acoustics, Speech, and Signal Processing*, vol. ASSP-28, pp. 110–111, February, 1980.

[Parikh and Ahmed, 1978]
Parikh, D., and N. Ahmed, "On an Adaptive Algorithm for IIR Filters," *Proc. IEEE*, vol. 65, pp. 585–587, May, 1978.

[Popov, 1973]
Popov, V. M., *Hyperstability of Control Systems*, Springer-Verlag, New York, 1973.

[Rabiner, and Gold, 1975]
Rabiner, L. R., and B. Gold, *Theory and Application of Digital Signal Processing*, Prentice-Hall, Englewood Cliffs, N.J., 1975.

[Rabiner and Schafer, 1978]
Rabiner, L. R., and R. W. Schafer, *Digital Processing of Speech Signals*, Prentice-Hall, Englewood Cliffs, N.J., 1978.

[Salz, 1973]
Salz, J., "Optimum Mean-Square Decision Feedback Equalization," *Bell System Tech. J.*, vol. 50, pp. 1341–1373, October, 1973.

[Satorius and Alexander, 1979]
Satorius, E. H., and S. T. Alexander, "Channel Equalization Using Adaptive Lattice Algorithms," *IEEE Trans. on Communications*, vol. COM-27, pp. 899–905, June, 1979.

[Shynk and Gooch, 1985]
Shynk, J. J., and R. P. Gooch, "Frequency-Domain Adaptive Pole-Zero Filtering," *Proc. IEEE*, vol. 73, pp. 1526–1528, October, 1985.

[Söderström et al., 1978]
Söderström, T., L. Ljung, and I. Gustavsson, "A Theoretical Analysis of Recursive Identification Methods," *Automatica*, vol. 14, pp. 231–244, May, 1978.

[Solo, 1979]
Solo, V., "The Convergence of AML," *IEEE Trans. on Automat. Control*, vol. AC-24, pp. 958–962, December, 1979.

[Stearns, 1981]
Stearns, S. D., "Error Surfaces of Recursive Adaptive Filters," *IEEE Trans. on Acoustics, Speech, and Signal Processing*, vol. ASSP-29, pp. 763–766, June, 1981.

[Stearns et al., 1976]
Stearns, S. D., G. R. Elliott, and N. Ahmed, "On Adaptive Recursive Filtering," *Proc. 10th Asilomar Conf. on Circuits, Sys., and Computers*, Pacific Grove, Calif., pp. 5–10, November, 1976.

[Tant, 1974]
Tant, M. J., *The White Noise Book*, White Crescent Press, Luton, England, 1974.

[Taylor and Johnson, 1982]
Taylor, T., and C. R. Johnson, Jr., "A Problem with Effectively Overspecified Adaptive Identifiers," *Systems and Control Letters*, vol. 1, pp. 227–231, January, 1982.

[Tomizuka, 1982]
Tomizuka, M., "Parallel MRAS without Compensation Block," *IEEE Trans. on Automat. Control*, vol. AC-27, pp. 505–506, April, 1982.

[Treichler, 1979]
Treichler, J. R., "Transient and Convergent Behavior of the Adaptive Line Enhancer," *IEEE Trans. on Acoustics, Speech, and Signal Processing*, vol. ASSP-27, pp. 53–62, February, 1979.

[Treichler, 1980]
Treichler, J. R., "Observability and Its Effect on the Design of ML and MAP Joint Estimators," *IEEE Trans. on Information Theory*, vol. IT-26, pp. 498–503, July, 1980.

[Treichler, 1981]
Treichler, J. R., "The Use of Digital Processors for Wideband Analog Communications," *Proc. 15th Asilomar Conf. on Circuits, Sys., and Computers*, Pacific Grove, Calif., November, 1981.

[Treichler and Agee, 1983]
Treichler, J. R., and B. G. Agee, "A New Approach to Multipath Correction of Constant Modulus Signals," *IEEE Trans. on Acoustics, Speech, and Signal Processing*, vol. ASSP-31, pp. 459–472, April, 1983.

[Treichler and Larimore, 1985a]
Treichler, J. R., and M. G. Larimore, "New Processing Techniques Based on the Constant Modulus Adaptive Algorithm," *IEEE Trans. on Acoustics, Speech, and Signal Processing*, vol. ASSP-33, pp. 420–431, April, 1985.

[Treichler and Larimore, 1985b]
Treichler, J. R. and M. G. Larimore, "The Tone Capture Properties of CMA-Based Interference Suppressors," *IEEE Trans. on Acoustics, Speech, and Signal Processing*, vol. ASSP-33, pp. 946–958, August, 1985.

[Tretter, 1976]
Tretter, S. A., *Introduction to Discrete Time Signal Processing*, Wiley, New York, chap. 2, 1976.

[White, 1975]

White, S. A., "An Adaptive Recursive Filter," *Proc. 9th Asilomar Conf. on Circuits, Sys., and Computers,* Pacific Grove, Calif., pp. 21–25, November, 1975.

[Widrow et al., 1975*a*]

Widrow, B., J. M. McCool, and M. Ball, 1975, "The Complex LMS Algorithm," *Proc. IEEE,* vol. 63, pp. 719–720, April, 1975.

[Widrow, et. al., 1975*b*]

Widrow, B., J.R. Glover, J. M. McCool, J. Kaunitz, C. S. Williams, R. H. Hearn, J. R. Zeidler, E. Dong, Jr., and R. C. Goodlin, "Adaptive Noise Cancelling: Principles and Applications," *Proc IEEE,* vol. 63, pp. 1692–1716, December, 1975.

[Widrow and Hoff, 1960]

Widrow, B., and M. Hoff, Jr., "Adaptive Switching Circuits," *IRE WESCON Conv. Record,* pt. 4, pp. 96–104, 1960.

[Widrow and McCool, 1976]

Widrow, B., and J. M. McCool, "A Comparison of Adaptive Algorithms Based on the Methods of Steepest Descent and Random Search," *IEEE Trans. of Antennas and Propagation,* vol. AP-24, pp. 615–636, September, 1976.

[Widrow et al., 1976]

Widrow, B., J. M. McCool, M. G. Larimore, and C. R. Johnson, Jr., "Stationary and Nonstationary Learning Characteristics of the LMS Adaptive Filter," *Proc. IEEE,* vol. 64, pp. 1151–1162, August, 1976.

[Yam and Redman, 1983]

Yam, E. S., and M. D. Redman, "Development of a 60-Channel FDM-TDM Transmultiplexer," *COMSAT Tech. Rev.,* vol. 13, pp. 1–56, Spring, 1983.

[Zahm, 1973]

Zahm, C. L., "Application of Adaptive Arrays to Suppress Strong Jammers in the Presence of Weak Signals," *IEEE Trans. on Aerospace and Electronic Systems,* vol. AES-9, pp. 260–271, March, 1973.

Index

339